VOLUME ONE HUNDRED AND THIRTEEN

SEMICONDUCTORS AND SEMIMETALS
Non-layered 2D materials

SERIES EDITORS

ZETIAN MI
Professor
Department of Electrical Engineering and Computer Science
University of Michigan
1310 Beal Avenue
Ann Arbor MI48109
United States of America

HARK HOE TAN
The Australian National University, Australia

VOLUME ONE HUNDRED AND THIRTEEN

SEMICONDUCTORS AND SEMIMETALS
Non-layered 2D materials

Edited by

IRFAN HAIDER ABIDI
Royal Melbourne Institute of Technology, Melbourne, Australia

ACADEMIC PRESS
An imprint of Elsevier

Academic Press is an imprint of Elsevier
50 Hampshire Street, 5th Floor, Cambridge, MA 02139, United States
525 B Street, Suite 1650, San Diego, CA 92101, United States
The Boulevard, Langford Lane, Kidlington, Oxford OX5 1GB, United Kingdom
125 London Wall, London, EC2Y 5AS, United Kingdom

First edition 2023

Copyright © 2023 Elsevier Inc. All rights reserved.

No part of this publication may be reproduced or transmitted in any form or by any means, electronic or mechanical, including photocopying, recording, or any information storage and retrieval system, without permission in writing from the publisher. Details on how to seek permission, further information about the Publisher's permissions policies and our arrangements with organizations such as the Copyright Clearance Center and the Copyright Licensing Agency, can be found at our website: www.elsevier.com/permissions.

This book and the individual contributions contained in it are protected under copyright by the Publisher (other than as may be noted herein).

Notices
Knowledge and best practice in this field are constantly changing. As new research and experience broaden our understanding, changes in research methods, professional practices, or medical treatment may become necessary.

Practitioners and researchers must always rely on their own experience and knowledge in evaluating and using any information, methods, compounds, or experiments described herein. In using such information or methods they should be mindful of their own safety and the safety of others, including parties for whom they have a professional responsibility.

To the fullest extent of the law, neither the Publisher nor the authors, contributors, or editors, assume any liability for any injury and/or damage to persons or property as a matter of products liability, negligence or otherwise, or from any use or operation of any methods, products, instructions, or ideas contained in the material herein.

ISBN: 978-0-443-19394-1
ISSN: 0080-8784

For information on all Academic Press publications
visit our website at https://www.elsevier.com/books-and-journals

Publisher: Zoe Kruze
Acquisitions Editor: Jason Mitchell
Editorial Project Manager: Sneha Apar
Production Project Manager: Abdulla Sait
Cover Designer: Victoria Pearson

Typeset by MPS Limited, India

Contents

Contributors	*xi*
Preface	*xv*

1. Non-layered two-dimensional metalloids   **1**

Rahul Mitra, Ajay Kumar Verma, Unnikrishnan Manju, and Yongxiang Li

1.	Introduction	2
2.	Boron-based non-layered 2D materials	4
3.	Si-based non-layered 2D materials	7
4.	Germanium-based non-layered 2D materials	10
5.	As-based non-layered 2D materials	13
6.	Sb-based non-layered 2D materials	14
7.	Te-based non-layered 2D materials	17
8.	Conclusions and future perspectives	19
	References	20

2. 2D metal oxides   **27**

Vahid Khorramshahi and Fatemeh Safari

1.	Introduction	28
2.	Structural characteristics of non-layered 2D metal oxides	29
3.	Ultrathin sheets with various bonding patterns	29
4.	Structural arrangements and their influence on properties	35
	4.1 Thickness control	35
	4.2 Vacancy manipulation	36
	4.3 Elemental doping	37
5.	Properties of non-layered 2D metal oxides	38
6.	Fabrication approaches for non-layered 2D metal oxides	43
7.	Liquid metal approach	44
8.	Hydrothermal approach	48
9.	Chemical vapor deposition (CVD) approach	50
10.	Template-assisted approach	52
11.	Challenges and future perspectives	54
	References	56

3. 2D non-layered metal dichalcogenides 63

Mostafa M.H. Khalil, Abdelrahman M. Ishmael, and
Islam M. El-Sewify

1.	Introduction	64
	1.1 Definition of non-layered 2D metal dichalcogenides	64
	1.2 Importance of studying non-layered 2D metal dichalcogenides	66
2.	Structure of different types of non-layered 2D metal dichalcogenides	67
	2.1 Non-layered transition metal dichalcogenides (TMDCs)	67
	2.2 Non-layered metal dichalcogenides beyond TMDCs	71
	2.3 Mixed metal dichalcogenides	74
3.	Structural characteristics of non-layered 2D metal dichalcogenides	74
	3.1 Monolayer and few-layer structures	74
	3.2 Defects and grain boundaries	75
	3.3 Structural phase transitions	75
4.	Properties of non-layered 2D metal dichalcogenides	76
	4.1 Electronic properties	76
	4.2 Mechanical properties	77
	4.3 Chemical properties non-layered 2D metal dichalcogenides	78
5.	Synthesis of non-layered 2D metal dichalcogenides	78
	5.1 Dry methods	79
	5.2 Wet chemistry methods	83
6.	Applications of non-layered 2D metal dichalcogenides	84
	6.1 Electronics	85
	6.2 Optoelectronics	85
	6.3 Energy conversion and storage	86
	6.4 Catalysis	88
	6.5 Biomedical applications	90
7.	Challenges and future directions	91
	7.1 Current challenges in the research of 2D metal dichalcogenides	91
	7.2 Future research directions in the field of 2D metal dichalcogenides	92
	References	93

4. 2D III-V semiconductors 101

Sattar Mirzakuchaki and Atefeh Nazary

1.	Introduction	102
2.	Non-layered III-V compounds	103
	2.1 Boron nitride (BN)	103
	2.2 Boron phosphide (BP)	107
	2.3 Boron arsenide (BAs)	110

2.4	Aluminum nitride (AlN)	113
2.5	Aluminum antimonide (AlSb)	116
2.6	Indium arsenide (InAs)	119
2.7	Indium phosphide (InP)	123
2.8	Indium antimonide (InSb)	126
2.9	Gallium nitride (GaN)	129
2.10	Gallium arsenide (GaAs)	134
2.11	Gallium arsenide (GaSb)	138
References		141

5. Two dimensional perovskites — **145**

Memoona Qammar and Faiza

1.	Introduction	145
2.	Synthesis of 2D perovskites	149
	2.1 Hot injection (HI)	150
	2.2 Ligand assisted reprecipitation (LARP)	151
	2.3 Other methods	154
3.	Structure and physical properties	155
4.	Applications	158
5.	Conclusions and perspective	163
References		164

6. CVD growth of 2D non layered materials — **169**

Shumaila Karamat and Shabeya Kanwal

1.	Introduction	170
2.	Growth of 2D non-layered materials	170
3.	Chemical vapor deposition (CVD)	171
	3.1 Basic overview: How does CVD work?	171
4.	Growth of phosphorene	172
5.	Two-dimensional selenium nanoflakes (SeNFs)	173
6.	Growth of di-indium tri-sulfide (In_2S_3)	175
7.	Growth of β-Ga_2O_3	176
8.	Growth of cadmium sulfide (CdS)	177
9.	Growth of hematite (α-Fe_2O_3)	179
10.	Growth of ε-Fe_2O_3	180
11.	Growth of lead sulfide (PbS)	181
12.	Growth of titanium dioxide (TiO_2)	183
13.	Conclusion	184
References		184

7. 2D non-layered materials for energy applications 189

Harish Somala, Muzammil Mushtaq, and Uma Sathyakam Piratla

1. Introduction	189
2. 2D materials	190
2.1 2D layered materials	191
2.2 2D non-layered materials	191
3. 2D non layered materials for energy application	193
4. Supercapacitors	193
4.1 Electric double-layer capacitor	193
4.2 Pseudocapacitor	195
4.3 Hybrid capacitor	196
5. Lithium-ion batteries	200
6. Sodium-ion batteries	205
7. Fuel cells	209
8. Other energy storage	213
9. Conclusions	213
References	213

8. Sensing applications of non-layered 2D materials 217

Tuan Sang Tran and Dzung Viet Dao

1. Introduction	218
2. Structure and synthesis of non-layered 2D materials	219
2.1 Structure	219
2.2 Synthesis	220
2.3 Wet chemical synthesis	221
2.4 Optimised chemical vapour deposition	223
2.5 Liquid metals synthesis	225
3. Sensing mechanisms of non-layered 2D materials	227
3.1 Piezoresistive mechanism	227
3.2 Piezoelectric mechanism	229
3.3 Electrochemical mechanism	229
3.4 Optoelectronic mechanism	230
3.5 Thermoresistive mechanism	231
3.6 Fluorescence resonance mechanism	231
4. Sensing applications of non-layered 2D materials	232
4.1 Nanoarchitecture designs for sensing applications	232
4.2 Non-layered 2D materials for solute sensing	235
4.3 Non-layered 2D materials for gas sensing	238

Contents ix

4.4 Non-layered 2D materials for mechanical sensing 242
4.5 Photodetectors and radiation sensing 243
5. Conclusions 245
References 246

9. 2D-non-layered materials: Advancement and application in biosensors, memristors, and energy storage 253

Zina Fredj and Mohamad Sawan

1. Introduction 254
2. Non-layered 2D materials based biosensing platforms 255
   2.1 Electrochemical biosensors 255
   2.2 Optical biosensors 262
   2.3 Nano-FET based biosensor 264
3. Memristors based 2D non layered material 267
4. 2D non-layered materials in energy storage 270
5. Challenges and future perspectives 272
References 273

10. Environment applications of non-layered 2D materials 277

Mohamed Bahri and Peiwu Qin

1. Introduction 278
2. Assessing environmental quality and remediation applications 279
   2.1 NH_3 and H_2S quantification 279
   2.2 Heavy metals monitoring 282
   2.3 pH sensing 283
3. Catalysis 285
   3.1 Catalysis of water splitting 285
   3.2 Catalysis of CO conversion and organic reactions 287
4. Summary and outlooks 289
Acknowledgments 291
References 292

11. Biomedical applications of non-layered 2DMs 297

Seyedeh Nooshin Banitaba, Abeer Ahmed Qaed Ahmed,
Mohammad-Reza Norouzi, and Sanaz Khademolqorani

1. Introduction 298
2. The recent advancement in the synthesis and structure of NL2DMs 299
3. Biosensors devices based on 2D non-layered composition 304

4.	Antibacterial activity of the NL2DMs	307
5.	The application of non-layered 2D arrays in controlled drug delivery systems	311
6.	The use of NL2DMs 2D non-layered structures in tissue engineering	315
7.	Current challenges	319
	References	320

12. Thermoelectric applications of non-layered 2-D materials   323

Ajay Kumar Verma, Rahul Mitra, Bhasker Gahtori, and
Sumeet Walia

1.	Introduction	324
	1.1 Thermoelectric effect	324
	1.2 Thermoelectric devices and materials	325
2.	Non-layered 2-D materials	327
	2.1 Metalloids	328
	2.2 Metal oxides	329
	2.3 Metal chalcogenides	330
	2.4 III–V Semiconductors	331
3.	Thermoelectric applications	332
	3.1 Power generation	332
	3.2 Cooling applications	333
	3.3 Space technologies	333
	3.4 Wearable medical applications	333
	3.5 Self-powered sensors	334
	3.6 Other applications	334
	References	335

Index   *339*

Contributors

Faiza
School of Natural Sciences, National University of Science and Technology, Islamabad, Pakistan; Department of Macromolecular Engineering, Case Western Reserve University, Cleveland, OH, United States

Abeer Ahmed Qaed Ahmed
Department of Environmental Sciences, School of Agriculture and Environmental Sciences, University of South Africa, Florida, Johannesburg, South Africa

Mohamed Bahri
Center of Precision Medicine and Healthcare, Tsinghua-Berkeley Shenzhen Institute, Shenzhen, Guangdong Province; Institute of Biopharmaceutical and Health Engineering, Tsinghua Shenzhen International Graduate School, Tsinghua University, Shenzhen, P.R. China

Dzung Viet Dao
School of Engineering and Built Environment, Griffith University, Gold Coast; Queensland Micro, and Nanotechnology Centre, Griffith University, Brisbane, QLD, Australia

Islam M. El-Sewify
Chemistry Department, Faculty of Science, Ain Shams University, Abbassia, Cairo, Egypt

Zina Fredj
CenBRAIN Neurotech, School of Engineering, Westlake University, Hangzhou, P.R. China

Bhasker Gahtori
CSIR-National Physical Laboratory, Dr. K.S. Krishnan Marg, New Delhi; Academy of Scientific & Innovative Research (AcSIR), Ghaziabad, India

Abdelrahman M. Ishmael
Nanomaterials Science Program, Faculty of Science, Benha University, Cairo, Egypt

Shabeya Kanwal
Department of Physics, COMSATS University, Islamabad, Pakistan

Shumaila Karamat
Department of Physics, COMSATS University, Islamabad, Pakistan

Sanaz Khademolqorani
Emerald Experts Laboratory, Isfahan Science and Technology Town; Department of Textile Engineering, Isfahan University of Technology, Isfahan, Iran

Mostafa M.H. Khalil
Chemistry Department, Faculty of Science, Ain Shams University, Abbassia, Cairo, Egypt

Vahid Khorramshahi
Materials and Energy Research Center, Dezful Branch, Islamic Azad University, Dezful, Iran

Yongxiang Li
School of Engineering, RMIT University, Melbourne, Victoria, Australia

Unnikrishnan Manju
Materials Chemistry Department, CSIR-Institute of Minerals and Materials Technology, Bhubaneswar, Odisha; Academy of Scientific and Innovative Research (AcSIR), Ghaziabad, India

Sattar Mirzakuchaki
School of Electrical Engineering, Iran University of Science and Technology, Tehran, Iran

Rahul Mitra
School of Engineering, RMIT University, Melbourne, Victoria, Australia; Materials Chemistry Department, CSIR-Institute of Minerals and Materials Technology, Bhubaneswar, Odisha; Academy of Scientific and Innovative Research (AcSIR), Ghaziabad, India

Muzammil Mushtaq
School of Electrical Engineering, Vellore Institute of Technology, Vellore, Tamil Nadu, India

Atefeh Nazary
Department of Electrical, Biomedical and Mechatronics Engineering, Qazvin Branch, Islamic Azad University, Qazvin, Iran

Seyedeh Nooshin Banitaba
Department of Textile Engineering, Amirkabir University of Technology, Tehran; Emerald Experts Laboratory, Isfahan Science and Technology Town, Isfahan, Iran

Mohammad-Reza Norouzi
Department of Textile Engineering, Isfahan University of Technology, Isfahan, Iran

Uma Sathyakam Piratla
School of Electrical Engineering, Vellore Institute of Technology, Vellore, Tamil Nadu, India

Memoona Qammar
Department of Chemistry, The Hong Kong University of Science and Technology (HKUST), Kowloon, Hong Kong (SAR), P.R. China; School of Natural Sciences, National University of Science and Technology, Islamabad, Pakistan

Peiwu Qin
Center of Precision Medicine and Healthcare, Tsinghua-Berkeley Shenzhen Institute, Shenzhen, Guangdong Province; Institute of Biopharmaceutical and Health Engineering, Tsinghua Shenzhen International Graduate School, Tsinghua University, Shenzhen, P.R. China

Fatemeh Safari
Materials and Energy Research Center, Dezful Branch, Islamic Azad University, Dezful, Iran

Mohamad Sawan
CenBRAIN Neurotech, School of Engineering, Westlake University, Hangzhou, P.R. China

Harish Somala
School of Electrical Engineering, Vellore Institute of Technology, Vellore, Tamil Nadu, India

Tuan Sang Tran
School of Engineering, RMIT University, Melbourne, Victoria; School of Chemical Engineering, The University of New South Wales, Sydney, NSW, Australia

Ajay Kumar Verma
School of Engineering, RMIT University, Melbourne, Victoria, Australia; CSIR–National Physical Laboratory, Dr. K.S. Krishnan Marg, New Delhi; Academy of Scientific and Innovative Research (AcSIR), Ghaziabad, India

Sumeet Walia
School of Engineering, RMIT University, Melbourne, Victoria, Australia

Preface

Welcome to the intriguing world of "Non-Layered 2D Materials." Within the pages of this book, we invite you to embark on a fascinating journey to explore atomically thin materials with exceptional properties to solve numerous technological challenges. Since the discovery of graphene, past couple of decades have witnessed an immense surge of interest in the realm of 2D materials; however, the majority of this enthusiasm has been directed towards "layered" 2D materials such as graphene and transition metal dichalcogenides. Remarkably, within this wave of research and discovery, the universe of atomically thin "non-layered" 2D materials has somewhat lingered in the background, awaiting its well-deserved spotlight. These non-layered materials are bestowed with captivating chemical and physical attributes, each holding immense potential across a vast spectrum of applications, spanning electronics, optoelectronics, energy storage, bio-sensing, catalysis, and environmental solutions. This book serves as your gateway to a comprehensive exploration of non-layered 2D materials. Our journey unfolds in three distinct sections, each designed to provide a deeper understanding of this captivating field.

In the first section, we delve into the abstract landscapes of structures and properties found within various classes of non-layered 2D materials. This section elucidates the correlation between their atomic structure and the remarkable properties they exhibit. In the second section, our focus turns to the synthesis techniques that bring these materials to life. We pay particular attention to the chemical vapor deposition technique, which stands as a scalable and industrially viable method for the production of a wide array of 2D materials, encompassing both layered and non-layered 2D materials. Finally, we venture into the most exciting aspect of this exploration—the practical applications of non-layered 2D materials. We explore how these materials can revolutionize the world of technology and problem-solving, from enhancing energy storage devices and enabling precise sensing to preserving our environment and advancing healthcare through innovative drug delivery and biomedical applications.

The existence of this book is a testament to the dedication and expertise of our esteemed authors and contributors from across the globe, all of whom are at the forefront of this rapidly evolving field. Their boundless passion for pushing the boundaries of science and technology shines through on every page, and we are deeply indebted to them for their invaluable contributions. Whether you are a student, a scientist, a curious reader, or someone with a thirst for knowledge, this book offers something for everyone. We aspire for this book to not only expand your understanding but also spark your creativity and curiosity.

We hope this book not only teaches you something new but also sparks your creativity and curiosity. We invite you to dream big and be a part of the exciting world of non-layered 2D materials.

Thank you for joining us on this adventure.

IRFAN H. ABIDI
Royal Melbourne Institute of Technology,
Melbourne, Australia

CHAPTER ONE

Non-layered two-dimensional metalloids

Rahul Mitra[a,b,c,*], Ajay Kumar Verma[a,c,d], Unnikrishnan Manju[b,c], and Yongxiang Li[a]

[a]School of Engineering, RMIT University, Melbourne, Victoria, Australia
[b]Materials Chemistry Department, CSIR-Institute of Minerals and Materials Technology, Bhubaneswar, Odisha, India
[c]Academy of Scientific and Innovative Research (AcSIR), Ghaziabad, India
[d]CSIR-National Physical Laboratory, Dr. K.S. Krishnan Marg, New Delhi, India
*Corresponding author. e-mail address: S3989184@student.rmit.edu.au

Contents

1. Introduction	2
2. Boron-based non-layered 2D materials	4
3. Si-based non-layered 2D materials	7
4. Germanium-based non-layered 2D materials	10
5. As-based non-layered 2D materials	13
6. Sb-based non-layered 2D materials	14
7. Te-based non-layered 2D materials	17
8. Conclusions and future perspectives	19
References	20

Abstract

The emergence of innovative two-dimensional (2D) materials that are not layered has ignited a global surge of interest in both fundamental research and practical applications, particularly in the field of optoelectronics. In this chapter, we aim to provide a comprehensive summary of recent advancements in materials and systems centered around non-layered 2D metalloids, which bring fresh and exciting additions to the ever-expanding 2D material family. We initiate our discussion by delving into the common methodologies used to create 2D structures from non-layered materials, with a specific focus on semimetal elements such as Germanium (Ge), Antimony (Sb), Tellurium (Te), among others. Following this, we delve into an exploration of the electronic properties, transport characteristics, and other pivotal attributes associated with these non-layered 2D materials. Lastly, we emphasize the significant challenges that lie ahead in the continuous development of this rapidly advancing research field. We also propose potential strategies to tackle these challenges, aiming to drive further progress and innovation in the study and application of non-layered 2D metalloids.

1. Introduction

Two-dimensional (2D) materials have experienced rapid development due to their fascinating properties as atomic-thin crystals with a unique flat structure (Schaibley et al., 2016; Mas-Ballesté et al., 2011). They typically have lateral dimensions ranging from 100 nm to several micrometers and a thickness of only a few atoms, usually less than 5 nm (Gupta et al., 2015). The emergence of 2D materials began in 2004 when graphene was successfully isolated from graphite using a micromechanical cleavage technique, marking the start of the 2D materials era (Novoselov et al., 2016). Graphene, a single-atom-thick crystalline carbon film, exhibits remarkable physical properties and has spurred the exploration and development of other ultrathin 2D materials (Gupta et al., 2015; Zheng et al., 2020). These newly developed 2D materials display extraordinary physical, chemical, optical, and electronic characteristics, offering promising applications in nanoelectronics, optoelectronics, catalysis, energy storage, and sensing (You et al., 2018; Androulidakis et al., 2018; Ma et al., 2021; Mitra et al., 2023; Singh et al., 2015; Akinwande et al., 2017). Due to electron confinement in two dimensions, they possess distinct electrical properties suitable for electronic device applications and condensed matter research. The ultrathin thickness and strong in-plane covalent bonding provide optical transparency, flexibility, and mechanical strength, making them attractive for flexible device fabrication among 2D materials (Das et al., 2014). Large lateral dimensions and ultrahigh specific surface areas make them ideal for surface-dependent applications and their high proportion of surface atoms enables properties modulation and functionalization through techniques like constructing heterojunctions, doping, introducing defects, and inducing phase engineering (Dai et al., 2019; Schulman et al., 2018; Yu et al., 2021; Su et al., 2020). Additionally, 2D materials exhibit a broad spectrum response, strong interactions with light, and low dark currents, making them well-suited for constructing high-performance functional devices (Pan et al., 2021).

From a crystal structure perspective, 2D materials can be broadly categorized into two types: Layered Categories: Examples include transition metal dichalcogenides and black phosphorus, where in-plane atoms are connected by strong chemical bonds, and stacking layers are held together by weak van der Waals interactions (Su et al., 2021; Turunen et al., 2022).

Non-layered Categories: These include materials which are formed through chemical bonding in three-dimensional directions, resulting in unsaturated dangling bonds on the surface and creating highly active and high-energy surfaces (Zhou et al., 2019).

In the realm of 3D materials, it's worth noting that only a small fraction exhibit layered structures, while the majority fall into the category of non-layered materials. These non-layered materials offer a wide range of tunable properties in areas such as electronics, optoelectronics, magnetism, and catalysis, making them highly promising for applications at the 2D scale (Zhou et al., 2019; Wang et al., 2017a). Compared to their layered counterparts, non-layered 2D materials bring together the advantages of atomic thinness and high-surface reactivity, resulting in a host of fascinating properties. Firstly, the influence of multibody interactions (such as electrons-electrons, electrons-phonons, electrons-plasma) on electronic transport, thermal conductivity, heat capacity, quasiparticle dynamics, and other properties gives rise to remarkable characteristics in 2D non-layered materials (You et al., 2018; Tan and Zhang, 2015). Moreover, these materials possess both bulk and 2D structural attributes, leading to a wealth of distinct physical phenomena. For instance, consider the III-V semiconductor GaN. Bulk GaN is widely used in modern optoelectronic devices due to its direct wide bandgap, rapid saturated migration rate, and high critical breakdown electric field. However, 2D GaN exhibits an enlarged bandgap, a blue-shifted photoluminescence emission peak, and enhanced internal quantum efficiency, making it a promising candidate for future nanoelectronics and optoelectronics (Lee et al., 2017; Camacho-Mojica and López-Urías, 2015). Secondly, as thickness decreases, lattice structure distortion occurs, reducing surface energy and stabilizing the crystal structure. This distortion also leads to changes in the electronic states compared to bulk materials, resulting in variations in energy band structure, electrical conductivity, carrier mobility, and even ferromagnetism (Schulman et al., 2018; Turunen et al., 2022; Das et al., 2019; Gibertini et al., 2019). Thirdly, the high-activity surface, driven by unsaturated dangling bonds and the unique electronic structure of 2D non-layered materials, translates into highly efficient catalytic capabilities and improved energy storage performance. Fourthly, the 2D planar structure of these materials aligns well with microelectronic processing technology, making them compatible with electronic device applications. Consequently, these materials are expected to advance our fundamental understanding of physical phenomena in 2D non-layered materials and promote the development of novel devices (Yu et al., 2021).

Two-dimensional (2D) metalloids, which are materials with properties intermediate between metals and nonmetals, hold significant importance in the realm of materials science and nanotechnology. These materials exhibit unique characteristics that make them valuable for various applications. 2D metalloids, such as arsenene and antimonene, can be easily tuned to exhibit a wide range of electrical properties. By adjusting the number of layers or introducing defects, their electrical conductivity can be finely controlled (Fortin-Deschênes and Moutanabbir, 2018; Gablech et al., 2018). This tunability is crucial for designing electronic components with tailored conductivity, making them vital in the development of next-generation electronic devices. Many 2D metalloids act as semiconductors, with bandgaps that can be modified to suit specific applications. This semiconductor behavior is fundamental for the creation of transistors, diodes, and other electronic components. The semiconducting properties of 2D metalloids also make them suitable for optoelectronic applications. They can be employed in the development of photodetectors, light-emitting diodes (LEDs), and solar cells (Zhou et al., 2019; Wang et al., 2017a). Their ability to efficiently absorb and emit light in the visible and infrared spectra is crucial for advancing these technologies. Non-layered 2D metalloids are exceptionally thin and exhibit mechanical flexibility. This property is advantageous for creating flexible and bendable electronic devices. They can be integrated into wearable technology, flexible displays, and other innovative applications where traditional bulk materials would be impractical (Zheng et al., 2020). Due to these myriad attributes, non-layered 2D metalloids are vital materials in the fields of electronics, optoelectronics, catalysis, and nanotechnology due to their tunable properties, semiconductor behavior, mechanical flexibility, surface reactivity, and compatibility with nanoscale applications (Zheng et al., 2020; Zhou et al., 2019; Tan and Zhang, 2015). Their unique characteristics open up new possibilities for the development of advanced technologies and hold the potential to revolutionize various industries in the near future.

2. Boron-based non-layered 2D materials

Extensive research in the field of non-van der Waals (non-vdW) crystals has typically commenced with layered materials (Su et al., 2021; Turunen et al., 2022). However, recent work by Backes et al. has demonstrated that the aspect ratio (length compared to thickness) of liquid-exfoliated nanosheets can be controlled by considering the ratio between the energy associated with in-plane tearing and that associated with out-of-plane peeling (Backes et al., 2017). This finding indicates that layered materials characterized by significant mechanical anisotropy tend to yield nanosheets with larger aspect ratios. This analysis also suggests that non-layered materials with anisotropic bonding arrangements can also undergo exfoliation, resulting in non-layered quasi-2D materials. Consequently, the production of nanosheets and nanoplatelets through the exfoliation of non-layered materials has the potential to significantly expand the family of 2D materials. This approach builds upon the exfoliation of non-layered materials like Fe_2O_3, WO_3, Se, pyrite, and various metal oxides (Yang et al., 2016; Masoumi et al., 2021; Zhou et al., 2015; Kaur et al., 2020). Boron carbide (B_4C) has been a subject of extensive study over the past decades due to its unique properties (Chang et al., 2020; Ojalvo et al., 2020). It is renowned as the third hardest material known to date, ranking only behind diamond and boron nitride. B_4C has two primary crystal structures: one consisting of B_{12} icosahedra with C-C-C chains and the other featuring $B_{11}C$ icosahedra with C-B-C chains (Guo et al., 2021).

In the study conducted by Guo et al., non-layered B_4C nanosheets were fabricated through a gentle processing method involving the tip sonication of B_4C bulk powder in various organic solvents (Guo et al., 2021). The confirmation of nanosheet structures was achieved through atomic force microscopy (AFM) and transmission electron microscopy (TEM). The selection of these nanosheet planes was based on an initial observation of the B_4C crystal structure, which indicated more favorable cleavage without cutting through icosahedral structures. These planes were further analyzed using density functional theory (DFT). The computational results demonstrated exfoliation energies for four chosen planes – (001), (012), (101), and (300) – with binding energies of 0.056, 0.105, 0.116, and 0.113 eV per atom, respectively. Importantly, these energies were found to be below the thermodynamic stability threshold of 0.2 eV per atom for a free-standing single-layer or few-layer nanosheet.

DFT computations were also employed to calculate the formation energies of various B_4C slabs cleaved along four different planes: (300), (001), (101), and (012). While the majority of experimental results in the study focused on a B_4C structure with C-B-C chains, the computations were carried out with C-C-C chains for two primary reasons. Firstly, structures with C-B-C chains demanded extensive computational resources due to the multitude of possible surface terminations, each containing between 90 and 225 atoms. In contrast, simulating B_4C slabs with C-C-C chains required only 29 terminations, each comprising 46–176 atoms. Secondly, the simulations revealed that structures with C-B-C chains exhibited energy trends similar to those of C-C-C chains, both in the bulk and for comparable terminations in the slabs. In the bulk, the total energy of structures with C-B-C chains was found to be lower than those with C-C-C chains by 0.075 eV per atom. In the case of the (001) slab, for terminations 2 and 6, slabs with C-B-C bonds displayed formation energies lower than those with C-C-C bonds by 0.042 eV per atom and 0.071 eV per atom, respectively (Gupta et al., 2022; Xie et al., 2015).

The formation energies of various terminations along these directions are depicted in Fig. 1A–D. Each panel included different terminations at

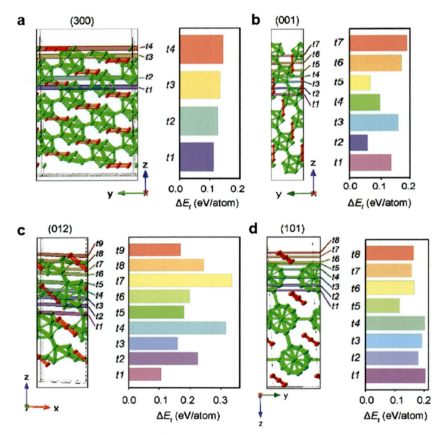

Fig. 1 The figures illustrate the formation energies of various terminations when cleaving along different planes of non-layered 2D boron carbide: (A) (300), (B) (001), (C) (012), and (D) (101). In these diagrams, boron atoms are represented in green, while carbon atoms are depicted in red. Each cleavage plane is labeled with terminations numbered as tn, with n incrementing from the bottom to the top (Guo et al., 2021).

slightly shifted positions, labeled as t1, t2, and so on. The bar graphs represented the formation energies (ΔE_f) for each termination in each plane. The minimum ΔE_f for the (300) and (001) planes were 0.113 and 0.056 eV per atom, respectively, while for the (101) and (012) planes, they were 0.116 and 0.105 eV per atom, respectively. Remarkably, the proximity of these energies suggested that there was no significant preference for a particular cleavage plane.

Yang et al. have developed an innovative method for the rapid synthesis of a sandwiched Ni-B_i/graphene/Ni-B_i hybrid composite at room temperature (Yang et al., 2017). In this process, ultrathin non-layered Ni-B_i nanosheets are vertically arranged on both sides of the conductive graphene, creating an open and exposed heterostructure framework. This unique configuration effectively prevents aggregation and enhances the exposure of active sites within the 2D nanosheets in the composite heterostructure. Simultaneously, the in-situ integration of these ultrathin nanosheets with graphene facilitates rapid charge transport in the thickness dimension due to intimate interfacial contact. Consequently, the Ni-Bi/G heterostructure exhibits exceptional photoelectrocatalytic activity and stability, coupled with outstanding structural integrity. The study showcases a straightforward and scalable approach for producing large quantities of ^{2}D–^{2}D layered heterostructures, which are crucial for various technological applications.

3. Si-based non-layered 2D materials

In a study by Lalmi et al., they claimed the synthesis of 2D silicene on Ag(111), based solely on observations made through scanning tunneling microscopy (STM) (Lalmi et al., 2010). However, since no additional complementary experimental evidence was presented, the conclusion regarding the synthesis of silicene remained somewhat speculative. Furthermore, the Si-Si distance reported by Lalmi et al. was 0.19 nm, a value significantly smaller than the expected range of 0.22–0.24 nm based on both theoretical density functional theory (DFT) calculations and general expectations.

In contrast, Vogt et al. conducted a study focusing on one-atom-thick Si sheets formed on the Ag(111) surface (Vogt et al., 2012). They meticulously prepared clean and well-ordered Ag(111) surfaces using Ar^+ bombardment and subsequent annealing of (111)-oriented Ag single crystals under ultrahigh vacuum conditions. Si deposition was carried out by evaporating Si from a heated source, while the Ag sample was maintained at temperatures between 220 and 260 °C. To ascertain the energetic stability of the proposed silicene structure, the researchers performed ab initio calculations based on DFT, incorporating complete structural relaxations. These calculations yielded a total energy value for the structure model, indicating a negative value and an adhesion energy of 0.48 eV for each Si atom. This outcome demonstrated that the arrangement of silicene on Ag(111) is indeed energetically stable, as represented in Fig. 2.

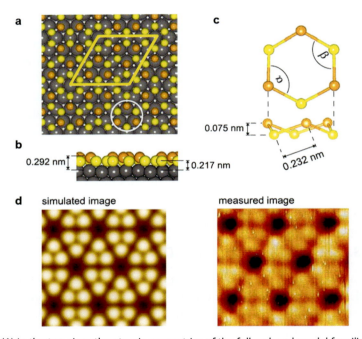

Fig. 2 (A) In the top view, the atomic geometries of the fully relaxed model for silicene on the Ag(111) surface are depicted. (B) The side view offers a complementary perspective, showing the vertical arrangement of atoms in the same model as depicted in (A). It illustrates the relative positions of the silicene layer and the Ag(111) substrate. (C) An enlarged image specifically focuses on a hexagonal silicene ring within the structure. This region is indicated by a white circle in the top view (A), allowing for a detailed examination of the atomic arrangement within the hexagonal ring. (D) A simulated STM (Scanning Tunneling Microscopy) image is presented on the left, corresponding to the structure shown in (A). This simulated image faithfully reproduces the structural features observed in the experimental STM image shown on the right. These features include a hexagonal arrangement of the triangular structures surrounding dark centers, aligning with the experimental observations (Vogt et al., 2012).

Upon closer examination, it was observed that the Si atoms in the silicene layer, directly above the underlying Ag atoms, displayed slight displacement in the z direction concerning the bottom Si atoms. The distance between the top and bottom Si atoms and the average height of the first Ag layer were found to be 0.292 ± 0.002 nm and 0.217 ± 0.003 nm, respectively. This displacement resulted in the formation of triangular structures in the top silicene layer, each consisting of three Si atoms separated by 0.38 nm. These findings were in excellent agreement with the experimental STM observations, which also revealed a triangular structure with a separation of approximately 0.40 nm (Lalmi et al., 2010). In addition, a hexagonal silicene ring was examined, and the theoretically determined Si–Si distance was found to be 0.232 nm, aligning closely with the experimentally determined value of 0.22 nm. Furthermore, the top and bottom Si atoms within this structure were separated by 0.075 nm.

To further understand the bonding in the silicene layer, the bond angles of the Si atoms were determined, shedding light on their hybridization states. The results showed that the six top Si atoms exhibited bond angles of approximately $110°$, closely resembling the ideal sp^3-hybridized Si atom angle of $109.5°$. Among the remaining 12 bottom Si atoms, six were purely sp^2 hybridized with bond angles of $120°$, while the other six had bond angles ranging from $112°$ to $118°$, indicating sp^3/sp^2 hybridization. These varying bond angles of Si atoms resulted from their displacement in the z direction, a consequence of their interaction with the Ag(111) substrate. However, the majority of the atoms exhibited sp^3/sp^2 hybridization, likely representing an equilibrium state, in agreement with calculations for freestanding Si (Cinquanta et al., 2013).

Density functional theory (DFT) calculations performed by Cai et al. have unveiled intriguing findings regarding atomically thin silicon nanosheets (Cai et al., 2023). These calculations have demonstrated that these ultrathin silicon nanosheets exhibit enhanced kinetics for hydrogen evolution when compared to their multilayer counterparts and bulk silicon. This phenomenon can be attributed to their notably small hydrogen adsorption free energy and a low energy barrier for the hydrogen evolution reaction. Building upon these theoretical insights, they successfully prepared both atomically thin and multilayer silicon nanosheets through a straightforward liquid-phase exfoliation method. These silicon nanosheets were subsequently employed as photocatalysts without the need for additional co-catalysts in the hydrogen production reaction. In direct comparison to thicker multilayer silicon nanosheets and bulk silicon, the atomically thin silicon nanosheets demonstrated a substantial improvement in their photocatalytic hydrogen production activity. These experimental results align with the predictions of the theoretical calculations. This combined approach of theoretical calculations and experimental validation underscores the effectiveness of atomically thin silicon nanosheets as co-catalyst-free photocatalysts. Furthermore, it highlights the critical role of controlling thickness in tailoring the electronic properties and unlocking novel attributes of non-layered 2D materials (Ma et al., 2021; Schulman et al., 2018; Vogt et al., 2012).

4. Germanium-based non-layered 2D materials

Following the successful synthesis of silicene in 2012, which ignited a wave of research on two-dimensional (2D) non-layered materials beyond graphene, there has been a persistent quest to obtain germanene, the germanium-based counterpart to graphene, which was initially predicted to exist in 2009 (Vogt et al., 2012; Houssa et al., 2010). While its fully hydrogenated form, known as germanane, was produced using wet chemistry methods in 2013, the elusive germanene remained undiscovered (Xu et al., 2013; Bianco et al., 2013). However, in the study by Dávila et al., they present compelling experimental and theoretical evidence confirming its synthesis through a dry epitaxial growth process on a gold (111) surface (Dávila et al., 2014).

The discovery of graphene sparked intensive research into 2D materials, especially those of elemental composition. In 2012, a silicon counterpart to graphene, called silicene, was successfully synthesized on two different metallic templates: a silver (111) surface and the zirconium diboride (0001) surface of a thin film grown on a silicon (111) substrate (Vogt et al., 2012). Subsequently, silicene was also grown on an iridium (111) surface (Meng et al., 2013). Germanene, another germanium-based analog of graphene, along with silicene, had been predicted by Cahangirov et al. to be stable as novel 2D allotropes of germanium and silicon, arranged in a low buckled honeycomb lattice (Cahangirov et al., 2017).

In the pursuit of germanene, its fully hydrogen-terminated counterpart, germanane (GeH), was first fabricated through the deintercalation of the layered van der Waals solid calcium digermanide ($CaGe_2$) (Liu et al., 2019). To improve the stability of germanane, researchers replaced the hydrogen termination with methyl groups (Dávila et al., 2014). For the synthesis of silicene, silicon was deposited onto silver (111) surfaces because the reverse system, namely silver grown on silicon (111) surfaces, was known to form atomically abrupt interfaces without intermixing. The choice of an Au (111) substrate in this study was guided by a similar strategy. Among the four noble metals studied on elemental semiconductor systems (Au, Ag/Ge, Si(111)), gold exhibited the most similarities, particularly in terms of growth mode characterized by the formation of a $\sqrt{3} \times \sqrt{3}$ R30° superstructure (wetting layer). This superstructure was associated with the formation of gold trimers on Ge(111) or silver ones on Si(111). Notably, a two-dimensional non-layered germanium with multiple phases was grown in-situ by depositing germanium onto the Au(111) surface, similar to the formation of silicene on Ag(111). One of these phases exhibited a distinct honeycomb structure with minimal corrugation in scanning tunneling microscopy (STM) imaging, as shown in Fig. 3A–C. Comprehensive core-level spectroscopy measurements and advanced density functional theory calculations provided conclusive evidence, identifying this phase as a $\sqrt{3} \times \sqrt{3}$ reconstructed germanene layer atop a $\sqrt{7} \times \sqrt{7}$ Au(111) surface.

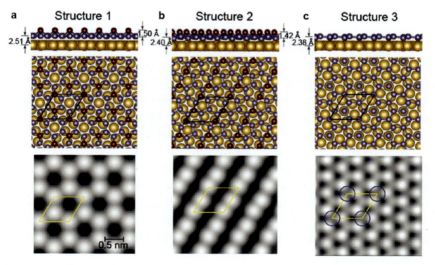

Fig. 3 (A–C) The provided visuals depict atomic structures in both side and top views, as well as simulated scanning tunneling microscopy (STM) images, illustrating three distinct models of non-layered 2D germanene positioned on the $\sqrt{7} \times \sqrt{7}$ Au(111) surface. Structures 1 and 2 exhibit a 2×2 periodic arrangements, while structure 3 showcases a $\sqrt{3} \times \sqrt{3}$ periodicity concerning the germanene layer. In these images, the Ge atoms that protrude are emphasized in a dark red shade. Additionally, the STM images feature highlighted supercells delineated by yellow lines for clarity (Dávila et al., 2014).

This discovery established the growth of nearly flat germanene, a synthetic germanium allotrope with graphene-like characteristics seldom found in nature.

The inherent indirect bandgap of bulk germanium (Ge), measured at 0.67 eV, has long posed a significant challenge for its practical application in optoelectronic devices, primarily due to its limited optical properties, including photoluminescence (Houssa et al., 2010; Dávila et al., 2014). Achieving the desired optical characteristics in Ge has been a formidable task, whether through the synthesis of ultrathin two-dimensional (2D) analogue of germanium on silicon-based substrates or by introducing substantial structural modifications to its crystal lattice.

A novel approach was developed by Hussain et al., involving a vacuum-tube hot-pressing technique to create strain-engineered two-dimensional germanium nanoplates (Ge-NPts) directly on a fused silica substrate (SiO_2) (Hussain et al., 2021). This innovative method capitalized on the significant mismatch in the coefficient of thermal expansion between Ge and SiO_2 substrates at elevated temperatures (700 °C), combined with hydrostatic pressure ($\sim$2 GPa). As a result, a biaxial compressive strain of approximately 1.23 ± 0.06% was induced within the Ge lattice. This strain-induced engineering led to a remarkable transformation from an indirect to a direct bandgap with an exceptionally wide bandgap of 2.91 eV. These strained Ge nanoplates exhibited an extraordinary 42-fold enhancement in blue photoluminescence at 300 K compared to bulk Ge and the enhanced photoluminescence was accompanied by significant quantum confinement effects, as evidenced by a quantum shift of approximately 114 meV observed as the thickness of the Ge nanoplates decreased.

Importantly, this method not only offered a means to engineer compressive strain in 2D materials but also enables the further realization of non-layered 2D materials from different groups across periodic table, that were previously inaccessible using conventional techniques (Pan et al., 2021; Mannix et al., 2017). Furthermore, investigations into the photoluminescence as a function of thickness revealed the presence of quantum confinement effects even at room temperature (300 K). The photoluminescence bands were deconvoluted to identify contributions from the band-edge, loosely bound oxide species, and luminescent defect centers in the strained Ge-NPts. These strain-engineered 2D Ge-NPts hold significant promise for monolithic integration into advanced photonic and optoelectronic devices, such as optical modulators, nano-LEDs, and nano-lasers, representing a substantial advancement in the field.

5. As-based non-layered 2D materials

GaAs, a typical III-V semiconductor material, boasts a direct bandgap of 1.42 eV and excellent carrier mobility (Stone et al., 2009; Lilly et al., 2003). It has found applications in the development of infrared optoelectronic devices, including solar cells, photodetectors, and micro-sensors (Wei et al., 2023; Fiederle et al., 2020; Ajayan et al., 2019). However, when bulk materials are reduced to just a few atomic layers, forming what's known as two-dimensional non-layered sheets (2DNLSs), they exhibit unique properties akin to two-dimensional materials, such as a high surface-to-volume ratio and strong quantum confinement. Nevertheless, 2DNLSs also come with certain drawbacks, distinct from two-dimensional materials, due to the presence of significant surface states resulting from dangling bonds or surface oxide layers. These surface states significantly limit the performance of 2DNLS materials, particularly in photodetectors, affecting responsivity and detectivity. Hence, there is a pressing need to find effective and convenient methods to mitigate surface states in 2DNLS materials and enhance photodetector performance. Surface treatment methods are commonly employed to modify the surface characteristics of nanomaterials and low-dimensional materials (Gupta et al., 2022; Cai et al., 2023; Zheng et al., 2021; Wang et al., 2017b; Priyadarshini et al., 2023). Two convenient techniques used for eliminating surface states are sulfur passivation and plasma treatment. Sulfur passivation involves the removal of the natural oxide layer on the surface and the formation of protective Ga–S bonds on the GaAs surface.

In a study conducted by Guo et al., they thoroughly investigated the impact of plasma treatment and sulfur passivation on the performance of GaAs 2DNLS photodetectors (Guo et al., 2022). Initially, the as-prepared GaAs 2DNLS photodetector exhibited a responsivity of 0.1 A/W and a detectivity of 9.6×10^{10} Jones at 8 V. However, after passivation with $(NH_4)_2S$, its responsivity saw a significant 14-fold increase, and its dark current notably decreased. This improvement was attributed to the reduction in surface state density. Furthermore, for photodetectors subjected to plasma treatment, their responsivities under various power densities closely matched those of the passivated device. The primary factor contributing to performance enhancement was the removal of the surface oxide layer, as confirmed through meticulous measurements and characterization. These results unequivocally demonstrate that plasma treatment and sulfur passivation are effective methods for enhancing the performance of two-dimensional non-layered materials and optoelectronic devices.

On the flip side, a monolayer of partially sp^3 hybridized group-V arsenic atoms arranged in a puckered or buckled honeycomb structure, has been gaining increasing attention due to its unique fundamental bandgap and novel carrier transport properties. This material, known as arsenene, has piqued interest for its potential applications in photovoltaic devices and nanotransistors (Fortin-Deschênes and Moutanabbir, 2018; Gablech et al., 2018). However, addressing the challenges related to its indirect bandgap properties is crucial as they can lead to reduced energy conversion efficiency and relatively low carrier mobility. These factors can result in increased Joule heat, ultimately degrading the reliability and performance of electronic devices (Yu et al., 2021).

To address these challenges without introducing additional complexity such as strain or external electric fields, which are often used to modulate arsenene's properties, chemical decoration to flatten the atomic structure and induce a transition from an indirect to a direct bandgap is an effective approach. By combining an element from group-III, which has s^2p^1 orbital electrons, with an element possessing s^2p^3 orbitals, it is possible to transform the buckled crystal structure of arsenene into a planar configuration with sp^2 hybridization similar to graphene. In a study conducted by Zhang et al., they utilized first-principles calculations, ab initio molecular dynamics simulations, and deformation potential theory to predict the high stability of 2D AsB with a planar honeycomb structure (Zhang et al., 2019). Electronic structure calculations revealed that the band extremum is primarily composed of the p_z orbitals of As or B atoms, as shown in Fig. 4. Due to its unfilled p_z orbitals and substantial elastic modulus, AsB exhibits high carrier mobility and a direct bandgap, making it a promising candidate for applications in nanoelectronics.

6. Sb-based non-layered 2D materials

In recent years, researchers have explored various low-dimensional materials, including one-dimensional (1D) nanostructures and two-dimensional (2D) nanostructures, as sensing materials for high-performance photodetectors (Peng et al., 2021; Fang et al., 2020; Shang et al., 2020). Among these materials, ultrathin 2D substances like graphene and transition metal dichalcogenides (TMDCs) have garnered significant attention due to their unique physical and chemical properties, making them

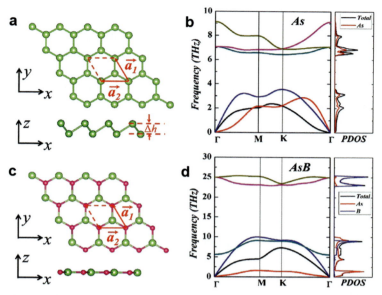

Fig. 4 Optimized structures of (A) pure arsenene and (C) arsenene doped with boron are depicted in both top and side views. The figures also include coordinate axes (x, y, z) and lattice vectors for reference. On the right side, the corresponding phonon spectra and phonon density of states (PDOS) are presented in (B) and (D), respectively (Zhang et al., 2019).

promising candidates for electronics and optoelectronics (Mawlong and Ahn, 2022; Özdemir et al., 2019). 2D materials with asymmetric lattices possess distinct periodic atomic arrangements and density variations within their planes. This results in highly asymmetric band structures and strong in-plane anisotropy, affecting properties such as Raman and photoluminescence spectra, electrical and thermal conductivity, and photoelectric response (Su et al., 2021; Hess, 2021). These anisotropic characteristics make 2D materials excellent candidates for polarization devices, particularly polarized photodetectors. However, the limited number of 2D materials suitable for infrared polarized photodetectors has been a challenge due to their intrinsic band gaps.

In the research conducted by Chai and colleagues, they employed a straightforward chemical vapor deposition method to synthesize 2D nanoplate structures of ZnSb, a compound belonging to the II-V group with a non-layered configuration (Chai et al., 2020). Unlike other non-layered 2D materials, ZnSb nanoplates possess an orthorhombic structure and are formed through covalent bonding. This unique structure is expected to give rise to distinct optical and electronic properties not typically observed in conventional 2D layered materials.

The as-synthesized ZnSb nanoplates exhibited robust in-plane electrical and optical anisotropy, owing to their low-symmetry crystal structure. Leveraging these ZnSb nanoplates as the sensing components, the researchers developed polarized infrared photodetectors, showcasing exceptional optoelectronic performance across a wide spectral range. Moreover, due to the asymmetric lattice structure of ZnSb, the photodetectors demonstrated in-plane angle-resolved conductance and anisotropic photocurrent when exposed to incident polarized laser light, with anisotropic factors reaching values of 1.58 and 1.28. The band structure calculations for orthorhombic ZnSb revealed an indirect band gap of 0.46 eV between the Z and Y points in the Brillouin zone, making it suitable for mid-infrared photodetection. The band structure further exhibited highly anisotropic characteristics, with varying slopes and band curvatures near the bottom of the conduction band and the top of the valence band, as shown in Fig. 5. Analysis of the density of states (DOS) indicated that the conduction bands predominantly originated from Zn s, p, and Sb s, p states, while the valence bands were primarily formed by Zn-p and Sb-p hybridized states. The fabricated photodetector device demonstrated a broad spectral response, excellent stability in ambient conditions, high specific detectivity, and responsivity at room temperature.

Additionally, theoretical calculations provided insights into the origin of the material's anisotropic properties, corroborating experimental findings. The angle-resolved electronic measurements and incident polarization photoelectric measurements confirmed the material's high anisotropic conductivity and linear polarization sensitivity.

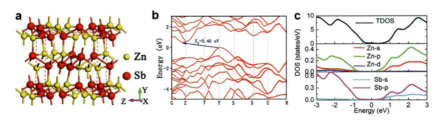

Fig. 5 (A) The depiction of the crystal structure of non-layered ZnSb in its orthorhombic form. (B) The presentation of ZnSb electronic energy band structure. (C) The graphical representation of the total and partial electron density of states for ZnSb (Chai et al., 2020).

This study opens up opportunities for the synthesis of non-layered 2D compounds belonging to group II-V elements, paving the way for their application in infrared polarization photodetectors (Takagi et al., 2019; Toko and Suemasu, 2020).

7. Te-based non-layered 2D materials

Van der Waals epitaxy (vdWE) has recently emerged as a straightforward method for synthesizing ultrathin two-dimensional (2D) layered and non-layered materials, as well as their vertical hetero-structures (Zhang et al., 2022; Yu et al., 2020; Manna et al., 2018; Neumayer et al., 2020; Mitra and Manju, 2023). In contrast to traditional heteroepitaxy, vdWE relies on substrates with chemically inert surfaces, lacking surface dangling bonds like fluorophlogopite mica (Chang et al., 2022). In vdWE growth, the connection between the overlayer and substrate primarily involves weak van der Waals interactions instead of strong chemical bonds. Consequently, vdWE eliminates the stringent requirement for lattice matching, enabling the growth of defect-free overlayers with different crystalline symmetries than the substrate. Moreover, vdWE allows the overlayer to remain perfectly relaxed, avoiding excessive strain at the heterointerface. These advantages make vdWE a powerful technique for producing various 2D layered materials with exceptionally high crystalline quality.

Chalcogens, specifically selenium (Se) and tellurium (Te), are p-type narrow band gap elemental semiconductors (with Se having a band gap of 1.6 eV and Te of 0.34 eV) (Scheiner, 2021; Huang et al., 2021). Chalcogens possess a highly anisotropic crystal structure, with Se and Te atoms initially forming helical chains along the c-axis through covalent bonding. These helical chains can be easily arranged into a hexagonal crystal structure via van der Waals interactions. Chalcogens are of significant interest due to their remarkable physical and chemical properties, including high photoconductivity, thermal conductivity anisotropy, hydration, oxidation catalytic activity, high piezoelectricity, thermoelectricity, and nonlinear optical responses (Kim et al., 2021; Mohl et al., 2020; Buffiere et al., 2019; Gao and Chelikowsky, 2020). Various one-dimensional (1D) chalcogen nanostructures have been successfully synthesized using methods such as microwave-assisted synthesis in ionic liquids, vapor phase growth, and solution-phase approaches (Pal et al., 2023; Jha and Bhat, 2020).

Wang et al. conducted research wherein they demonstrated the vapor-phase deposition of 2D hexagonal tellurium (Te) nanoplates on flexible and transparent fluorophlogopite mica sheets using van der Waals epitaxy (vdWE) (Wang et al., 2014). The resulting 2D hexagonal Te nanoplates exhibited highly single crystallinity, large lateral dimensions measuring 6–10 micrometers, and thin thickness ranging from 30 to 80 nanometers, as shown in Fig. 6. The absence of surface dangling bonds on the mica substrate allowed Te adatoms to readily migrate along the mica surface and quickly reach the growth sites. This rapid migration facilitated the lateral growth of 2D hexagonal Te nanoplates, eliminating the need for strict lattice matching, as compared to growth on Si substrate, where under the same experimental conditions, a multitude of irregular microcrystals formed. In particular, the researchers created a photodetector using a single 2D Te hexagonal nanoplate directly on a flexible mica growth substrate. The resulting device exhibited a high photoresponse, even after the flexible mica sheet was bent 100 times. While the primary focus of this work centered on 2D Te hexagonal nanoplates, the researchers suggested that the insights gained could be applied to explore 2D nanostructures of other

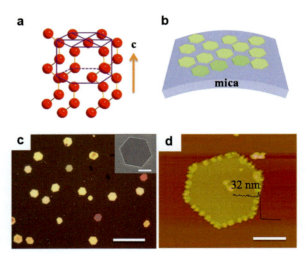

Fig. 6 (A) The crystal structure of Se/Te resembles helical chains. (B) A schematic representation of the 2D Te hexagonal nanoplates on flexible and transparent mica substrates using vdWE. (C) A representative optical microscope (OM) image displaying 2D Te hexagonal nanoplates. Inset: Scanning electron microscope (SEM) image of an individual 2D Te hexagonal nanoplate, with a scale bar of 4 μm. (D) Atomic force microscopy (AFM) image of a Te hexagonal nanoplate with a thickness of 32 nm (Wang et al., 2014).

non-layered materials, such as selenium (Se) and Se/Te alloy, which share the same crystal structure as Te. Additionally, the findings may open avenues for utilizing 2D Te hexagonal nanoplates as templates for generating various Te-based 2D functional nanomaterials like zinc telluride (ZnTe) and cadmium telluride (CdTe) through solution-phase methods (Gorai et al., 2023; Tarasov et al., 2021; Wei et al., 2019).

Furthermore, Hu et al. have successfully synthesized hexagonal tellurium (Te) nanosheets through a rapid solution-based method (Hu et al., 2021). Despite Te's natural inclination to form one-dimensional nanostructures, the introduction of polyvinylpyrrolidone (PVP) as a surfactant, along with a swift reaction process, significantly enhanced the lateral growth rate of these non-layered nanosheets. The growth rate, size, and quality of Te nanosheets were also influenced by factors such as reaction temperature, the quantities of sodium hydrogen telluride (NaHTe) and PVP. Consequently, they efficiently produced stable hexagonal Te nanosheets characterized by substantial dimensions (ranging from 5 to 11 μm) and minimal thickness (ranging from 50 to 170 nm) in an ethanol solution, employing 620 mg of PVP and 0.3 mL of NaHTe at 50 °C.

These Te nanosheets displayed a robust light absorption range spanning from 350 to 1100 nm in UV–vis–NIR spectrophotometry. Furthermore, they exhibited commendable photoresponse and long-term stability when employed as a photoanode material in a photoelectrochemical (PEC)-type photodetector under simulated light. Notably, the nanosheets achieved a substantial modulation depth (29%) in Z-scan experiments. In addition, they were effectively employed as a saturable absorber material in an erbium-doped fiber ring laser, successfully achieving mode-locking operation at 1550 nm. This research underscores the potential of hexagonal Te nanosheets in the development of optoelectronic and ultrafast photonic devices. Moreover, this swift and convenient solution-based approach holds promise for the synthesis of other non-layered 2D materials, including selenium (Se) and tellurium/selenium (Te/Se) alloys (Pal et al., 2023; Yang et al., 2019; Yang and Lee, 2020).

8. Conclusions and future perspectives

2D non-layered metalloid-based materials offer unique physical properties, flexibility, and compatibility with Si-based electronic devices, making them promising candidates for applications in optoelectronics, catalysis, energy conversion and storage, and topological crystalline insulators. While the past decade has witnessed extensive research on 2D layered materials, including metallic, semi-metallic, semiconducting, and insulating materials, the attention has increasingly shifted towards 2D non-layered metalloid-based materials. This comprehensive commentary and review, in the form of a book chapter, provides an overview of recent advancements in the field, encompassing synthesis methods, properties, and applications. Synthesis control is a pivotal initial step in working with novel materials. Dry synthesis techniques such as chemical vapor deposition (CVD) and van der Waals epitaxy (vdWE) have proven effective in producing 2D non-layered materials characterized by high crystalline quality, contamination-free composition, and ultrathin thickness, often below 10 nm. Nevertheless, achieving large-scale growth of 2D materials with precise control over thickness, especially reducing it to a few atoms and obtaining uniform, wafer-scale 2D non-layered materials, presents a significant challenge. Moreover, exploring methods for integrating 2D materials with Si-complementary metal-oxide-semiconductor (CMOS) devices, such as in situ growth on SiO_2/Si substrates via a vdWE buffer layer, remains unexplored and requires further investigation. On the other hand, wet chemical techniques like solution-based exfoliation, transfer and deposition methods, chemical solution synthesis, and chemical transformation methods offer better thickness control and scalability, making them suitable for applications in catalysis, energy conversion, and storage. Nonetheless, new synthesis approaches are needed to produce larger-sized 2D non-layered materials, and there is a multitude of ultrathin materials that remain unexplored.

The field of 2D non-layered materials is still in its early stages, with much of the focus placed on materials synthesis. However, there is a growing need to shift attention towards functional devices. Consequently, the challenge lies in designing and realizing functional devices based on these materials. Furthermore, leveraging the 2D configuration, combining layered and non-layered materials, and developing hybrid devices may unlock new capabilities. Given the intriguing nature of 2D geometry, particularly in ultrathin non-layered materials, there is much anticipation for the surprises they may bring to the field.

References

Ajayan, J., Nirmal, D., Mohankumar, P., Kuriyan, D., Fletcher, A.S.A., Arivazhagan, L., et al., 2019. GaAs metamorphic high electron mobility transistors for future deep space-biomedical–millitary and communication system applications: a review. Microelectron. J. 92, 104604. https://doi.org/10.1016/j.mejo.2019.104604.

Akinwande, D., Brennan, C.J., Bunch, J.S., Egberts, P., Felts, J.R., Gao, H., et al., 2017. A review on mechanics and mechanical properties of 2D materials—graphene and beyond. Extreme Mech. Lett. 13, 42–77. https://doi.org/10.1016/j.eml.2017.01.008.

Androulidakis, C., Zhang, K., Robertson, M., Tawfick, S., 2018. Tailoring the mechanical properties of 2D materials and heterostructures. 2D Mater. 5, 032005. https://doi.org/10.1088/2053-1583/aac764.

Backes, C., Higgins, T.M., Kelly, A., Boland, C., Harvey, A., Hanlon, D., et al., 2017. Guidelines for exfoliation, characterization and processing of layered materials produced by liquid exfoliation. Chem. Mater. 29, 243–255. https://doi.org/10.1021/acs.chemmater.6b03335.

Bianco, E., Butler, S., Jiang, S., Restrepo, O.D., Windl, W., Goldberger, J.E., 2013. Stability and exfoliation of germanane: a germanium graphane analogue. ACS Nano 7, 4414–4421. https://doi.org/10.1021/nn4009406.

Buffiere, M., Dhawale, D.S., El-Mellouhi, F., 2019. Chalcogenide materials and derivatives for photovoltaic applications. Energy Technol. 7. https://doi.org/10.1002/ente.201900819.

Cahangirov, S., Sahin, H., Le Lay, G., Rubio, A., 2017. Germanene, Stanene and Other 2D Materials. pp. 63–85. ⟨https://doi.org/10.1007/978-3-319-46572-2_5⟩.

Cai, R., Wang, Y., Wang, J., Zhang, J., Yu, C., Qin, Y., et al., 2023. Accelerated hydrogen production on atomically thin silicon nanosheets photocatalyst with unique surface adsorption chemistry. Int. J. Hydrog. Energy. https://doi.org/10.1016/j.ijhydene.2023.09.064.

Camacho-Mojica, D.C., López-Urías, F., 2015. GaN haeckelite single-layered nanostructures: monolayer and nanotubes. Sci. Rep. 5, 17902. https://doi.org/10.1038/srep17902.

Chai, R., Chen, Y., Zhong, M., Yang, H., Yan, F., Peng, M., et al., 2020. Non-layered ZnSb nanoplates for room temperature infrared polarized photodetectors. J. 8, 6388–6395. https://doi.org/10.1039/d0tc00755b.

Chang, Y., Sun, X., Ma, M., Mu, C., Li, P., Li, L., et al., 2020. Application of hard ceramic materials B4C in energy storage: design B4C@C core-shell nanoparticles as electrodes for flexible all-solid-state micro-supercapacitors with ultrahigh cyclability. Nano Energy 75, 104947. https://doi.org/10.1016/j.nanoen.2020.104947.

Chang, Y.-W., Yang, W.-C., Lo, W.-R., Lo, Z.-X., Ma, C.-H., Chu, Y.-H., et al., 2022. Direct growth of flexible GaN film via van der Waals epitaxy on mica. Mater. Today Chem. 26, 101243. https://doi.org/10.1016/j.mtchem.2022.101243.

Cinquanta, E., Scalise, E., Chiappe, D., Grazianetti, C., van den Broek, B., Houssa, M., et al., 2013. Getting through the nature of silicene: An sp2–sp3 two-dimensional silicon nanosheet. J. Phys. Chem. C 117, 16719–16724. https://doi.org/10.1021/jp405642g.

Dai, Z., Liu, L., Zhang, Z., 2019. Strain engineering of 2D materials: issues and opportunities at the interface. Adv. Mater. 31. https://doi.org/10.1002/adma.201805417.

Das, S., Kim, M., Lee, J., Choi, W., 2014. Synthesis, properties, and applications of 2-D materials: a comprehensive review. Crit. Rev. Solid. State Mater. Sci. 39, 231–252. https://doi.org/10.1080/10408436.2013.836075.

Das, S., Pandey, D., Thomas, J., Roy, T., 2019. The role of graphene and other 2D materials in solar photovoltaics. Adv. Mater. 31. https://doi.org/10.1002/adma.201802722.

Dávila, M.E., Xian, L., Cahangirov, S., Rubio, A., Le Lay, G., 2014. Germanene: A novel two-dimensional germanium allotrope akin to graphene and silicene. N. J. Phys. 16. https://doi.org/10.1088/1367-2630/16/9/095002.

Fang, J., Zhou, Z., Xiao, M., Lou, Z., Wei, Z., Shen, G., 2020. Recent advances in low-dimensional semiconductor nanomaterials and their applications in high-performance photodetectors. InfoMat 2, 291–317. https://doi.org/10.1002/inf2.12067.

Fiederle, M., Procz, S., Hamann, E., Fauler, A., Fröjdh, C., 2020. Overview of GaAs und CdTe pixel detectors using medipix electronics. Cryst. Res. Technol. 55. https://doi.org/10.1002/crat.202000021.

Fortin-Deschênes, M., Moutanabbir, O., 2018. Recovering the semiconductor properties of the epitaxial group V 2D materials antimonene and arsenene. J. Phys. Chem. C 122, 9162–9168. https://doi.org/10.1021/acs.jpcc.8b00044.

Gablech, I., Pekárek, J., Klempa, J., Svatoš, V., Sajedi-Moghaddam, A., Neužil, P., et al., 2018. Monoelemental 2D materials-based field effect transistors for sensing and biosensing: phosphorene, antimonene, arsenene, silicene, and germanene go beyond graphene. TrAC. Trends Anal. Chem. 105, 251–262. https://doi.org/10.1016/j.trac.2018.05.008.

Gao, W., Chelikowsky, J.R., 2020. Prediction of intrinsic ferroelectricity and large piezoelectricity in monolayer arsenic chalcogenides.

Gibertini, M., Koperski, M., Morpurgo, A.F., Novoselov, K.S., 2019. Magnetic 2D materials and heterostructures. Nat. Nanotechnol. 14, 408–419. https://doi.org/10.1038/s41565-019-0438-6.

Gorai, P., Krasikov, D., Grover, S., Xiong, G., Metzger, W.K., Stevanović, V., 2023. A search for new back contacts for CdTe solar cells. Sci. Adv. 9. https://doi.org/10.1126/sciadv.ade3761.

Guo, Y., Gupta, A., Gilliam, M.S., Debnath, A., Yousaf, A., Saha, S., et al., 2021. Exfoliation of boron carbide into ultrathin nanosheets. Nanoscale 13, 1652–1662. https://doi.org/10.1039/d0nr07971e.

Guo, S., Yang, D., Wang, D., Fang, X., Fang, D., Chu, X., et al., 2022. Response improvement of GaAs two-dimensional non-layered sheet photodetector through sulfur passivation and plasma treatment. Vacuum 197. https://doi.org/10.1016/j.vacuum.2021.110792.

Gupta, A., Biswas, T., Singh, A.K., 2022. Anomalous stability of non-van der Waals bonded B4C nanosheets through surface reconstruction. J. Appl. Phys. 132. https://doi.org/10.1063/5.0123687.

Gupta, A., Sakthivel, T., Seal, S., 2015. Recent development in 2D materials beyond graphene. Prog. Mater. Sci. 73, 44–126. https://doi.org/10.1016/j.pmatsci.2015.02.002.

Hess, P., 2021. Bonding, structure, and mechanical stability of 2D materials: the predictive power of the periodic table. Nanoscale Horiz. 6, 856–892. https://doi.org/10.1039/D1NH00113B.

Houssa, M., Pourtois, G., Afanas'ev, V.V., Stesmans, A., 2010. Electronic properties of two-dimensional hexagonal germanium. Appl. Phys. Lett. 96. https://doi.org/10.1063/1.3332588.

Huang, X., Sun, J., Wang, L., Tong, X., Dou, S.X., Wang, Z.M., 2021. Advanced high-performance potassium–chalcogen (S, Se, Te) batteries. Small 17. https://doi.org/10.1002/smll.202004369.

Hussain, N., Yisen, Y., Ur Rehman Sagar, R., Anwar, T., Murtaza, M., Huang, K., et al., 2021. Quantum-confined blue photoemission in strain-engineered few-atomic-layer 2D germanium. Nano Energy 83. https://doi.org/10.1016/j.nanoen.2021.105790.

Hu, H., Zeng, Y., Gao, S., Wang, R., Zhao, J., You, K., et al., 2021. Fast solution method to prepare hexagonal tellurium nanosheets for optoelectronic and ultrafast photonic applications. J. Mater. Chem. C Mater. 9, 508–516. https://doi.org/10.1039/d0tc04106h.

Jha, R.K., Bhat, N., 2020. Recent progress in chemiresistive gas sensing technology based on molybdenum and tungsten chalcogenide nanostructures. Adv. Mater. Interfaces 7. https://doi.org/10.1002/admi.201901992.

Kaur, H., Tian, R., Roy, A., McCrystall, M., Horvath, D.V., Lozano Onrubia, G., et al., 2020. Production of quasi-2D platelets of nonlayered iron pyrite (FeS2) by liquid-phase exfoliation for high performance battery electrodes. ACS Nano 14, 13418–13432. https://doi.org/10.1021/acsnano.0c05292.

Kim, Y., Woo, W.J., Kim, D., Lee, S., Chung, S., Park, J., et al., 2021. Atomic-layer-deposition-based 2D transition metal chalcogenides: synthesis, modulation, and applications. Adv. Mater. 33. https://doi.org/10.1002/adma.202005907.

Lalmi, B., Oughaddou, H., Enriquez, H., Kara, A., Vizzini, S., Ealet, B., et al., 2010. Epitaxial growth of a silicene sheet. Appl. Phys. Lett. 97. https://doi.org/10.1063/1.3524215.

Lee, C.H., Krishnamoorthy, S., O'Hara, D.J., Brenner, M.R., Johnson, J.M., Jamison, J.S., et al., 2017. Molecular beam epitaxy of 2D-layered gallium selenide on GaN substrates. J. Appl. Phys. 121. https://doi.org/10.1063/1.4977697.

Lilly, M.P., Reno, J.L., Simmons, J.A., Spielman, I.B., Eisenstein, J.P., Pfeiffer, L.N., et al., 2003. Resistivity of dilute 2D electrons in an undoped GaAs heterostructure. Phys. Rev. Lett. 90, 056806. https://doi.org/10.1103/PhysRevLett.90.056806.

Liu, N., Bo, G., Liu, Y., Xu, X., Du, Y., Dou, S.X., 2019. Recent progress on germanene and functionalized germanene: preparation, characterizations, applications, and challenges. Small 15. https://doi.org/10.1002/smll.201805147.

Manna, S., Gorai, P., Brennecka, G.L., Ciobanu, C.V., Stevanović, V., 2018. Large piezoelectric response of van der Waals layered solids. J. Mater. Chem. C Mater. 6, 11035–11044. https://doi.org/10.1039/C8TC02560F.

Mannix, A.J., Kiraly, B., Hersam, M.C., Guisinger, N.P., 2017. Synthesis and chemistry of elemental 2D materials. Nat. Rev. Chem. 1, 0014. https://doi.org/10.1038/s41570-016-0014.

Mas-Ballesté, R., Gómez-Navarro, C., Gómez-Herrero, J., Zamora, F., 2011. 2D materials: to graphene and beyond. Nanoscale 3, 20–30. https://doi.org/10.1039/C0NR00323A.

Masoumi, Z., Tayebi, M., Lee, B.-K., 2021. Ultrasonication-assisted liquid-phase exfoliation enhances photoelectrochemical performance in α-Fe2O3/MoS2 photoanode. Ultrason. Sonochem. 72, 105403. https://doi.org/10.1016/j.ultsonch.2020.105403.

Mawlong, L.P.L., Ahn, J.-H., 2022. 3D-structured photodetectors based on 2D transition-metal dichalcogenide. Small Struct. 3. https://doi.org/10.1002/sstr.202100149.

Ma, Q., Ren, G., Xu, K., Ou, J.Z., 2021. Tunable optical properties of 2D materials and their applications. Adv. Opt. Mater. 9. https://doi.org/10.1002/adom.202001313.

Meng, L., Wang, Y., Zhang, L., Du, S., Wu, R., Li, L., et al., 2013. Buckled silicene formation on Ir(111). Nano Lett. 13, 685–690. https://doi.org/10.1021/nl304347w.

Mitra, R., Manju, U., 2023. Negative capacitance and intrinsic ferroelectric behavior in α-MoO3 culminating as a robust piezoelectric energy harvester. ACS Appl. Electron. Mater. 5, 3130–3143. https://doi.org/10.1021/acsaelm.3c00203.

Mitra, R., Prusty, A., Manju, U., 2023. Ferroelectric perovskites as electro-optic switching devices, modulators and optical memory. Perovskite Metal OxidesElsevier, pp. 617–643. ⟨https://doi.org/10.1016/B978-0-323-99529-0.00022-9⟩.

Mohl, M., Rautio, A., Asres, G.A., Wasala, M., Patil, P.D., Talapatra, S., et al., 2020. 2D tungsten chalcogenides: synthesis, properties and applications. Adv. Mater. Interfaces 7. https://doi.org/10.1002/admi.202000002.

Neumayer, S.M., Tao, L., O'Hara, A., Susner, M.A., McGuire, M.A., Maksymovych, P., et al., 2020. The concept of negative capacitance in ionically conductive Van der Waals ferroelectrics. Adv. Energy Mater. 10, 2001726. https://doi.org/10.1002/aenm.202001726.

Novoselov, K.S., Mishchenko, A., Carvalho, A., Castro Neto, A.H., 2016. 2D materials and van der Waals heterostructures. Science 353, 1979. https://doi.org/10.1126/science.aac9439.

Ojalvo, C., Moreno, R., Guiberteau, F., Ortiz, A.L., 2020. Processing of orthotropic and isotropic superhard B4C composites reinforced with reduced graphene oxide. J. Eur. Ceram. Soc. 40, 3406–3413. https://doi.org/10.1016/j.jeurceramsoc.2020.02.027.

Özdemir, O., Ramiro, I., Gupta, S., Konstantatos, G., 2019. High sensitivity hybrid PbS CQD-TMDC photodetectors up to 2 μm. ACS Photonics 6, 2381–2386. https://doi.org/10.1021/acsphotonics.9b00870.

Pal, M.K., Karmakar, G., Shah, A.Y., Tyagi, A., Bhuvanesh, N., Dey, S., 2023. Accessing copper selenide nanostructures through a 1D coordination polymer of copper ($<$scp$>$ii$<$/scp$>$) with 4,4′-dipyridyldiselenide as a molecular precursor. N. J. Chem. 47, 16954–16963. https://doi.org/10.1039/D3NJ02062B.

Pan, Y., Abazari, R., Yao, J., Gao, J., 2021. Recent progress in 2D metal-organic framework photocatalysts: synthesis, photocatalytic mechanism and applications. J. Phys. Energy 3. https://doi.org/10.1088/2515-7655/abf721.

Peng, M., Ma, Y., Zhang, L., Cong, S., Hong, X., Gu, Y., et al., 2021. All-inorganic $CsPbBr_3$ perovskite nanocrystals/2D non-layered cadmium sulfide selenide for high-performance photodetectors by energy band alignment engineering. Adv. Funct. Mater. 31. https://doi.org/10.1002/adfm.202105051.

Priyadarshini, B.S., Mitra, R., Manju, U., 2023. Titania nanoparticle-stimulated ultralow frequency detection and high-pass filter behavior of a flexible piezoelectric nanogenerator: a self-sustaining energy harvester for active motion tracking. ACS Appl. Mater. Interfaces 15, 45812–45822. https://doi.org/10.1021/acsami.3c07413.

Schaibley, J.R., Yu, H., Clark, G., Rivera, P., Ross, J.S., Seyler, K.L., et al., 2016. Valleytronics in 2D materials. Nat. Rev. Mater. 1, 16055. https://doi.org/10.1038/natrevmats.2016.55.

Scheiner, S., 2021. Participation of S and Se in hydrogen and chalcogen bonds. CrystEngComm 23, 6821–6837. https://doi.org/10.1039/D1CE01046H.

Schulman, D.S., Arnold, A.J., Das, S., 2018. Contact engineering for 2D materials and devices. Chem. Soc. Rev. 47, 3037–3058. https://doi.org/10.1039/C7CS00828G.

Shang, H., Chen, H., Dai, M., Hu, Y., Gao, F., Yang, H., et al., 2020. A mixed-dimensional 1D Se–2D InSe van der Waals heterojunction for high responsivity self-powered photodetectors. Nanoscale Horiz. 5, 564–572. https://doi.org/10.1039/C9NH00705A.

Singh, A.K., Mathew, K., Zhuang, H.L., Hennig, R.G., 2015. Computational screening of 2D materials for photocatalysis. J. Phys. Chem. Lett. 6, 1087–1098. https://doi.org/10.1021/jz502646d.

Stone, K.W., Gundogdu, K., Turner, D.B., Li, X., Cundiff, S.T., Nelson, K.A., 2009. Two-quantum 2D FT electronic spectroscopy of biexcitons in GaAs quantum wells. Science 324 (1979), 1169–1173. https://doi.org/10.1126/science.1170274.

Su, Q., Li, Y., Hu, R., Song, F., Liu, S., Guo, C., et al., 2020. Heterojunction photocatalysts based on 2D materials: the role of configuration. Adv. Sustain. Syst. 4. https://doi.org/10.1002/adsu.202000130.

Su, S.-K., Chuu, C.-P., Li, M.-Y., Cheng, C.-C., Wong, H.-S.P., Li, L.-J., 2021. Layered semiconducting 2D materials for future transistor applications. Small Struct. 2. https://doi.org/10.1002/sstr.202000103.

Takagi, S., Kato, K., Takenaka, M., 2019. Group IV based bi-layer tunneling field effect transistor. ECS Trans. 93, 23–27. https://doi.org/10.1149/09301.0023ecst.

Tan, C., Zhang, H., 2015. Wet-chemical synthesis and applications of non-layer structured two-dimensional nanomaterials. Nat. Commun. 6. https://doi.org/10.1038/ncomms8873.

Tarasov, A.S., Mikhailov, N.N., Dvoretsky, S.A., Menshchikov, R.V., Uzhakov, I.N., Kozhukhov, A.S., et al., 2021. Preparation of atomically clean and structurally ordered surfaces of epitaxial CdTe films for subsequent epitaxy. Semiconductors 55, S62–S66. https://doi.org/10.1134/S1063782621090220.

Toko, K., Suemasu, T., 2020. Metal-induced layer exchange of group IV materials. J. Phys. D Appl. Phys. 53, 373002. https://doi.org/10.1088/1361-6463/ab91ec.

Turunen, M., Brotons-Gisbert, M., Dai, Y., Wang, Y., Scerri, E., Bonato, C., et al., 2022. Quantum photonics with layered 2D materials. Nat. Rev. Phys. 4, 219–236. https://doi.org/10.1038/s42254-021-00408-0.

Vogt, P., De Padova, P., Quaresima, C., Avila, J., Frantzeskakis, E., Asensio, M.C., et al., 2012. Silicene: compelling experimental evidence for graphenelike two-dimensional silicon. Phys. Rev. Lett. 108. https://doi.org/10.1103/PhysRevLett.108.155501.

Wang, Q., Safdar, M., Xu, K., Mirza, M., Wang, Z., He, J., 2014. Van der Waals epitaxy and photoresponse of hexagonal tellurium nanoplates on flexible mica sheets. ACS Nano 8, 7497–7505. https://doi.org/10.1021/nn5028104.

Wang, F., Wang, Z., Shifa, T.A., Wen, Y., Wang, F., Zhan, X., et al., 2017a. Two-dimensional non-layered materials: synthesis, properties and applications. Adv. Funct. Mater. 27. https://doi.org/10.1002/adfm.201603254.

Wang, Z., Błaszczyk, A., Fuhr, O., Heissler, S., Wöll, C., Mayor, M., 2017b. Molecular weaving via surface-templated epitaxy of crystalline coordination networks. Nat. Commun. 8, 14442. https://doi.org/10.1038/ncomms14442.

Wei, Y., Xie, P., Lei, H., Lu, Z., Liu, C., Zhang, B.-B., et al., 2019. Luminescence and optical properties of Fe2+:ZnTe crystal grown by temperature gradient solution method. J. Alloy. Compd. 805, 774–778. https://doi.org/10.1016/j.jallcom.2019.07.124.

Wei, B., Mao, X., Liu, W., Ji, C., Yang, G., Bao, Y., et al., 2023. Recent progress of surface plasmon–enhanced light trapping in GaAs thin-film solar cells. Plasmonics. https://doi.org/10.1007/s11468-023-01902-0.

Xie, K.Y., An, Q., Toksoy, M.F., McCauley, J.W., Haber, R.A., Goddard, W.A., et al., 2015. Atomic-level understanding of "asymmetric twins" in boron carbide. Phys. Rev. Lett. 115, 175501. https://doi.org/10.1103/PhysRevLett.115.175501.

Xu, M., Liang, T., Shi, M., Chen, H., 2013. Graphene-like two-dimensional materials. Chem. Rev. 113, 3766–3798. https://doi.org/10.1021/cr300263a.

Yang, P.-K., Lee, C.-P., 2020. 2D-layered nanomaterials for energy harvesting and sensing applications. Appl. Electromech. Dev. Mach. Electr. Mobil. Solut. 1–14. https://doi.org/10.5772/intechopen.85791.

Yang, Z., Liang, H., Wang, X., Ma, X., Zhang, T., Yang, Y., et al., 2016. Atom-thin SnS2–xSex with adjustable compositions by direct liquid exfoliation from single crystals. ACS Nano 10, 755–762. https://doi.org/10.1021/acsnano.5b05823.

Yang, M.Q., Dan, J., Pennycook, S.J., Lu, X., Zhu, H., Xu, Q.H., et al., 2017. Ultrathin nickel boron oxide nanosheets assembled vertically on graphene: a new hybrid 2D material for enhanced photo/electro-catalysis. Mater. Horiz. 4, 885–894. https://doi.org/10.1039/c7mh00314e.

Yang, M., Zhu, H., Yi, W., Li, S., Hu, M., Hu, Q., et al., 2019. Electrical transport and thermoelectric properties of Te–Se solid solutions. Phys. Lett. A 383, 2615–2620. https://doi.org/10.1016/j.physleta.2019.05.019.

You, J.W., Bongu, S.R., Bao, Q., Panoiu, N.C., 2018. Nonlinear optical properties and applications of 2D materials: theoretical and experimental aspects. Nanophotonics 8, 63–97. https://doi.org/10.1515/nanoph-2018-0106.

Yu, J., Wang, L., Hao, Z., Luo, Y., Sun, C., Wang, J., et al., 2020. Van der Waals epitaxy of III-nitride semiconductors based on 2D materials for flexible applications. Adv. Mater. 32. https://doi.org/10.1002/adma.201903407.

Yu, X., Wang, X., Zhou, F., Qu, J., Song, J., 2021. 2D van der Waals heterojunction nanophotonic devices: from fabrication to performance. Adv. Funct. Mater. 31. https://doi.org/10.1002/adfm.202104260.

Zhang, J., Liu, H., Gao, Y., Xia, X., Huang, Z., 2019. The sp2 character of new two-dimensional AsB with tunable electronic properties predicted by theoretical studies. Phys. Chem. Chem. Phys. 21, 20981–20987. https://doi.org/10.1039/c9cp03385h.

Zhang, Z., Yang, X., Liu, K., Wang, R., 2022. Epitaxy of 2D materials toward single crystals. Adv. Sci. 9. https://doi.org/10.1002/advs.202105201.

Zheng, G., Xing, Z., Gao, X., Nie, C., Xu, Z., Ju, Z., 2021. Fabrication of 2D Cu-BDC MOF and its derived porous carbon as anode material for high-performance Li/K-ion batteries. Appl. Surf. Sci. 559, 149701. https://doi.org/10.1016/j.apsusc.2021.149701.

Zheng, Z., Yao, J., Li, J., Yang, G., 2020. Non-layered 2D materials toward advanced photoelectric devices: progress and prospects. Mater. Horiz. 7, 2185–2207. https://doi.org/10.1039/d0mh00599a.

Zhou, P., Xu, Q., Li, H., Wang, Y., Yan, B., Zhou, Y., et al., 2015. Fabrication of two-dimensional lateral heterostructures of WS2/WO3·H2O through selective oxidation of monolayer WS2. Angew. Chem. 127, 15441–15445. https://doi.org/10.1002/ange.201508216.

Zhou, N., Yang, R., Zhai, T., 2019. Two-dimensional non-layered materials. Mater. Today Nano 8. https://doi.org/10.1016/j.mtnano.2019.100051.

CHAPTER TWO

2D metal oxides

Vahid Khorramshahi* and Fatemeh Safari

Materials and Energy Research Center, Dezful Branch, Islamic Azad University, Dezful, Iran
*Corresponding author. e-mail address: va.khoramshahi@iau.ac.ir

Contents

1. Introduction	28
2. Structural characteristics of non-layered 2D metal oxides	29
3. Ultrathin sheets with various bonding patterns	29
4. Structural arrangements and their influence on properties	35
4.1 Thickness control	35
4.2 Vacancy manipulation	36
4.3 Elemental doping	37
5. Properties of non-layered 2D metal oxides	38
6. Fabrication approaches for non-layered 2D metal oxides	43
7. Liquid metal approach	44
8. Hydrothermal approach	48
9. Chemical vapor deposition (CVD) approach	50
10. Template-assisted approach	52
11. Challenges and future perspectives	54
References	56

Abstract

Two-dimensional metal oxides can be classified into two groups based on their crystal structure: layered and non-layered. Non-layered two-dimensional (2D) metal oxides have emerged as a fascinating class of materials with unique structures, properties, and diverse applications. This book chapter provides a comprehensive overview of the fundamental aspects of non-layered 2D metal oxides, including their structures, properties, fabrication techniques, and prospects. The chapter begins by discussing the structural characteristics of non-layered 2D metal oxides. Unlike traditional layered materials, these oxides exhibit atomically thin sheets with various bonding patterns. The absence of well-defined layers leads to intriguing structural arrangements, influencing their properties. The properties of non-layered 2D metal oxides are explored next. The synthesis of 2D non-layered metal oxides (MOXs) is a promising field with diverse applications and intriguing properties. Various synthesis strategies have evolved rapidly to address challenges in thickness control, cost, and quality. The liquid metal approach involves stripping oxide skin on metal surfaces to create ultrathin 2D MOXs. Bottom-up methods like hydrothermal, template-assisted, and chemical vapor deposition enable precise control over atom or molecule arrangement. Template-assisted synthesis offers control over morphology using templates. Nevertheless,

---

Semiconductors and Semimetals, Volume 113
ISSN 0080-8784, https://doi.org/10.1016/bs.semsem.2023.09.011

Copyright © 2023 Elsevier Inc.
All rights reserved.

scalability, stability, and cost-effectiveness remain challenges. Future research directions involve theoretical calculations, composite materials, and exploring diverse applications to unleash the full potential of 2D non-layered metal oxides.

1. Introduction

Over the past few years, there has been a remarkable surge in research interest surrounding two-dimensional (2D) nanomaterials, and they have now assumed a pivotal role in driving advancements in materials and their properties. Numerous layered materials have undergone extensive study, including (1) Group IVA elements like graphene (Novoselov et al., 2004), silicene (Benasutti et al., 2012), and germanene (Ni et al., 2012), (2) black phosphorene (Neal et al., 2014), (3) h-BN (Dean et al., 2010), (4) transition metal dichalcogenides such as MoS_2, $MoSe_2$, $MoTe_2$, WS_2, WSe_2, and WTe_2 (Tabatabaei et al., 2020), (5) III–VI compounds like In_2Se_3, GaS, GaSe, and GaTe (Liu et al., 2020; Liu et al., 2019), (6) V–VI compounds like Bi_2Se_3 and Bi_2Te_3 (Butler et al., 2013; Wang et al., 2015), and (7) metal oxides (MOXs) (Mahmood et al., 2019).

With a thickness of merely one or a few atomic layers, these materials exhibit extraordinary optical, electrical, chemical, and mechanical properties. Consequently, they hold immense potential for a wide range of applications, including photovoltaics, catalysts, sensors, and thermoelectrics. The fascinating properties exhibited by these 2D nanomaterials have opened up exciting avenues for innovation and practical use in various fields.

Two-dimensional metal oxides (MOXs) have garnered significant global attention due to their exceptional performance in a wide range of applications. These applications include electronics, optoelectronics, electrochemistry, sensors, energy storage, catalysis, and more. As a result, there has been extensive research and interest in these materials worldwide. In general, 2D MOXs can be classified into two groups based on their crystal structure: layered and non-layered. Non-layered two-dimensional (2D) MOXs have emerged as a fascinating class of materials with unique structures, properties, and diverse applications. This book chapter provides a comprehensive overview of the fundamental aspects of non-layered 2D MOXs, including their structures, properties, fabrication techniques, and their prospects. The chapter begins by discussing the structural characteristics of non-layered 2D MOXs. Unlike traditional layered materials, these oxides exhibit atomically thin sheets with various bonding patterns. The absence of well-defined layers leads to intriguing structural arrangements, influencing their properties. The properties of non-layered 2D MOXs are explored next. These materials display exceptional electrical conductivity, distinct optical properties, high surface area, and remarkable catalytic activity. The chapter delves into the underlying mechanisms behind these properties, highlighting their significance in electronics, energy storage, catalysis, and sensing applications. Fabrication techniques for non-layered 2D MOXs are then presented. Various methods such as chemical vapor deposition, liquid metal, and template-assisted approaches are discussed, providing insights into the synthesis and control of the 2D structures. Finally, we will discuss challenges and future prospectives.

2. Structural characteristics of non-layered 2D metal oxides

In this section, we explore metals transforming into oxides in the form of 2D structures. The transformation of a metal surface into an oxide phase to create a metal-oxide hybrid material, exhibiting unique properties, has captured the attention of researchers in their quest to explore this material in its ultrathin 2D configuration. The crystal structure of these MOXs can vary from cubic to triclinic symmetry, adding to their intriguing nature. Several advanced computational techniques are employed to gain precise insights into the electronic density changes occurring in these 2D MOX structures. Notably, density functional theory (DFT) and molecular dynamics (MD) simulations play crucial roles in making accurate predictions, aiding researchers in better understanding the characteristics and behavior of these fascinating materials (Gangopadhyay et al., 2014; Hoang and Johannes, 2014; Richter et al., 2013).

3. Ultrathin sheets with various bonding patterns

Bulk NiO is widely recognized as a Mott insulator, characterized by an odd number of electrons and an antiferromagnetic spin arrangement. Its crystal structure follows a rock salt pattern with the cubic $F\bar{m}3m$ space group, featuring octahedral Ni^{2+} and O^{2-} sites, as shown in Fig. 1. This material acts as a p-type wide band gap semiconductor, with a band gap ranging from approximately 3.6 to 4 eV. Interestingly, its magnetic and

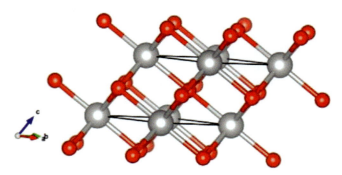

Fig. 1 Crystal structures of NiO. The Gray and red spheres are Ni, and O atoms, respectively (Kumbhakar et al., 2021). *Reprinted from Kumbhakar, P., Chowde Gowda, C., Mahapatra, P.L., Mukherjee, M., Malviya, K.D., Chaker, M., Chandra, A., Lahiri, B., Ajayan, P.M., Jariwala, D., Singh, A., Tiwary, C.S., 2021. Emerging 2D metal oxides and their applications. Mater. Today 45, 142–168, Copyright (2021), with permission from Elsevier.*

electrical properties exhibit notable variations depending on the size of the particles (Dubal et al., 2015).

CuO, another rare earth metal exhibiting antiferromagnetic spin ordering, has garnered significant interest in recent years (Majumdar and Ghosh, 2021). It stands as the most stable form of oxidized copper, mainly due to its unique characteristics as a p-type semiconductor with a relatively narrow band gap, ranging from 1.2 to 1.8 eV. The crystal structure of CuO can be found in both the cubic $F\bar{m}3m$ space group and the monoclinic C2/c arrangement (see Fig. 2a and b) (Pawar et al., 2017). The favorable state of 2D CuO was found to be an antiferromagnetic rectangular atomic structure in the ground state (Kvashnin et al., 2019).

Cobalt monoxide (CoO) is another fascinating example of a rock salt-type MOX (see Fig. 3), featuring two interpenetrating face-centered cubic (fcc) sublattices composed of Co^{2+} and O^{2-} ions. At the bulk level, CoO exhibits an antiferromagnetic character, with a Neel temperature occurring at approximately 298 K (Pawar et al., 2017).

Bulk TiO_2 is a straightforward inorganic compound that exists in four fundamental crystal forms. As a wide band-gap semiconductor, it possesses a bandgap of approximately 3 eV. Interestingly, when TiO_2 takes on the form of 2D nanosheets, the bandgap increases to about 3.65 eV, indicating a shift in its optical properties. The transformation from a 3D structure to a 2D configuration causes a reduction in thickness, leading to a "blue-shift"

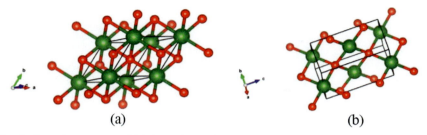

Fig. 2 Crystal structures of (a) the cubic and (b) the monoclinic of CuO. The Green and red spheres are Cu, and O atoms, respectively (Kumbhakar et al., 2021). *Reprinted from Kumbhakar, P., Chowde Gowda, C., Mahapatra, P.L., Mukherjee, M., Malviya, K.D., Chaker, M., Chandra, A., Lahiri, B., Ajayan, P.M., Jariwala, D., Singh, A., Tiwary, C.S., 2021. Emerging 2D metal oxides and their applications. Mater. Today 45, 142–168, Copyright (2021), with permission from Elsevier.*

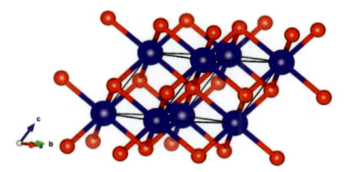

Fig. 3 Crystal structures of CoO. The blue and red spheres are Co, and O atoms, respectively (Kumbhakar et al., 2021). *Reprinted from Kumbhakar, P., Chowde Gowda, C., Mahapatra, P.L., Mukherjee, M., Malviya, K.D., Chaker, M., Chandra, A., Lahiri, B., Ajayan, P.M., Jariwala, D., Singh, A., Tiwary, C.S., 2021. Emerging 2D metal oxides and their applications. Mater. Today 45, 142–168, Copyright (2021), with permission from Elsevier.*

in the UV–Vis absorption spectrum of these ultrathin 2D nanosheets. Extensive research has been conducted on the phase control and phase transformation of anatase-to-rutile and brookite over the past few decades (Alkathiri et al., 2020; Hwang et al., 2023). Regarding its crystal structure, TiO_2 adopts various arrangements, such as the Tetragonal $I4_1/amd$ and $P4_2/mnm$ space groups, as presented in Fig. 4. These findings provide valuable insights into the properties and potential applications of TiO_2 in different forms and structures (Sun et al., 2014).

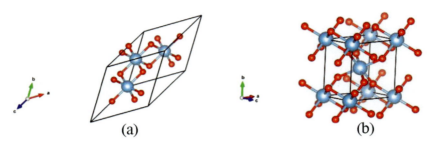

Fig. 4 Crystal structures of (a) the Tetragonal I41/amd and (b) P42/mnm of TiO$_2$. The light blue and red spheres are Ti, and O atoms, respectively (Kumbhakar et al., 2021). *Reprinted from Kumbhakar, P., Chowde Gowda, C., Mahapatra, P.L., Mukherjee, M., Malviya, K.D., Chaker, M., Chandra, A., Lahiri, B., Ajayan, P.M., Jariwala, D., Singh, A., Tiwary, C.S., 2021. Emerging 2D metal oxides and their applications. Mater. Today 45, 142–168, Copyright (2021), with permission from Elsevier.*

Zinc oxide (ZnO) adopts a Wurtzite crystal structure and arranges itself in the hexagonal P6$_3$mc space group (see Fig. 5). In this structure, Zn^{2+} ions are bonded to four equivalent O^{2-} atoms, forming corner-sharing ZnO$_4$ tetrahedra. These tetrahedra consist of three shorter Zn–O bond lengths (1.97 Å) and one longer Zn–O bond length (1.98 Å). On the other hand, O^{2-} ions are bonded to four equivalent Zn^{2+} atoms, creating corner-sharing OZn$_4$ tetrahedra. ZnO behaves as an n-type wide band-gap semiconductor with a bandgap of approximately 3.37 eV. What makes it particularly intriguing is its tunable properties, making it suitable for gas sensing applications and catalytic performances. The unique characteristics of ZnO open up exciting possibilities for various practical uses in these domains (Sun et al., 2014; Wang et al., 2016).

SnO$_2$ is one of the most extensively employed materials in commercial sensors. It adopts a Rutile crystal structure and crystallizes in the tetragonal P4$_2$/mnm space group. Within this structure, Sn^{4+} ions are bonded to six equivalent O^{2-} atoms, giving rise to a combination of corner-sharing and edge-sharing SnO$_6$ octahedra. Notably, the corner-sharing octahedral tilt angles measure 51°. The Sn–O bond lengths in this compound exhibit interesting variations. Specifically, there are two shorter Sn–O bond lengths measuring 2.06 Å and four longer Sn–O bond lengths measuring 2.07 Å. This unique arrangement and bonding configuration contribute to the remarkable properties that make SnO$_2$ an indispensable material in various commercial sensor applications (Li et al., 2017; Sun et al., 2013).

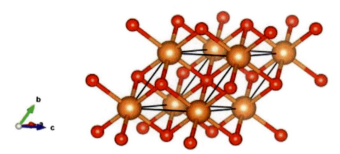

Fig. 5 Crystal structures of ZnO. The orange and red spheres are Zn, and O atoms, respectively (Kumbhakar et al., 2021). *Reprinted from Kumbhakar, P., Chowde Gowda, C., Mahapatra, P.L., Mukherjee, M., Malviya, K.D., Chaker, M., Chandra, A., Lahiri, B., Ajayan, P.M., Jariwala, D., Singh, A., Tiwary, C.S., 2021. Emerging 2D metal oxides and their applications. Mater. Today 45, 142–168, Copyright (2021), with permission from Elsevier.*

MnO_2 is a MOX that has garnered considerable attention in recent years due to its exceptional electrochemical properties. It possesses a Rutile-like structure and forms a tetragonal I4/m crystal arrangement (see Fig. 6). Within this structure, Mn^{4+} ions are bonded to six O^{2-} atoms, resulting in a combination of edge-sharing and corner-sharing MnO_6 octahedra. Interestingly, the corner-sharing octahedral tilt angles are measured at 50°. A noteworthy feature of MnO_2 is the variety of Mn–O bond distances, spanning from 1.88 to 1.91 Å. Moreover, there are two distinct O^{2-} sites within the compound. In the first O^{2-} site, O^{2-} forms a bond in a distorted trigonal planar geometry with three equivalent Mn^{4+} atoms. On the other hand, in the second O^{2-} site, O^{2-} is bonded in a distorted trigonal non-coplanar geometry with three equivalent Mn^{4+} atoms. This combination of unique structural features and bonding arrangements contributes to the exceptional properties exhibited by MnO_2, making it a promising material for various electrochemical applications (Sun et al., 2014; Zeng et al., 2021).

The metastable phase of VO_2 (B) features a monoclinic layered structure, similar to V_6O_{13}, and belongs to the C2/m (12) space group. At room temperature, the stable form is VO_2 (M), which exhibits a monoclinic structure with the P21/C (14) space group. However, upon undergoing a first-order transition, it transforms into the rutile structure of VO_2 (R) with the P42/mmm (136) space group. During the annealing process, the nanosheets experience an irreversible conversion to the VO_2 (R) phase.

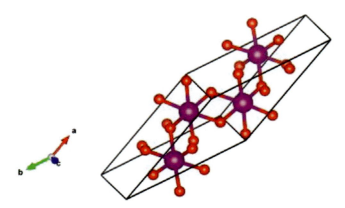

Fig. 6 Crystal structures of MnO$_2$. The purple and red spheres are Mn, and O atoms, respectively (Kumbhakar et al., 2021). *Reprinted from Kumbhakar, P., Chowde Gowda, C., Mahapatra, P.L., Mukherjee, M., Malviya, K.D., Chaker, M., Chandra, A., Lahiri, B., Ajayan, P.M., Jariwala, D., Singh, A., Tiwary, C.S., 2021. Emerging 2D metal oxides and their applications. Mater. Today 45, 142–168, Copyright (2021), with permission from Elsevier.*

Subsequently, when cooled back down to room temperature, the VO$_2$ (R) phase can be reversibly transformed back into the monoclinic VO$_2$ (M) phase. This transition between different phases and structures in VO$_2$ showcases its intriguing behavior, making it a promising material for various applications and further research (Morin, 1959; Wu et al., 2013).

Co$_3$O$_4$ has garnered significant attention among other MOXs due to its wide range of applications in energy storage, photocatalysis, and electrochromic devices (Kang et al., 2021; Zhong et al., 2022). It stands as one of the most stable cobalt oxide structures, exhibiting a normal spinel structure with cubic closed packing and an Fd3m space group (see Fig. 7). Bulk Co$_3$O$_4$ possesses antiferromagnetic spin ordering.

Complex structures like porous Co$_3$O$_4$ nanosheets have been observed to undergo thermal recrystallization, transforming into a single crystal framework, which enhances the performance of the electrode material. In the case of 2D structures of Co$_3$O$_4$, the hexagonal sheets exhibit mesopores resulting from the thermal decomposition reaction of Co(OH)$_2$ nanosheets. This structural transition from 3D to 2D leads to a larger exposed surface area, with more atoms located on the surface, thereby increasing the structural strain in the material. This phenomenon significantly impacts the electrochemical performance of Co$_3$O$_4$, further highlighting its potential for advanced applications in various fields (Zhan et al., 2009).

2D metal oxides

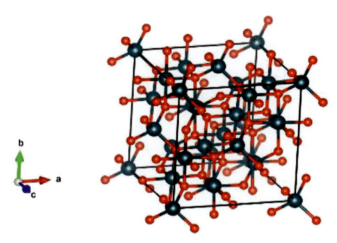

Fig. 7 Crystal structures of Co_3O_4. The dark green and red spheres are Co, and O atoms, respectively (Kumbhakar et al., 2021). *Reprinted from Kumbhakar, P., Chowde Gowda, C., Mahapatra, P.L., Mukherjee, M., Malviya, K.D., Chaker, M., Chandra, A., Lahiri, B., Ajayan, P.M., Jariwala, D., Singh, A., Tiwary, C.S., 2021. Emerging 2D metal oxides and their applications. Mater. Today 45, 142–168, Copyright (2021), with permission from Elsevier.*

4. Structural arrangements and their influence on properties

The remarkable benefits of 2D nanomaterials in photovoltaic, sensing, electrocatalysis, and optoelectronic applications stem from their exceptional morphology and atomic structures. In this section, we will explore the control of several crucial parameters in 2D non-layered materials, such as thickness, point defects (vacancies and doping), and heterogeneity, and examine their impact on electrocatalysis applications.

4.1 Thickness control

Controlling the thickness of materials at the nanometer scale is a highly effective method to manipulate their electronic structure and chemical reactivity. When a semiconductor material is reduced to just a few or even a single atomic layer, the bandgap widens due to the quantum confinement effect. This reduction in thickness also leads to a significant increase in the electronic density of states (DOS) at the material's surface compared to its bulk interior. Additionally, the thinner dimensions can induce surface lattice distortion and alter the electronic structure, including lowering the work function, thereby offering opportunities for tunable band alignment in electrocatalysis design.

Moreover, as the thickness decreases to the nanometer scale, the ratio of exposed surface atoms sharply rises, resulting in enhanced surface effects. The reduction in neighboring atoms leads to a surplus of low-coordination surface atoms with dangling bonds. These atoms are more reactive and tend to form bonds with other atoms or molecules, significantly improving the material's chemical activity. An illustrative example is the catalytic performance of 2D SnO_2 with varying thicknesses, where sub nanometer-thick 2D SnO_2 exhibited notably enhanced CO catalytic activity compared to thicker SnO_2 nanosheets and bulk SnO_2. The thinner SnO_2 displayed lower activation energy and a reduction in CO full-conversion temperature.

Controlling the thickness of 2D non-layered materials poses challenges, as they lack a significant driving force for anisotropic growth, unlike layered crystal structures. However, several strategies have been explored to achieve controlled thickness. For instance, a graphene oxide-templated synthesis strategy was employed to grow various binary oxides into 2D morphology with tunable thicknesses. Despite these efforts, achieving thickness down to a single atomic layer remained rare until the advent of ionic layer epitaxy (ILE). This cutting-edge technology enables precise unit-cell level thickness control and has proven effective for synthesizing promising electrocatalysts like ZnO and CoO. The self-limited thickness control offered by ILE opens up new possibilities for precisely tuning the thickness of 2D non-layered materials, facilitating more in-depth and quantitative studies on 2D electrocatalysis, such as exploring the direct relationship between the thickness of the Stern layer (Zn^{2+} concentrated zone) and the thickness of the 2D materials.

4.2 Vacancy manipulation

Cation and anion vacancies are well-known factors that play a crucial role in determining the physical and catalytic properties of materials. These vacancies influence various characteristics, such as the electronic structure, carrier concentration, electrical conductivity, and atom coordination. In 2D non-layered materials, vacancies are commonly found due to the intrinsic nature of their 3D crystal lattices, and their presence can significantly impact catalytic performance.

For instance, in 2D SnO_2, the Sn/O dual vacancies can transform into isolated Sn vacancies under a small electric field, leading to a reversible transition between a semiconductor and a half-metal (Li et al., 2017). This transition is accompanied by a drastic conductivity change of up to 10^3 times. DFT calculations further revealed that 2D SnO_2 with Sn/O dual vacancies exhibits semiconductive behavior, while isolated Sn vacancies induce a half-metallic characteristic, primarily originating from the O 2p state (Liu et al., 2018). Similarly, in 2D In_2O_3, the presence of O-vacancies increases the density of states (DOS) at the valence band edge and creates a new defect level in the forbidden band. This alteration in the electronic structure makes it easier for electrons to be excited into the conduction band, resulting in 2D In_2O_3 with O-vacancies having a higher carrier concentration compared to a perfect lattice.

Furthermore, the physicochemical properties of materials can be tailored by introducing and tuning vacancies. In bulk crystals, vacancies can be created and controlled through various methods, many of which can be adapted to 2D non-layered materials. Fast heating phase transformation is a powerful technique for engineering surface defects. For example, ultrathin $In(OH)_3$ can be transformed into 2D In_2O_3 with controlled O-vacancies by fast heating at 400 °C for 3 min, adjusting the oxygen partial pressure in the calcination atmosphere. The ILE technique is another versatile method for controlling defect evolution in quasi-2D crystal lattices. By introducing a water–oil interface, polycrystalline 2D ZnO with an unprecedented Zn vacancy concentration of approximately 33% can be synthesized. Stabilizing such a high Zn vacancy concentration is attributed to local charge balancing from the surfactants and fast growth kinetics in the ultrathin geometry. Additionally, plasma treatment has been shown to be an effective strategy for introducing surface vacancies. For example, Ar plasma treatment on 2D Co_3O_4 can partially reduce Co^{3+} to Co^{2+}, creating oxygen vacancies. The combination of surface oxygen vacancies and the high surface area of 2D Co_3O_4 leads to a substantial enhancement in electrocatalytic activity.

4.3 Elemental doping

Extrinsic point defects, which involve the introduction of elemental dopants, have the fascinating ability to bring about various physical and chemical changes in materials. These alterations include distortions in atomic arrangement, redistribution of electron density, an increase in delocalized electrons, and the exposure of more active sites. These changes present exciting opportunities to manipulate 2D materials and enhance their catalytic applications. Recently, confined doping within atomic layers has been highlighted as an effective approach to modulating the catalytic properties of 2D non-layered materials (Liu et al., 2021a). This technique strategically places dopant elements in the basal planes of the material while preserving the 2D atomic arrangement and electron-conjugated system. The result is an abundance of catalytically active sites, allowing for precise control over the electronic structure to optimize electrocatalytic dynamics.

In summary, this section discusses three crucial and distinctive structural factors in 2D non-layered materials: thickness, intrinsic defects (vacancies), and extrinsic defects (dopants). It explores how these factors are controlled and how they are fundamentally related to catalytic performance, shedding light on their potential in enhancing various catalytic applications.

5. Properties of non-layered 2D metal oxides

MOXs are widely used as catalysts for oxygen evolution reactions (OER) due to their stability and abundance of oxygen vacancies. In recent years, there has been extensive research on earth-rich transition MOXs (TMOs) like CoO (Wang et al., 2017), Co_3O_4 (Zhong et al., 2022), CuO (Kvashnin et al., 2019), NiO (Lin et al., 2018), La_2O_3 (Yan et al., 2020), and $NiCo_2O_4$ (Liu et al., 2020; Liu et al., 2019) as potential replacements for noble MOXs such as RuO_2 (Stoerzinger et al., 2017) and IrO_2 (Kuo et al., 2017). Among these, Co_3O_4 stands out for its exceptional performance and sustainability, being environmentally friendly and rich in reserves.

Researchers like Li et al. have successfully synthesized 2D Co_3O_4 through a surfactant-free cyanogel–$NaBH_4$ method (Li et al., 2018). This 2D Co_3O_4 material had a uniform thickness of approximately 1.5 nm and featured nanoscale pores, which contributed to its excellent OER performance. Compared to commercial RuO_2, the 2D Co_3O_4 showed a lower onset potential and overpotential, thanks to its high surface area and abundant defect sites at the pore edges, which facilitated mass transport.

To enhance the oxygen vacancy concentration, Bao et al. created bimetallic oxide ($NiCo_2O_4$) in a 2D morphology (Bao et al., 2015). The resulting 2D $NiCo_2O_4$ had a thickness of 1.6 nm and a rich presence of oxygen vacancies due to calcination in an oxygen-deficient atmosphere. These oxygen vacancies improved the reactivity of active sites and reduced

H_2O adsorption energy. The ultrathin structure also increased the number of active sites, thereby enhancing surface reactions and electrocatalytic performance.

Additionally, 2D MOXs like $NiCo_2O_4$ and $ZnCo_2O_4$ have been explored for oxygen reduction reactions (ORR) (Bao et al., 2018; Liu et al., 2020; Liu et al., 2019; Wang et al., 2021). These complex MOXs, with their spinel structure, offer higher ORR activity compared to binary MOXs. Liu et al. synthesized 2D $NiCo_2O_4$ with a thickness of 2 nm, which exhibited improved ORR electrocatalytic performance and performed comparably to commercial Pt/C in Zn–air batteries (Liu et al., 2020; Liu et al., 2019). The ultrathin structure and high concentration of oxygen vacancies contributed to the enhanced electrocatalytic activity.

Furthermore, 2D MOXs have been studied for CO_2 reduction electrocatalysis. Thinner 2D Co_3O_4 demonstrated higher CO_2 reduction activity compared to thicker versions, owing to a higher fraction of low-coordinated surface Co atoms. Oxygen vacancies in 2D ZnO also played a crucial role in promoting CO_2 activation and facilitating CO production.

NiO, when in its 2D form, exhibits distinct electronic and optical properties compared to its bulk counterpart, primarily due to the enhanced specific surface area and quantum confinement effect. Researchers, like Patil and his group, have delved into the impact of nanosized 2D NiO on charge-storing mechanisms. Their study demonstrated that the specific capacitance value increased to 746 F/g, which is three times larger than in bulk NiO. Thin-films of 2D NiO were obtained on indium–tin–oxide thin-films and proved to be highly suitable for supercapacitor applications due to their superior energy, power densities, and electrochemical stability (Patil et al., 2016).

In a different investigation, 2D nanosheets of NiO were arranged in a flower-like architecture through a solvothermal process, focusing on gas sensing performance. The goal was to reduce the operating temperature for gas sensors, as conventional bulk NiO gas sensors typically work above 200 °C. The sensor showed excellent results for formaldehyde gas sensing compared to other gases (San et al., 2017).

Furthermore, 2D NiO structures were created through laser ablation, demonstrating their potential as a plasmonic photocatalyst for solar H_2 evolution. The 2D effect in this case enhanced electron doping and suppressed carrier recombination, making it suitable for photocatalysis. The introduction of plasmonic effects provided active sites for light harvesting and boosted solar H_2 evolution (Lin et al., 2018).

Additionally, a thermal decomposition strategy was explored by Zheng et al. for fabricating 2D NiO porous nanosheets. These nanosheets showed improved lithium storage capability compared to microsized NiO polyhedrons, exhibiting high reversible capacity and rate capability (Zheng et al., 2019). The porous structure had a significant impact on electrochemical performance, offering a higher surface area for ions and promoting active interactions during intercalation/de-intercalation (Wang et al., 2010a).

Collectively, these observations from various 2D NiO experiments illustrate that reducing the material to a few layers enhances specific properties, such as specific capacitance value, low-temperature sensor operation, reduced carrier recombination in photocatalysis, and increased absorption in the visible range for solar photovoltaics applications. Overall, 2D NiO shows great promise for catalysis and semiconductor device applications, presenting exciting prospects for the future.

The antiferromagnetic ground state of 2D CuO is strikingly different from the paramagnetic behavior observed in bulk CuO. Moreover, 2D CuO displays a remarkable stiffness of $124\,N/m$ due to the strong covalent bonds between Cu and O in its 2D sheets (Kano et al., 2017). When we consider the above findings of 2D CuO, it becomes evident that while there are similarities in their overall applications compared to bulk CuO, the performance of 2D CuO is significantly enhanced on a broader scale. These 2D materials exhibit improved properties, such as enhanced temperature stability, increased optical band gap, and a change in magnetic state. As a result, these 2D materials are well-suited for electronic applications, as indicated by the study.

Similar to NiO and CuO, bulk ZnO is well-known for its excellent thermal and chemical stabilities. It functions as an n-type wide band gap semiconductor with properties that make it ideal for gas sensing and catalytic applications. When ZnO is reduced to its 2D form, several properties undergo modifications. For instance, 2D ZnO nanowalls, with a thickness of $60\,nm$, exhibit rapid response and recovery times, making them effective methane sensors. This is attributed to their high surface-to-volume ratio and n-type behavior, which leads to oxygen vacancy defects (Chen et al., 2013).

Recent research by Zhou et al. resulted in the creation of 2D ZnO nanosheets-regenerated cellulose (ZNSRC) photocatalytic composite thin films. These $80\,nm$ thick and $350–450\,nm$ transverse sized 2D ZNSRC showed remarkable photocatalytic activity and were effective in degrading organic dyes (Zhou et al., 2019).

In another study, 2D ZnO nanoplates were synthesized using an oxalic acid–assisted hydrothermal method. By adjusting calcination temperatures and additives, the morphology of ZnO can be easily controlled, resulting in uniform nanoparticles of less than 50 nm. ZnO demonstrated excellent sensing performance, particularly with selectivity towards triethylamine (TEA). Moreover, first-principles calculations indicated that 2D ZnO nanosheets, when supported by Au clusters, displayed high catalytic activity, showing promise for catalytic applications (Guo et al., 2014).

The low-dimensional structures of ZnO inherit the piezoelectricity from their bulk counterpart, especially with increasing thickness at about ten Zn–O double layers (Li et al., 2007). Furthermore, the substitution of Co atoms in a Zn site of ZnO introduces a Jahn–Teller distortion in the structure, resulting in a strong ferromagnetic coupling without any defect or doping. This opens up exciting possibilities for future device applications using 2D ZnO (Schmidt et al., 2010).

In summary, a comparison between bulk and 2D ZnO reveals improved catalytic activity in the 2D form, as well as increased sensitivity to pressure and gases, making it an efficient candidate for piezoelectric and gas sensor applications.

2D forms of cobalt monoxide (CoO) have garnered significant attention for their remarkable electron transport capability and reduced dimensions, making them highly valuable in electrochemical storage devices. Li et al. conducted a study on 2D CoO and demonstrated its effectiveness in lithium ion batteries. They achieved this by creating ultrathin and well-ordered nanosheet arrays using the galvanostatic electrodeposition technique. The porous and polycrystalline nature of CoO nanoarrays contributed to their enhanced functions and application performance (Li et al., 2014). Various research has highlighted the exceptional electrochemical performance of dimensionally reduced CoO compared to its bulk counterpart. This improvement is attributed to the enhanced ion transport capability and reduced resistance to ionic diffusion. Given these findings, it is evident that these 2D materials are well-suited for semiconductor device applications.

The preference for TiO_2 nanosheets as photoanodes is due to their high surface area, which allows for easy adsorption of dye molecules and increases the reaction interfaces between the semiconductor, dye, and electrolyte. These nanosheets promote efficient in-plane transport of electrons, as they possess non-equilibrium highly active surfaces, thereby enhancing the photochemical reactions. In solar cell applications, when used as an absorber layer, the TiO_2 nanosheets demonstrated solar conversion efficiencies of 6.63% and 8.28% for photoanodes with thicknesses of 4 mm and 8 mm, respectively (Sheng et al., 2017).

Functionalizing SnO_2 with carbon-based structures and incorporating additives have shown improved performance and higher sensitivity in sensing various gases. Researchers have explored various nanostructures to develop industrial-grade, high-performance sensors for detecting hydrogen gas at sub-ppm levels. Mesoporous 2D MOX structures have demonstrated better catalytic activities compared to their 3D/1D counterparts (Yin, Wang et al., 2019; Yin, Lv et al., 2019; Zhang et al., 2018). Wang et al. conducted research on highly porous SnO_2 nanometer-sized sheets, which exhibited an average pore diameter of 16.2 nm, a surface area of 180.3 m^2/g, and a pore volume of 1.028 cm^3/g (Wang et al., 2010b). However, SnO_2 experiences intrinsic volume changes due to pulverization, leading to rapid capacity deterioration. To address this issue, Zhu et al. synthesized 2D ultrathin SnO_2 nanosheets using ultra-rapid microwave-assisted solution-phase growth. The nanosheets demonstrated reversible capacities of 757 mA h/g (at 200 mA/g) and 571 mA h/g (at 300 mA/g) up to the 40th cycle (Zhu et al., 2015).

In gas sensing applications, SnO_2 nanosheets have proven favorable for detecting alkene gases compared to commercially available sensors. The electrical resistance of SnO_2 nanosheet gas sensors varies when exposed to different alkene gases at 50 ppm (Choi et al., 2019). The enhanced sensitivity of SnO_2 gas sensors is achieved when the grain size is less than twice the thickness of the surface charge layers, allowing the grain to be fully involved in the space-charge layer. By decreasing the depletion layer widths in adjacent grains, the tunneling probability through inter-grain barriers increases, leading to photoexcitation that affects the grain boundaries. The introduction of 2D materials has opened up new possibilities with a wide range of tunable conductive and electronic properties, making sensing a highly interesting area of research.

MnO_2 is a type of MOX with remarkable electrochemical properties, and in recent years, it has gained significant attention (Zeng et al., 2021). While Mn oxides are widely used in energy storage applications, they also show great potential in catalytic fields. The nanosheet structure of MnO_2 (manganese dioxide) exhibits strong electrocatalytic activity towards the hydrogen evolution reaction (HER) in alkaline environments. This activity is attributed to the presence of abundant oxygen vacancies and numerous active Mn^{3+} sites within the nanosheets (Bai et al., 2015).

In addition to its catalytic applications, 2D ultrathin sheets of MnO_2 have been utilized as electrode materials for supercapacitors and other energy storage devices. During the initial lithiation process, MnO_2 undergoes a 2D-confined conversion reaction, transforming into MnO and then forming a homogeneous layer of Mn dispersed between the bilayers of reduced graphene oxide (rGO). The performance of the material was tested for both lithium and sodium battery applications, and it exhibited excellent specific capacities. At a low current density of 0.1 A/g, the specific capacities achieved were 1325 mA h/g for lithium storage and 795 mA h/g for sodium storage. Even at a high current density of 12.8 A/g, the specific capacities were still impressive, measuring 370 mA h/g for lithium and 245 mA h/g for sodium storage. Moreover, the material demonstrated exceptional cyclability, with an ultra-low capacity decay rate of 0.004% and 0.0078% per cycle, respectively, for lithium and sodium storage, over the course of 5000 cycles (Kundu et al., 2013).

6. Fabrication approaches for non-layered 2D metal oxides

Due to the intriguing properties and potential applications of 2D MOXs, their synthesis procedures are rapidly evolving to cope with challenges such as thickness control, cost, and quality. The synthesis methods can be categorized into three main strategies: top-down, bottom-up, and liquid metals-assisted methods. The top-down approach in the synthesis of 2D materials involves starting with a larger bulk material and then selectively reducing its dimensions to obtain a thin, two-dimensional structure. Mechanical and liquid exfoliation are commonly used techniques in this approach. Mechanical exfoliation involves applying shear forces to separate thin layers from the bulk material. Liquid exfoliation, on the other hand, disperses the bulk material in a suitable solvent and subjects it to disruptive forces like sonication to obtain individual layers of desired thickness (Lee et al., 2019). This approach is often utilized for layered 2D MOXs. Conversely, the bottom-up approach involves building up the material from atomic or molecular components, leading to the formation of a thin, two-dimensional structure. The bottom-up approach allows precise control over the arrangement of atoms or molecules to create the desired

2D material. This approach is not limited to oxide crystal structures and offers promising prospects for the fabrication of various 2D materials with tailored properties (Li et al., 2022). Regarding the liquid metal approach, it is common for the oxide layer that develops on the liquid metal or alloy surface to be removed through techniques such as printing and other processes (Goff et al., 2021). In addition to liquid metal methods, this discussion will cover hydrothermal synthesis, template-assisted synthesis, and vapor-phase deposition synthesis as part of the bottom-up synthesis category.

7. Liquid metal approach

The formation of a self-limiting thin oxide layer at the metal–air interface is a common occurrence for most metals. Building on this principle, a technique for producing ultrathin 2D oxides was developed by stripping away the oxide skin from the metal (Daeneke et al., 2018). When the oxide skin forms on the surface of liquid metal, the interaction force between the liquid metal and the 2D oxide skin remains weak and localized due to the nonpolar nature of liquid metal (Akbari et al., 2021). This characteristic allows for the separation of the 2D oxide skin from the liquid metal, enabling the production of ultrathin layers. In recent times, the liquid metal method has garnered attention as a novel and efficient approach to producing 2D MOXs on a significant scale (Allioux et al., 2022; Cao et al., 2023; Petkov et al., 2023; Tang et al., 2022). This method not only overcomes obstacles related to cost-effectiveness and productivity but also exhibits remarkable maneuverability in tailoring the properties of the synthesized materials.

Mahmood and colleagues introduced an innovative approach for synthesizing ultrathin extra-large 2D sheets of non-layered functional materials. They employed a liquid metal-based route, showcasing the ability to achieve ZnO layers with remarkable thinness, down to the sub-nanometer scale (Mahmood et al., 2021). The experiment utilized pure zinc metal (99.8%, Roto Metals) that was melted on a heated glass slide at 500 °C within a controlled atmosphere. This temperature was chosen to facilitate the zinc melting process, which typically occurs at 420 °C, and to promote the interfacial reaction with ambient oxygen. As a result, a self-limiting Cabrera–Mott reaction took place, leading to the formation of an oxide layer on the outer surface of the liquid metal droplet.

To enable the synthesis of ZnO sheets, the oxide layer was removed by preconditioning the glass slides with the reflective Zn surface. Additionally, to create covalent interactions between the as-grown ZnO sheets and the substrate, a heated SiO_2/Si substrate was gently touched to the surface of the droplet. The researchers obtained sheets of varying thicknesses by conducting multiple consecutive exfoliations from the same area. Fig. 8a demonstrates the ZnO sheet exfoliation process, while Fig. 8b displays optical images of the synthesized ZnO nanosheets. Atomic force microscopy (AFM) analysis reveals that a single exfoliation step produces ultrathin wurtzite ZnO sheets, approximately one and a half unit-cell thick (3 Zn–O layers), as depicted in Fig. 8c–i. By extending the exfoliation time, these sheets can be made thicker, reaching around two and a half unit-cell thickness (5 Zn–O layers), as shown in Fig. 8c–ii. The TEM study on the prepared sample is also depicted in Fig. 8d–g.

Another approach to obtaining 2D MOX is through the blowing method, also known as the gas injection method, which shares similarities with the direct bonding technique. The main idea behind this method is that when oxygen bubbles come into contact with the liquid metal surface, the oxides on the surface will form bubbles and separate from the liquid metal. These oxides then appear in the upper layer of the solution due to their low buoyancy. The gas injection method is similar to the sticking method, which means adjusting the oxide type on the surface by modifying the molten metal. However, the gas injection method differs in that it allows for the production of precise two-dimensional ultrathin nanomaterials with higher yields. This is achieved by adjusting the solvent type and reaction temperature. (Liao et al., 2023).

Using this approach, Liu and colleagues showcased the ability to achieve a controllable synthesis of a centimeter-scale two-dimensional (2D) ZnO-doped Ga_2O_3 nanostructure layer. They employed a liquid Ga–Zn alloy printing strategy near room temperature for this achievement (Liu et al., 2023). They placed Ga and Zn metals on a high-temperature-resistant quartz substrate for melting to form liquid Ga–Zn alloys. To prevent thermal oxidation and pollution, they filled the tube furnace chamber with pure argon gas. The annealing temperature was set at 300 °C with a heating rate of 10 °C/min. The annealing process lasted for 4 h at 300 °C to ensure uniform composition in the liquid alloys. A polydimethylsiloxane (PDMS) intermediate substrate was

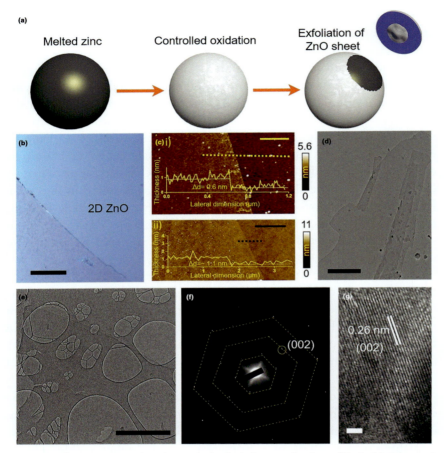

Fig. 8 Synthesis, structural and morphological characterization of ZnO: (a) Schematic illustration of growth of ZnO sheets and their exfoliation. (b) Optical image delineating homogenous millimeter sized ZnO sheet, scale bar is 100 lm. (c) AFM images (i) ~0.6 nm thick ZnO sheets at scale bar of 500 nm and (ii) ~1.1 nm thick ZnO sheet at scale bar of 4 lm (the insets show thickness profiles). (d) TEM image showing a folded edge to delineate the ultrathin features, scale bar is 500 nm. (e) A TEM image of a thicker ZnO nanosheet, scale bar is 1 lm. (f) SAED pattern indexed to (0 0 2) plane showing hexagonal ZnO. (g) HRTEM image showing preferred growth direction in wurtzite phase of ZnO, scale bar is 1 nm. *Reprinted from Mahmood, N., Khan, H., Tran, K., Kuppe, P., Zavabeti, A., Atkin, P., Ghasemian, M.B., Yang, J., Xu, C., Tawfik, S.A., 2021. Maximum piezoelectricity in a few unit-cell thick planar ZnO–a liquid metal-based synthesis approach. Mater. Today 44, 69–77, Copyright (2021), with permission from Elsevier.*

prepared by mixing the prepolymer and cross-linker in a 10:1 ratio. To prepare a 2D ZnO-doped Ga_2O_3 film, the liquid Ga–Zn alloy was heated and transferred to a tape. They brushed the liquid alloy on the tape over the smooth PDMS substrate. Then they could transfer a large-scale amorphous ZnO-doped Ga_2O_3 film with an atomically thin thickness from PDMS to Si, SiO_2, and quartz substrates. The printing preparation of ZnO-doped Ga_2O_3 films was conducted at room temperature without additional heating. However, they noted that heating the alloy might be necessary if the experiment is conducted at a low temperature. Table 1 presents a compilation of non-layered 2D metal oxides synthesized using the liquid metal approach in recent years.

Table 1 Non-layered 2D metal oxides synthesized using the liquid metal approach in recent years.

Material	Thickness (nm)	Ref.
ZnO	0.6–4	Mahmood et al. (2021)
ZnO-Doped Ga_2O_3	7.74–7.82	Liu et al. (2023)
PbO	1	Ghasemian et al. (2020)
In_2O_3	4.5	Alsaif et al. (2019)
MnO_2	1	Ghasemian et al. (2019)
Al_2O_3	1.1	Zavabeti et al. (2018)
TiO_2	3	Alkathiri et al. (2020)
TeO_2	1.5	Zavabeti et al. (2021)
ITO	1.5	Datta et al. (2020)
IZO	1.59	Jannat et al. (2021)
ZnO	5	Krishnamurthi et al. (2021)
$GaPO_4$	2.25–4.3	Syed et al. (2018)
Ga_2O_3	3	Xu et al. (2023)
In_2O_3	8	Liao et al. (2023)

8. Hydrothermal approach

Hydrothermal synthesis represents a wet chemical method employed for the production of diverse nanomaterials, including 2D metal oxides. In this technique, materials are dissolved in an aqueous solution under high vapor pressure at elevated temperatures, usually around 200–300 °C. The hydrothermal synthesis of 2D non-layered metal oxides presents numerous benefits, allowing precise control over surface morphology, crystallite size, and dopants. An additional benefit of the hydrothermal approach is its versatility in working with diverse substrates. It allows for the growth of 2D nanosheets on various substrates, including FTO glass, TiO_2 nanotubes, and 2D graphitic carbon nitride, by utilizing Na_2WO_4 as the precursor in the case of WO_3 (Li et al., 2016; Su et al., 2011). These advantageous characteristics make it well-suited for various applications, including catalysts, energy storage devices, and chemical/biological sensors (Choi et al., 2022).

Following this synthesis approach, researchers have successfully synthesized ultrathin 2D metal oxide materials such as ZnO (Zhu et al., 2023), TiO_2 (Hwang et al., 2023), SnO_2 (Hassan, 2023), and WO_3 (Wang et al., 2019).

Not only have binary metal oxides been prepared using hydrothermal methods, but composite metal oxides are also being synthesized through this approach. High-performance lithium storage materials were achieved through the construction of hierarchical $CoNiO_2@CeO_2$ nanosheets (Yi et al., 2021). A one-step hydrothermal process, followed by annealing, successfully synthesized these composite materials. Notably, the CeO_2 nanoparticles deposited on the $CoNiO_2$ nanosheet surface significantly improved the electrical contact between the two components, emphasizing their potential in lithium storage systems.

Moreover, Yang et al. have demonstrated the successful preparation of spinel-type 2D metal oxide nanosheets (including $ZnGa_2O_4$, g-Ga_2O_3, and $MnGa_2O_4$) using the hydrothermal reaction process (Yang et al., 2018). Each of the spinel-type nanosheets exhibited a characteristic hexagonal/triangular geometrical shape, with an extremely small thickness. The hydrothermal approach, employed in this research, holds great promise for the direct synthesis of various semiconductor nanosheets with a non-layered structure. With transitional metals incorporated into these 2D nanosheets, new possibilities emerge for integrating diverse optoelectronic nanodevices to serve specific functions across multiple fields.

Furthermore, ultrathin Co_3O_4 nanosheets@Ni foam was prepared through a facile hydrothermal method with the assistance of formamide and PVP (Kang et al., 2021). The Co_3O_4 nanosheets had a thickness of approximately 4 nm and were interconnected to form a highly hierarchical porous architecture. In the synthesis procedure, 0.3 g of $Co(NO_3)_2 \cdot 6H_2O$, 0.9 g of PVP, and 0.9 mL of formamide were added to 100 mL of ultrapure water. Subsequently, a piece of pretreated Ni foam was soaked into the as-formed light-pink solution, which was then hydrothermally treated at 140, 160, and 180 °C for 24 h. After the hydrothermal reaction, the products were washed with ultrapure water under ultrasonication, dried overnight, and thermally treated at 350 °C for 2 h in air. Fig. 9 depicts the process of forming nanosheet and nanowire-structured Co_3O_4 on Ni foam, including the chemical reactions occurring during the hydrothermal synthesis.

In addition to the hydrothermal method, the solvothermal synthesis technique has also proven effective in preparing 2D metal oxides. These two methods share similarities and can be grouped under the same synthesis category (Liang et al., 2019; Patil et al., 2023; Saikia et al., 2022). The hydrothermal/solvothermal synthesis method is a cost-effective and scalable approach for producing ultrathin 2D MOXs. However, understanding its growth mechanism is challenging due to the reactions occurring in a sealed autoclave, making it difficult to devise experimental schemes for alternative

Fig. 9 Highly hierarchical porous ultrathin Co_3O_4 nanosheets@Ni foam for high-performance supercapacitors. *Reprinted (adapted) with permission from Kang, M., Zhou, H., Wen, P., Zhao, N., 2021. Highly hierarchical porous ultrathin Co3O4 nanosheets@ Ni foam for high-performance supercapacitors. ACS Appl. Energy Mater. 4, 1619–1627.Copyright 2021 American Chemical Society.*

2D MOXs. Additionally, the method's sensitivity to various experimental conditions poses difficulties in achieving precise and reproducible control over the resulting structure or morphology (Xie et al., 2022).

9. Chemical vapor deposition (CVD) approach

Chemical vapor deposition (CVD) is a high-temperature synthesis process utilized for depositing desired materials onto substrates. This technique has found application in the synthesis of diverse 2D materials, including non-layered metal oxides (Zhang et al., 2023). The process involves introducing precursor materials into a heated reaction chamber, usually maintained at temperatures above the boiling point of the precursors. As the precursor vapor interacts with the substrate surface, it leads to the deposition of the desired metal oxide material in a non-layered structure. The CVD method presents numerous advantages in synthesizing 2D non-layered metal oxides. It allows the production of large-sized crystals with precise atomic layer thickness and minimal defects. This method has proven successful in synthesizing various 2D non-layered metal oxides, including MO_2 (Wu et al., 2018), VO_2 (Ran et al., 2022), α-MoO_3 (Ran et al., 2022), Fe_3O_4 (Yin et al., 2020), Bi_2O_3 (Messalea et al., 2018) and Sb_2O_3 (Liu et al., 2021b). The resulting materials find extensive applications in electronics, catalysis, and energy storage, among others.

Guo et al. successfully synthesized 2D vertical heterostructures of MoO_3–MoS_2 on SiO_2/Si substrates through a one-step CVD process (Guo et al., 2021). By maintaining a fixed reaction temperature at 720 °C and adjusting the location of the sulfur source, different products were obtained by varying the concentration of sulfur vapor. The vertical MoO_3–MoS_2 heterostructures exhibit an average size of ~20 μm and a thickness down to ~10 nm. The investigation also revealed a direct relationship between the synthesis temperature and the thickness of MoO_3 nanosheets. As the temperature increased, the thickness of the MoO_3 nanosheets also exhibited a corresponding increase.

Pu et al. demonstrated a significant advancement in the synthesis of ultrathin MoO_2 nanosheets through an enhanced chemical vapor deposition (CVD) technique, utilizing solely molybdenum trioxide as the precursor material (Pu et al., 2017). The resulting MoO_2 nanosheets exhibited remarkable monodispersity, with an average thickness ranging from approximately 5 to 10 nm, and displayed excellent crystal quality. Fig. 10 shows the TEM image of a single MoO$_2$ nanosheet.

In another study, researchers reported on the controlled synthesis of rectangular and hexagonal molybdenum oxide nanosheets from a single precursor in the form of powdered MoO$_3$ (Wazir et al., 2020). Remarkably, the synthesis process was accomplished without the need for any post-annealing treatment, utilizing chemical vapor deposition (CVD) methods. Hexagonal molybdenum dioxides (MoO$_2$) nanosheets were synthesized using MoO$_3$ powder placed in a quartz tube and heated in a two-zone furnace under a nitrogen atmosphere. SiO$_2$/Si substrates were cleaned and positioned near the precursor powders. After purging with N$_2$ gas, the temperature was raised to 480 °C in one zone and 780 °C in the other, and the reaction was allowed to proceed in the N$_2$ environment for 20 min. The furnace was then naturally cooled, resulting in the deposition of hexagonal MoO$_2$ nanosheets on the substrates. Following a similar approach as described earlier, the synthesis of rectangular molybdenum dioxide (MoO$_2$) nanosheets was conducted in a two-zone tube furnace under an N$_2$ atmosphere. The thermo-blocks were positioned near the middle of the tube, with the left zone set at room temperature. The remaining parameters were kept consistent with those utilized for the synthesis of hexagonal MoO$_2$ nanosheets. As a result, rectangular MoO$_2$ nanosheets were grown on the SiO$_2$/Si substrates. AFM analysis revealed that the thin rectangular nanosheets had an approximate thickness of 30 nm, while the hexagonal nanosheets had a thickness of approximately 25 nm.

Taking into account the thickness of nanosheets, thinner MoO$_2$ nanosheets have also been prepared. Luo et al. reported the synthesis of

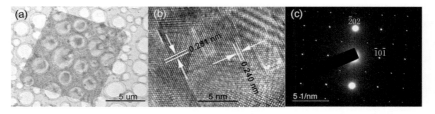

Fig. 10 (a) TEM image of a single MoO$_2$ nanosheet. (b) Corresponding HRTEM image of the sample. (c) SAED pattern taken from (b). Pu, E., Liu, D., Ren, P., Zhou, W., Tang, D., Xiang, B., Wang, Y., Miao, J., 2017. Ultrathin MoO2 nanosheets with good thermal stability and high conductivity. AIP Adv. 7, 025015; licensed under a Creative Commons Attribution (CC BY) license.

micrometer-scale 2D MoO_2 single crystals, showcasing an impressive thickness of only about 3.8 nm (Luo et al., 2020). The exceptional accomplishment was attained by reducing MoO_3 using hydrogen in the atmospheric pressure CVD method.

Despite its promise for synthesizing 2D non-layered metal oxides, the CVD method comes with several limitations. High temperatures are required to initiate the reaction, which can restrict substrate options and raise production costs. Moreover, the specialized equipment needed for CVD, such as a reaction chamber and gas delivery system, can be expensive and challenging to maintain. Additionally, the batch process of CVD makes scaling up for large-scale production difficult. The volatility requirement of precursor materials at high temperatures can also limit the range of precursors available for use. Furthermore, the resulting material may exhibit random crystal orientation, impacting its properties.

10. Template-assisted approach

The key element in template-assisted synthesis is the "template," which can be any entity with nanostructured characteristics (Poolakkandy and Menamparambath, 2020). The size, morphology, and charge dissemination of the template play a significant role in guiding the structure. The process begins with preparing the template, followed by fabricating the desired material using the template and, if necessary, removing the template. Fabrication methods can include physical techniques like surface coating or chemical methods such as addition, elimination, substitution, or isomerization reactions. After the reaction, the template can be removed through physical methods like dissolution or chemical methods like calcination. The major advantage of the template-assisted method lies in its effective control over the structure, dimension, and morphology of the final product.

There are three different methods for synthesizing metal oxides using a template-assisted approach. The first method is the hard-template method, which uses solid molds with well-defined designs to create the final product. The second method is the soft-template method, which utilizes flexible nanostructures composed of surfactants, organic molecules, or block copolymers to control the size and morphology of the product. The third method is the colloidal-template method, which combines the features of both soft and hard templates by using inorganic nanoparticles bound with flexible polymer tails (Alhalili, 2023; Kaur et al., 2022; Li et al., 2023).

Salt crystals can act as a hard template for the growth of 2D MOXs due to their larger surface area, which allows the oxide material to spread out and form a larger sheet (Xiao et al., 2016). This method is hypothesized to occur via a match between the crystal lattices of the salt and the growing oxide (Huang et al., 2020). A general, fast, and low-temperature ($\approx$100 °C) salt-templated method to prepare various 2D metal oxides on a large scale, such as MoO_3, SnO_2, and SiO_2, has been proposed (Gu et al., 2019). This method utilizes the $CoCl_2 \cdot 6H_2O$ salt template, which could exert a nano-confined effect during the synthesis. The resulting 2D materials exhibited ultrathin features (2–7 nm) and large aspect ratios. The synthesis process of 2D metal oxide mainly involves three steps: first, $CoCl_2 \cdot 6H_2O$ and MCl_x (M = Mo, Si, and Sn) mixed uniformity in a certain proportion ($CoCl_2 \cdot 6H_2O$: MCl_x = 10:1); After that, a molten salt was obtained by heating the mixture to 100 °C and maintaining for 5 min. In this step, the hydrolysis reaction was taken place between MCl_x (M = Mo, Si, and Sn) and water broken from $CoCl_2 \cdot 6H_2O$. At the same time, the relevant metal oxide was generated during the hydrolysis. Then, the resultant 2D metal oxide nanosheets were formed by cooling the molten salt to room temperature naturally. This is because the molten $CoCl_2$ can still form $CoCl_2 \cdot 6H_2O$ in the cooling process, which confined the resultant metal oxide into the interlamination of $CoCl_2 \cdot 6H_2O$. Finally, the 2D metal oxide was separated from the mixed salt by washing with water and centrifugation.

Besides salt templates, there are other 2D hard templates such as Li_2O_2 nanosheets (Xue et al., 2020), graphitic carbon nitride nanosheets (Xu et al., 2019), and graphene oxide (Lei et al., 2018), which have been utilized in developing 2D MOXs. Lei et al. successfully developed a modified evaporation-induced self-assembly (EISA) method to synthesize the graphene-like CeO_2, TiO_2, ZrO_2, CoO_x, NiO, Al_2O_3, and CuO nanosheets (Lei et al., 2018). In general, the synthesis of porous 2D nanosheets of various metal oxides involved a two-step process. Initially, the metal nitrate precursors were uniformly deposited onto the surface of graphene oxide (GO) using the modified EISA method. Subsequently, the GO template was eliminated through heat treatment, resulting in the formation of the 2D metal oxides.

Due to the challenging and time-consuming template removal process in hard-template-assisted synthesis, there has been a notable shift towards the widespread use of soft-templates. This method has proven highly effective in synthesizing various 2D metal oxides, such as TiO_2 (Sun et al., 2023), CuO (Jagadale et al., 2018), WO_3 (Hu et al., 2018), ZnO (Yu et al., 2022), Co_3O_4 (Zhong et al., 2022), and MnO_2 (Chen et al., 2019).

To control the surface morphology of CuO nanosheets, Jagadale et al. used dibenzo thiazolyl dibenzo-18-crown-6 (DDCE) as a template for synthesis in their research. The synthesis of DDCE involved two steps: first, a mixture of dibenzo-18-crown-6, trifluoroacetic acid, and hexamethylenetetramine was stirred for 24 h at 90 °C under a nitrogen atmosphere, resulting in diformyldibenzo-18-crown-6 crystals. In the second step, diformyldibenzo-18-crown-6 was reacted with 2-amino benzene thiol in methanol for 24 h to produce pure DDCE crystals. This prepared DDCE was then employed as a structure-directing agent for the synthesis of nanostructured CuO through hydrothermal treatment. By varying the concentration of DDCE, CuO nanosheets with different surface morphologies were obtained.

11. Challenges and future perspectives

The synthesis of 2D non-layered metal oxides presents several challenges. These challenges include the need to control the thickness, which sets them apart from layered 2D metal oxides that have a fixed thickness due to their crystal structure. Depending on the synthesis method used, non-layered 2D metal oxides can have varying thicknesses, making it crucial to regulate this aspect to achieve the desired properties and applications (Zhou et al., 2023). Achieving high quality is another significant challenge. Non-layered 2D metal oxides are often created through complex chemical reactions, which can introduce impurities and defects into the final product. Ensuring high-quality non-layered 2D metal oxides becomes essential to maintain consistency and reproducibility in their properties (Xie et al., 2022). Scalability poses a considerable obstacle as well. Many synthesis methods for non-layered 2D metal oxides are still in the experimental phase and may not be suitable for large-scale production. The development of scalable synthesis methods becomes vital to commercialize non-layered 2D metal oxides for diverse applications (Zhou et al., 2023). Stability adds another layer of complexity to this journey. Non-layered 2D metal oxides can be vulnerable to degradation and oxidation, which can impact their properties and applications over time. To ensure their long-term performance, stable versions must be meticulously developed (Hu et al., 2022). Finally, cost-effectiveness presents a challenge worth conquering. Some synthesis methods for non-layered 2D metal oxides can be costly and may require specialized equipment. By devising cost-effective synthesis methods, these remarkable materials can become more accessible for various applications (Xie et al., 2022).

Considering the limitations in synthesis approaches, selecting the most appropriate method for the desired application warrants careful consideration. To aid in this decision-making process, Table 2 presents an overview of the advantages and disadvantages of the discussed approaches.

The fascinating properties and potential applications of 2D non-layered materials have generated a lot of interest in exploring their unexplored possibilities. As researchers explore this promising field, they have discovered several potential research directions that offer exciting opportunities for progress. Firstly, theoretical calculations are set to play a key role in guiding the growth of 2D non-layered materials. Unraveling their growth mechanisms and gaining a deeper understanding of their unique properties through theoretical insights will pave the way for innovative

Table 2 Advantages and disadvantages of 2D non-layered synthesis approaches.

Approach	Advantages	Disadvantages	Possible Applications
liquid metal	Large-scale production Low cost High yield	Low crystallinity Structural defects Residual	Energy storage CO_2 reduction Photodegradation
Hydrothermal/ solvothermal	Low cost few-step synthetic procedure Scalable	Thickness control	Gas sensing Photocatalysis Biosensing
CVD	Uniformity Precise control High purity	High cost Complex Operation Slow growth rate	Energy storage
Template assisted	High yield Low cost controllable morphology	Residual Complexity High cost	Electronic devices

synthetic strategies. This will enable researchers to fine-tune the synthesis process, controlling properties such as thickness and composition to tailor materials for specific applications. Secondly, researchers are urged to focus on exploring composite materials or heterojunctions based on 2D non-layered materials. By combining these materials with other elements or compounds, entirely new functionalities can be unlocked, leading to a realm of transformative applications. Lastly, the extensive investigation of the wide application range of 2D non-layered materials is of utmost importance. Their exceptional physical, chemical, optical, and electrical properties offer an array of opportunities in diverse fields, promising breakthroughs in technologies that can address pressing challenges in catalysis, electronics, optoelectronics, and energy storage fields.

References

Akbari, M.K., Verpoort, F., Zhuiykov, S., 2021. State-of-the-art surface oxide semaiconductors of liquid metals: an emerging platform for development of multifunctional two-dimensional materials. J. Mater. Chem. A 9, 34–73.

Alhalili, Z., 2023. Metal oxides nanoparticles: general structural description, chemical, physical, and biological synthesis methods, role in pesticides and heavy metal removal through wastewater treatment. Molecules 28, 3086.

Alkathiri, T., Dhar, N., Jannat, A., Syed, N., Mohiuddin, M., Alsaif, M.M.Y.A., et al., 2020. Atomically thin TiO2 nanosheets synthesized using liquid metal chemistry. Chem. Commun. 56, 4914–4917.

Allioux, F.-M., Ghasemian, M.B., Xie, W., O'Mullane, A.P., Daeneke, T., Dickey, M.D., et al., 2022. Applications of liquid metals in nanotechnology. Nanoscale Horiz. 7, 141–167.

Alsaif, M.M.Y.A., Kuriakose, S., Walia, S., Syed, N., Jannat, A., Zhang, B.Y., et al., 2019. 2D SnO/In2O3 van der Waals heterostructure photodetector based on printed oxide skin of liquid metals. Adv. Mater. Interfaces 6, 1900007.

Bai, B., Li, J., Hao, J., 2015. 1D-MnO2, 2D-MnO2 and 3D-MnO2 for low-temperature oxidation of ethanol. Appl. Catal. B Environ. 164, 241–250.

Bao, J., Wang, Z., Liu, W., Xu, L., Lei, F., Xie, J., et al., 2018. ZnCo2O4 ultrathin nanosheets towards the high performance of flexible supercapacitors and bifunctional electrocatalysis. J. Alloy. Compd. 764, 565–573.

Bao, J., Zhang, X., Fan, B., Zhang, J., Zhou, M., Yang, W., et al., 2015. Ultrathin spinel-structured nanosheets rich in oxygen deficiencies for enhanced electrocatalytic water oxidation. Angew. Chem. - Int. Ed. 54, 7399–7404.

Benasutti, P.B., Reserve, C.W., Jesuit, W., 2012. Electronic and Structural Properties of Silicene and Graphene Layered Structures.

Butler, S.Z., Hollen, S.M., Cao, L., Cui, Y., Gupta, J.A., Gutiérrez, H.R., et al., 2013. Progress, challenges, and opportunities in two-dimensional materials beyond graphene. ACS Nano 7, 2898–2926.

Cao, J., Li, X., Liu, Y., Zhu, G., Li, R.-W., 2023. Liquid metal-based electronics for on-skin healthcare. Biosensors 13, 84.

Chen, J., Meng, H., Tian, Y., Yang, R., Du, D., Li, Z., et al., 2019. Recent advances in functionalized MnO 2 nanosheets for biosensing and biomedicine applications. Nanoscale Horiz. 4, 321–338.

Chen, T.-P., Chang, S.-P., Hung, F.-Y., Chang, S.-J., Hu, Z.-S., Chen, K.-J., 2013. Simple fabrication process for 2D ZnO nanowalls and their potential application as a methane sensor. Sensors 13, 3941–3950.

Choi, P.G., Izu, N., Shirahata, N., Masuda, Y., 2019. SnO2 nanosheets for selective alkene gas sensing. ACS Appl. Nano Mater. 2, 1820–1827.

Choi, S.H., Yun, S.J., Won, Y.S., Oh, C.S., Kim, S.M., Kim, K.K., et al., 2022. Large-scale synthesis of graphene and other 2D materials towards industrialization. Nat. Commun. 13, 1484.

Daeneke, T., Khoshmanesh, K., Mahmood, N., De Castro, I.A., Esrafilzadeh, D., Barrow, S.J., et al., 2018. Liquid metals: fundamentals and applications in chemistry. Chem. Soc. Rev. 47, 4073–4111.

Datta, R.S., Syed, N., Zavabeti, A., Jannat, A., Mohiuddin, M., Rokunuzzaman, M., et al., 2020. Flexible two-dimensional indium tin oxide fabricated using a liquid metal printing technique. Nat. Electron. 3, 51–58.

Dean, C.R., Young, A.F., Meric, I., Lee, C., Wang, L., Sorgenfrei, S., et al., 2010. Boron nitride substrates for high-quality graphene electronics. Nat. Nanotechnol. 5, 722.

Dubal, D.P., Gomez-Romero, P., Sankapal, B.R., Holze, R., 2015. Nickel cobaltite as an emerging material for supercapacitors: an overview. Nano Energy 11, 377–399.

Gangopadhyay, S., Frolov, D.D., Masunov, A.E., Seal, S., 2014. Structure and properties of cerium oxides in bulk and nanoparticulate forms. J. Alloy. Compd. 584, 199–208.

Ghasemian, M.B., Mayyas, M., Idrus-Saidi, S.A., Jamal, M.A., Yang, J., Mofarah, S.S., et al., 2019. Self-limiting galvanic growth of MnO2 monolayers on a liquid metal—applied to photocatalysis. Adv. Funct. Mater. 29, 1901649.

Ghasemian, M.B., Zavabeti, A., Abbasi, R., Kumar, P.V., Syed, N., Yao, Y., et al., 2020. Ultra-thin lead oxide piezoelectric layers for reduced environmental contamination using a liquid metal-based process. J. Mater. Chem. A 8, 19434–19443.

Goff, A., Aukarasereenont, P., Nguyen, C.K., Grant, R., Syed, N., Zavabeti, A., et al., 2021. An exploration into two-dimensional metal oxides, and other 2D materials, synthesised via liquid metal printing and transfer techniques. Dalton Trans. 50, 7513–7526.

Gu, J., Zhang, C., Du, Z., Yang, S., 2019. Rapid and low-temperature salt-templated production of 2D metal oxide/oxychloride/hydroxide. Small 15, 1904587.

Guo, N., Lu, R., Liu, S., Ho, G.W., Zhang, C., 2014. High catalytic activity of Au clusters supported on ZnO nanosheets. J. Phys. Chem. C 118, 21038–21041.

Guo, Y., Kang, L., Song, P., Zeng, Q., Tang, B., Yang, J., et al., 2021. MoO3–MoS2 vertical heterostructures synthesized via one-step CVD process for optoelectronics. 2D Mater. 8, 35036.

Hassan, J.J., 2023. Hydrothermal synthesis and characterization of SnO2 nanosheets. Adv. Mater. Res. 1175, 47–53.

Hoang, K., Johannes, M.D., 2014. Defect chemistry in layered transition-metal oxides from screened hybrid density functional calculations. J. Mater. Chem. A 2, 5224–5235.

Hu, L., Hu, P., Chen, Y., Lin, Z., Qiu, C., 2018. Synthesis and gas-sensing property of highly self-assembled tungsten oxide nanosheets. Front. Chem. 6, 452.

Hu, X., Liu, K., Cai, Y., Zang, S.-Q., Zhai, T., 2022. 2D oxides for electronics and optoelectronics. Small Sci. 2, 2200008.

Huang, L., Hu, Z., Jin, H., Wu, J., Liu, K., Xu, Z., et al., 2020. Salt-assisted synthesis of 2D materials. Adv. Funct. Mater. 30, 1908486.

Hwang, J.Y., Lee, Y., Lee, G.H., Lee, S.Y., Kim, H.-S., Kim, S., et al., 2023. Room-temperature ammonia gas sensing via Au nanoparticle-decorated TiO2 nanosheets. Discov. Nano 18, 47.

Jagadale, S.D., Teli, A.M., Kalake, S.V., Sawant, A.D., Yadav, A.A., Patil, P.S., 2018. Functionalized crown ether assisted morphological tuning of CuO nanosheets for electrochemical supercapacitors. J. Electroanal. Chem. 816, 99–106.

Jannat, A., Syed, N., Xu, K., Rahman, M.A., Talukder, M.M.M., Messalea, K.A., et al., 2021. Printable single-unit-cell-thick transparent zinc-doped indium oxides with efficient electron transport properties. ACS Nano 15, 4045–4053.

Kang, M., Zhou, H., Wen, P., Zhao, N., 2021. Highly hierarchical porous ultrathin Co3O4 nanosheets@ Ni foam for high-performance supercapacitors. ACS Appl. Energy Mater. 4, 1619–1627.

Kano, E., Kvashnin, D.G., Sakai, S., Chernozatonskii, L.A., Sorokin, P.B., Hashimoto, A., et al., 2017. One-atom-thick 2D copper oxide clusters on graphene. Nanoscale 9, 3980–3985.

Kaur, A., Bajaj, B., Kaushik, A., Saini, A., Sud, D., 2022. A review on template assisted synthesis of multi-functional metal oxide nanostructures: status and prospects. Mater. Sci. Eng. B 286, 116005.

Krishnamurthi, V., Ahmed, T., Mohiuddin, M., Zavabeti, A., Pillai, N., McConville, C.F., et al., 2021. A visible-blind photodetector and artificial optoelectronic synapse using liquid-metal exfoliated ZnO nanosheets. Adv. Opt. Mater. 9, 2100449.

Kumbhakar, P., Chowde, G.C., Mahapatra, P.L., Mukherjee, M., Malviya, K.D., Chaker, M., et al., 2021. Emerging 2D metal oxides and their applications. Mater. Today 45, 142–168.

Kundu, M., Ng, C.C.A., Petrovykh, D.Y., Liu, L., 2013. Nickel foam supported mesoporous MnO2 nanosheet arrays with superior lithium storage performance. Chem. Commun. 49, 8459–8461.

Kuo, D.-Y., Kawasaki, J.K., Nelson, J.N., Kloppenburg, J., Hautier, G., Shen, K.M., et al., 2017. Influence of surface adsorption on the oxygen evolution reaction on IrO2(110). J. Am. Chem. Soc. 139, 3473–3479.

Kvashnin, D.G., Kvashnin, A.G., Kano, E., Hashimoto, A., Takeguchi, M., Naramoto, H., et al., 2019. Two-dimensional CuO inside the supportive bilayer graphene matrix. J. Phys. Chem. C 123, 17459–17465.

Lee, J.M., Kang, B., Jo, Y.K., Hwang, S.-J., 2019. Organic intercalant-free liquid exfoliation route to layered metal-oxide nanosheets via the control of electrostatic interlayer interaction. ACS Appl. Mater. Interfaces 11, 12121–12132.

Lei, L., Wu, Z., Liu, H., Qin, Z., Chen, C., Luo, L., et al., 2018. A facile method for the synthesis of graphene-like 2D metal oxides and their excellent catalytic application in the hydrogenation of nitroarenes. J. Mater. Chem. A 6, 9948–9961.

Li, C., Guo, W., Kong, Y., Gao, H., 2007. Size-dependent piezoelectricity in zinc oxide nanofilms from first-principles calculations. Appl. Phys. Lett. 90, 33108.

Li, D., Ding, L.-X., Wang, S., Cai, D., Wang, H., 2014. Ultrathin and highly-ordered CoO nanosheet arrays for lithium-ion batteries with high cycle stability and rate capability. J. Mater. Chem. A 2, 5625–5630.

Li, F., Chen, L., Knowles, G.P., MacFarlane, D.R., Zhang, J., 2017. Hierarchical mesoporous SnO2 nanosheets on carbon cloth: a robust and flexible electrocatalyst for CO_2 reduction with high efficiency and selectivity. Angew. Chem. Int. Ed. 56, 505–509.

Li, R., Rao, Y., Huang, Y., 2022. Advances in catalytic elimination of atmospheric pollutants by two-dimensional transition metal oxides. Chin. Chem. Lett. 34, 108000.

Li, J., Zhu, J., Dong, Z., Wu, Q., 2023. Nanomaterials derived from a template method for supercapacitor applications. ChemistrySelect 8, e202204487.

Li, Y., Li, F.M., Meng, X.Y., Li, S.N., Zeng, J.H., Chen, Y., 2018. Ultrathin Co3O4 nanomeshes for the oxygen evolution reaction. ACS Catal. 8, 1913–1920.

Li, Y., Wei, X., Yan, X., Cai, J., Zhou, A., Yang, M., et al., 2016. Construction of inorganic–organic 2D/2D WO 3/gC 3 N 4 nanosheet arrays toward efficient photoelectrochemical splitting of natural seawater. Phys. Chem. Chem. Phys. 18, 10255–10261.

Liang, Y., Yang, Y., Zou, C., Xu, K., Luo, X., Luo, T., et al., 2019. 2D ultra-thin WO3 nanosheets with dominant {002} crystal facets for high-performance xylene sensing and methyl orange photocatalytic degradation. J. Alloy. Compd. 783, 848–854.

Liao, B., Shen, H., Zhu, H., Gao, B., Wang, Z., Zhang, J., et al., 2023. Synthesis and selenization of thickness-controllable In2O3 films by printed oxide skin of liquid metals. ACS Appl. Electron. Mater. 5, 1088–1096.

Liao, G., Ren, L., Guo, Z., Qiao, H., Huang, Z., Wang, Z., et al., 2023. Synthesis and application of liquid metal based-2D nanomaterials: a perspective view for sustainable energy. Molecules 28, 524.

Lin, Z., Du, C., Yan, B., Wang, C., Yang, G., 2018. Two-dimensional amorphous NiO as a plasmonic photocatalyst for solar H2 evolution. Nat. Commun. 9, 4036.

Liu, W., Bao, J., Xu, L., Guan, M., Wang, Z., Qiu, J., et al., 2019a. NiCo 2 O 4 ultrathin nanosheets with oxygen vacancies as bifunctional electrocatalysts for Zn-air battery. Appl. Surf. Sci. 478, 552–559.

Liu, B., Tang, B., Lv, F., Zeng, Y., Liao, J., Wang, S., et al., 2020. Photodetector based on heterostructure of two-dimensional WSe2/In2Se3. Nanotechnology 31, 065203.

Liu, D., Barbar, A., Najam, T., Javed, M.S., Shen, J., Tsiakaras, P., et al., 2021a. Single noble metal atoms doped 2D materials for catalysis. Appl. Catal. B Environ. 297, 120389.

Liu, K., Jin, B., Han, W., Chen, X., Gong, P., Huang, L., et al., 2021b. A wafer-scale van der Waals dielectric made from an inorganic molecular crystal film. Nat. Electron. 4, 906–913.

Liu, Q., Guo, J., Li, J., Feng, L., Chen, L., Hua, Z., et al., 2023. Room-temperature preparation of large-area transparent two-dimensional ZnO-doped Ga2O3 nanostructure-based layers: implications for optoelectronic nanodevices. ACS Appl. Nano Mater. 6, 3027–3035.

Liu, Y., Xiao, C., Huang, P., Cheng, M., Xie, Y., 2018. Regulating the charge and spin ordering of two-dimensional ultrathin solids for electrocatalytic water splitting. Chem 4, 1263–1283.

Luo, J., Chen, H., Wang, J., Xia, F., Huang, X., 2020. Direct growth of 2D MoO2 single crystal on SiO2/Si substrate by atmospheric pressure chemical vapor deposition. Mater. Chem. Phys. 251, 123166.

Mahmood, N., Castro, De, Pramoda, I.A., Khoshmanesh, K., Bhargava, K., Kalantar-Zadeh, K, S.K., 2019. Atomically thin two-dimensional metal oxide nanosheets and their heterostructures for energy storage. Energy Storage Mater. 16, 455–480.

Mahmood, N., Khan, H., Tran, K., Kuppe, P., Zavabeti, A., Atkin, P., et al., 2021. Maximum piezoelectricity in a few unit-cell thick planar ZnO—a liquid metal-based synthesis approach. Mater. Today 44, 69–77.

Majumdar, D., Ghosh, S., 2021. Recent advancements of copper oxide based nanomaterials for supercapacitor applications. J. Energy Storage 34, 101995.

Messalea, K.A., Carey, B.J., Jannat, A., Syed, N., Mohiuddin, M., Zhang, B.Y., et al., 2018. Bi 2 O 3 monolayers from elemental liquid bismuth. Nanoscale 10, 15615–15623.

Morin, F.J., 1959. Oxides which show a metal-to-insulator transition at the Neel temperature. Phys. Rev. Lett. 3, 34–36.

Neal, A.T., Liu, H., Neal, A.T., Zhu, Z., Luo, Z., Xu, X., et al., 2014. Phosphorene: an unexplored 2D semiconductor with a high hole mobility. ACS Nano 8, 4033–4041.

Ni, Z., Liu, Q., Tang, K., Zheng, J., Zhou, J., Qin, R., et al., 2012. Tunable bandgap in silicene and germanene. Nano Lett. 12, 113–118.

Novoselov, K.S., Geim, A.K., Morozov, S.V., Jiang, D.A., Zhang, Y., Dubonos, S.V., et al., 2004. Electric field effect in atomically thin carbon films. Sciences (80-) 306, 666–669.

Patil, R.A., Chang, C.-P., Devan, R.S., Liou, Y., Ma, Y.-R., 2016. Impact of nanosize on supercapacitance: study of 1D nanorods and 2D thin-films of nickel oxide. ACS Appl. Mater. Interfaces 8, 9872–9880.

Patil, S.A., Jagdale, P.B., Singh, A., Singh, R.V., Khan, Z., Samal, A.K., et al., 2023. 2D zinc oxide—synthesis, methodologies, reaction mechanism, and applications. Small 19, 2206063.

Pawar, S.M., Pawar, B.S., Hou, B., Kim, J., Aqueel Ahmed, A.T., Chavan, H.S., et al., 2017. Self-assembled two-dimensional copper oxide nanosheet bundles as an efficient oxygen evolution reaction (OER) electrocatalyst for water splitting applications. J. Mater. Chem. A 5, 12747–12751.

Petkov, A., Mishra, A., Cattelan, M., Field, D., Pomeroy, J., Kuball, M., 2023. Electrical and thermal characterisation of liquid metal thin-film Ga 2 O 3–SiO 2 heterostructures. Sci. Rep. 13, 3437.

Poolakkandy, R.R., Menamparambath, M.M., 2020. Soft-template-assisted synthesis: a promising approach for the fabrication of transition metal oxides. Nanoscale Adv. 2, 5015–5045.

Pu, E., Liu, D., Ren, P., Zhou, W., Tang, D., Xiang, B., et al., 2017. Ultrathin MoO_2 nanosheets with good thermal stability and high conductivity. Aip Adv. 7, 025015.

Ran, M., Zhao, C., Xu, X., Kong, X., Lee, Y., Cui, W., et al., 2022. Boosting in-plane anisotropy by periodic phase engineering in two-dimensional VO_2 single crystals. Fundam. Res. 2, 456–461.

Richter, N.A., Sicolo, S., Levchenko, S.V., Sauer, J., Scheffler, M., 2013. Concentration of vacancies at metal-oxide surfaces: case study of MgO(100). Phys. Rev. Lett. 111, 45502.

Saikia, S., Devi, R., Gogoi, P., Saikia, L., Choudary, B.M., Raja, T., et al., 2022. Regioselective Friedel–Crafts acylation reaction using single crystalline and ultrathin nanosheet assembly of scrutinyite-SnO2. ACS Omega 7, 32225–32237.

Schmidt, T.M., Miwa, R.H., Fazzio, A., 2010. Ferromagnetic coupling in a Co-doped graphenelike ZnO sheet. Phys. Rev. B 81, 195413.

Sheng, L., Liao, T., Kou, L., Sun, Z., 2017. Single-crystalline ultrathin 2D TiO_2 nanosheets: a bridge towards superior photovoltaic devices. Mater. Today Energy 3, 32–39.

Stoerzinger, K.A., Diaz-Morales, O., Kolb, M., Rao, R.R., Frydendal, R., Qiao, L., et al., 2017. Orientation-dependent oxygen evolution on RuO2 without lattice exchange. ACS Energy Lett. 2, 876–881.

Su, J., Feng, X., Sloppy, J.D., Guo, L., Grimes, C.A., 2011. Vertically aligned WO3 nanowire arrays grown directly on transparent conducting oxide coated glass: synthesis and photoelectrochemical properties. Nano Lett. 11, 203–208.

Sun, J., Zhen, W., Xue, C., 2023. Magnetic template-assisted construction of 2D PCN/TiO2 heterostructures for efficient photocatalytic hydrogen generation. Appl. Surf. Sci. 623, 157131.

Sun, Y., Lei, F., Gao, S., Pan, B., Zhou, J., Xie, Y., 2013. Atomically thin tin dioxide sheets for efficient catalytic oxidation of carbon monoxide. Angew. Chem. 125, 10763–10766.

Sun, Z., Liao, T., Dou, Y., Hwang, S.M., Park, M.-S., Jiang, L., et al., 2014. Generalized self-assembly of scalable two-dimensional transition metal oxide nanosheets. Nat. Commun. 5, 3813.

Syed, N., Zavabeti, A., Ou, J.Z., Mohiuddin, M., Pillai, N., Carey, B.J., et al., 2018. Printing two-dimensional gallium phosphate out of liquid metal. Nat. Commun. 9, 3618.

Tabatabaei, S.-M., Honari, N., Farshchi-Heydari, M.-J., Rastgoo, M., Fathipour, M., 2020. A comparative study on the application of single-layer MoS2 and WS2 for probing methylated and mutated nucleobases: a vdW-DFT study. Appl. Surf. Sci. 501, 143892.

Tang, Y., Huang, C.-H., Nomura, K., 2022. Vacuum-free liquid-metal-printed 2D indium–tin oxide thin-film transistor for oxide inverters. ACS Nano 16, 3280–3289.

Wang, D., Huang, S., Li, H., Chen, A., Wang, P., Yang, J., et al., 2019. Ultrathin WO3 nanosheets modified by g-C3N4 for highly efficient acetone vapor detection. Sens. Actuators B Chem. 282, 961–971.

Wang, D., Ni, W., Pang, H., Lu, Q., Huang, Z., Zhao, J., 2010a. Preparation of mesoporous NiO with a bimodal pore size distribution and application in electrochemical capacitors. Electrochim. Acta 55, 6830–6835.

Wang, C., Zhou, Y., Ge, M., Xu, X., Zhang, Z., Jiang, J.Z., 2010b. Large-scale synthesis of SnO2 nanosheets with high lithium storage capacity. J. Am. Chem. Soc. 132, 46–47.

Wang, F., Seo, J.-H., Luo, G., Starr, M.B., Li, Z., Geng, D., et al., 2016. Nanometre-thick single-crystalline nanosheets grown at the water–air interface. Nat. Commun. 7, 10444.

Wang, F., Yu, Y., Yin, X., Tian, P., Wang, X., 2017. Wafer-scale synthesis of ultrathin CoO nanosheets with enhanced electrochemical catalytic properties. J. Mater. Chem. A 5, 9060–9066.

Wang, F., Wang, Z., Wang, Q., Wang, F., Yin, L., Xu, K., et al., 2015. Synthesis, properties and applications of 2D non-graphene materials. Nanotechnology 26, 292001.

Wang, Y., Ren, B., Zhen Ou, J., Xu, K., Yang, C., Li, Y., et al., 2021. Engineering two-dimensional metal oxides and chalcogenides for enhanced electro- and photocatalysis. Sci. Bull. 66, 1228–1252.

Wazir, N., Ding, C., Wang, X., Ye, X., Lingling, X., Lu, T., et al., 2020. Comparative studies on two-dimensional (2D) rectangular and hexagonal molybdenum dioxide nanosheets with different thickness. Nanoscale Res. Lett. 15, 1–9.

Wu, C., Feng, F., Xie, Y., 2013. Design of vanadium oxide structures with controllable electrical properties for energy applications. Chem. Soc. Rev. 42, 5157–5183.

Wu, H., Zhou, X., Li, J., Li, X., Li, B., Fei, W., et al., 2018. Ultrathin molybdenum dioxide nanosheets as uniform and reusable surface-enhanced Raman spectroscopy substrates with high sensitivity. Small 14, 1802276.

Xiao, X., Song, H., Lin, S., Zhou, Y., Zhan, X., Hu, Z., et al., 2016. Scalable salt-templated synthesis of two-dimensional transition metal oxides. Nat. Commun. 7, 11296.

Xie, H., Li, Z., Cheng, L., Haidry, A.A., Tao, J., Xu, Y., et al., 2022. Recent advances in the fabrication of 2D metal oxides. Iscience 25, 103598.

Xu, C., Qiu, P., Chen, H., Jiang, F., Wang, X., 2019. Fabrication of two-dimensional indium oxide nanosheets with graphitic carbon nitride nanosheets as sacrificial templates. Mater. Lett. 242, 24–27.

Xu, Y., Zhang, J., Han, X., Wang, X., Ye, C., Mu, W., et al., 2023. Squeeze-printing ultrathin 2D gallium oxide out of liquid metal for forming-free neuromorphic memristors. ACS Appl. Mater. Interfaces 15, 25831–25837.

Xue, L., Li, S., Shen, T., Ni, M., Qiu, C., Sun, S., et al., 2020. Two-dimensional metal (oxy) hydroxide and oxide ultrathin nanosheets via liquid phase epitaxy. Energy Storage Mater. 32, 272–280.

Yan, G., Wang, Y., Zhang, Z., Dong, Y., Wang, J., Carlos, C., et al., 2020. Nanoparticle-decorated ultrathin La2O3 nanosheets as an efficient electrocatalysis for oxygen evolution reactions. Nano-Micro Lett. 12, 49.

Yang, W., Li, J., Zhang, X., Zhang, C., Jiang, X., Liu, B., 2018. Hydrothermal approach to spinel-type 2D metal oxide nanosheets. Inorg. Chem. 58, 549–556.

Yi, T.F., Shi, L., Han, X., Wang, F., Zhu, Y., Xie, Y., 2021. Approaching high-performance lithium storage materials by constructing hierarchical CoNiO2@ CeO2 nanosheets. Energy Environ. Mater. 4, 586–595.

Yin, C., Gong, C., Chu, J., Wang, X., Yan, C., Qian, S., et al., 2020. Ultrabroadband photodetectors up to 10.6 µm based on 2D Fe3O4 nanosheets. Adv. Mater. 32, 2002237.

Yin, X.-T., Zhou, W.-D., Li, J., Wang, Q., Wu, F.-Y., Dastan, D., et al., 2019. A highly sensitivity and selectivity Pt-SnO2 nanoparticles for sensing applications at extremely low level hydrogen gas detection. J. Alloy. Compd. 805, 229–236.

Yin, X.T., Zhou, W.D., Li, J., Lv, P., Wang, Q., Wang, D., et al., 2019. Tin dioxide nanoparticles with high sensitivity and selectivity for gas sensors at sub-ppm level of hydrogen gas detection. J. Mater. Sci. Mater. Electron. 30, 14687–14694.

Yu, H., Yu, X., Liu, C., Zhang, Y., 2022. Significant effect of SDS on the optimum operating temperature of ZnO nanosheets gas-sensitive materials. Eur. Phys. J. Appl. Phys. 97, 3.

Zavabeti, A., Aukarasereenont, P., Tuohey, H., Syed, N., Jannat, A., Elbourne, A., et al., 2021. High-mobility p-type semiconducting two-dimensional β-TeO2. Nat. Electron. 4, 277–283.

Zavabeti, A., Zhang, B.Y., de Castro, I.A., Ou, J.Z., Carey, B.J., Mohiuddin, M., et al., 2018. Green synthesis of low-dimensional aluminum oxide hydroxide and oxide using liquid metal reaction media: ultrahigh flux membranes. Adv. Funct. Mater. 28, 1804057.

Zeng, L., Zhang, G., Huang, X., Wang, H., Zhou, T., Xie, H., 2021. Tuning crystal structure of MnO2 during different hydrothermal synthesis temperature and its electrochemical performance as cathode material for zinc ion battery. Vacuum 192, 110398.

Zhan, F., Geng, B., Guo, Y., 2009. Porous Co3O4 nanosheets with extraordinarily high discharge capacity for lithium batteries. Chemistry - Eur. J. 15, 6169–6174.

Zhang, R., Liu, X., Zhou, T., Wang, L., Zhang, T., 2018. Carbon materials-functionalized tin dioxide nanoparticles toward robust, high-performance nitrogen dioxide gas sensor. J. Colloid Interface Sci. 524, 76–83.

Zhang, X., Lai, J., Gray, T., 2023. Recent progress in low-temperature CVD growth of 2D materials. Oxford Open Mater. Sci. 3, itad010.

Zheng, Y., Li, Y., Huang, R., Huang, Y., Yao, J., Huang, B., et al., 2019. Fabrication of 2D NiO porous nanosheets with superior lithium storage performance via a facile thermal-decomposition method. ACS Appl. Energy Mater. 2, 8262–8273.

Zhong, R., Wang, Q., Du, L., Pu, Y., Ye, S., Gu, M., et al., 2022. Ultrathin polycrystalline Co3O4 nanosheets with enriched oxygen vacancies for efficient electrochemical oxygen evolution and 5-hydroxymethylfurfural oxidation. Appl. Surf. Sci. 584, 152553.

Zhou, K., Shang, G., Hsu, H., Han, S., Roy, V.A.L., Zhou, Y., 2023. Emerging 2D metal oxides: from synthesis to device integration. Adv. Mater. 35, 2207774.

Zhou, X., Li, X., Gao, Y., Li, L., Huang, L., Ye, J., 2019. Preparation and characterization of 2D ZnO nanosheets/regenerated cellulose photocatalytic composite thin films by a two-step synthesis method. Mater. Lett. 234, 26–29.

Zhu, H., Yuan, Z., Shen, Y., Gao, H., Meng, F., 2023. Highly selective and ppb-level butanone sensors based on SnO2/NiO heterojunction-modified ZnO nanosheets with electron polarity transport properties. ACS Sens. 8, 2635–2645.

Zhu, Y., Guo, H., Zhai, H., Cao, C., 2015. Microwave-assisted and gram-scale synthesis of ultrathin SnO2 nanosheets with enhanced lithium storage properties. ACS Appl. Mater. Interfaces 7, 2745–2753.

CHAPTER THREE

2D non-layered metal dichalcogenides

Mostafa M.H. Khalil[a,*], Abdelrahman M. Ishmael[b], and Islam M. El-Sewify[a]

[a]Chemistry Department, Faculty of Science, Ain Shams University, Abbassia, Cairo, Egypt
[b]Nanomaterials Science Program, Faculty of Science, Benha University, Cairo, Egypt
*Corresponding author. e-mail address: mostafa_khalil@sci.asu.edu.eg

Contents

1. Introduction	64
1.1 Definition of non-layered 2D metal dichalcogenides	64
1.2 Importance of studying non-layered 2D metal dichalcogenides	66
2. Structure of different types of non-layered 2D metal dichalcogenides	67
2.1 Non-layered transition metal dichalcogenides (TMDCs)	67
2.2 Non-layered metal dichalcogenides beyond TMDCs	71
2.3 Mixed metal dichalcogenides	74
3. Structural characteristics of non-layered 2D metal dichalcogenides	74
3.1 Monolayer and few-layer structures	74
3.2 Defects and grain boundaries	75
3.3 Structural phase transitions	75
4. Properties of non-layered 2D metal dichalcogenides	76
4.1 Electronic properties	76
4.2 Mechanical properties	77
4.3 Chemical properties non-layered 2D metal dichalcogenides	78
5. Synthesis of non-layered 2D metal dichalcogenides	78
5.1 Dry methods	79
5.2 Wet chemistry methods	83
6. Applications of non-layered 2D metal dichalcogenides	84
6.1 Electronics	85
6.2 Optoelectronics	85
6.3 Energy conversion and storage	86
6.4 Catalysis	88
6.5 Biomedical applications	90
7. Challenges and future directions	91
7.1 Current challenges in the research of 2D metal dichalcogenides	91
7.2 Future research directions in the field of 2D metal dichalcogenides	92
References	93

Semiconductors and Semimetals, Volume 113
ISSN 0080-8784
https://doi.org/10.1016/bs.semsem.2023.09.010

Copyright © 2023 Elsevier Inc.
All rights reserved.

Abstract

Two-dimensional (2D) materials have recently received a lot of interest due to their unusual electrical, optical, and mechanical capabilities. Metal dichalcogenides (MX_2), in particular, have emerged as a potential family of materials for a variety of applications. There is significant interest in investigating non-layered 2D metal dichalcogenides with various structural types, even though most investigations have concentrated on the layered structure of MX_2 materials. Owing to the fascinating properties, 2D non-layered materials are attracting dramatically increasing interests, and great progress have been made in the synthetic strategies and applications of 2D non-layered materials over the past few years. In this chapter, we outline the recent progress of 2D non-layered chalcogenides and advances in their synthetic methods and applications.

1. Introduction

2D metal dichalcogenides (MDCs) are a class of materials that exhibit interesting properties due to their unique atomic structure and two-dimensional nature (Chen et al., 2020; Mu and Sun, 2020; Susarla et al., 2018; Ye et al., 2019). They belong to the larger family of 2D materials, which includes graphene and other atomically thin materials. Thus, study of 2D nanostructures is largely limited to naturally layered materials, in which the in-plane atoms are connected via strong chemical bonding, and the stacking layers are combined via weak van der Waals interaction. Metal dichalcogenides are composed of a transition metal (such as molybdenum or tungsten) and a chalcogen element (such as sulfur or selenium) (Eftekhari, 2017). When these materials are thinned down to a single layer or a few layers, they exhibit remarkable electronic, optical, and mechanical properties, making them attractive for various applications (Lembke et al., 2015; Shim et al., 2017; Wen et al., 2019; Yang et al., 2018; Yang and Hao, 2019; Yeh et al., 2017; Yu et al., 2017a).

1.1 Definition of non-layered 2D metal dichalcogenides

Non-layered 2D metal dichalcogenides (MDs) refer to a class of materials that possess a two-dimensional (2D) structure but lack the typical layered crystal structure commonly associated with traditional layered MDs (Wang et al., 2017a, 2021a; Peng et al., 2021). Unlike layered MDs such as molybdenum disulfide (MoS_2) or tungsten diselenide (WSe_2), where the atoms are arranged in stacked layers held together by weak van der Waals forces, non-layered 2D MDs exhibit different crystal structures and bonding arrangements (Balendhran et al., 2013). In non-layered 2D MDs, the atoms are still arranged in a planar fashion, forming a two-dimensional sheet, but without the layered stacking as shown in Fig. 1. These materials can have a variety of crystal structures, including hexagonal, cubic, or other arrangements. The absence of layered stacking can result in unique properties and behaviors that differ from their layered counterparts (Wang et al., 2021b).

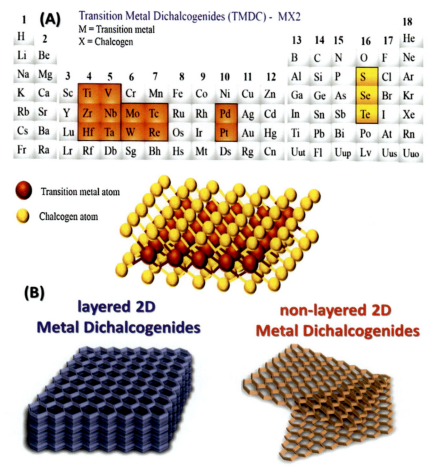

Fig. 1 (A) Transition metal dichalcogenides are layered materials and an example of a 2D material. The layered transition metal dichalcogenides material's crystalline structure is shown schematically. (B) Schematic investigation of layered 2D metal dichalcogenides and non-layered 2D metal dichalcogenides. *Reproduced with permission from Kumar, R., Goel, N., Hojamberdiev, M., Kumar, M., 2020. Transition metal dichalcogenides-based flexible gas sensors. Sensors and Actuators A: Physical, 303, 111875.*

Non-layered 2D MDs have gained significant attention due to their distinct structural features, which can lead to enhanced properties such as improved mechanical strength, novel electronic and optical properties, and unique catalytic activity (Ren et al., 2020). The non-layered 2D MDs represent a fascinating and rapidly growing class of materials with a diverse range of structures and properties, offering exciting possibilities for scientific research and technological advancements (Kumar et al., 2020).

Thus, study of 2D nanostructures is largely limited to naturally layered materials, in which the in-plane atoms are connected via strong chemical bonding, and the stacking layers are combined via weak van der Waals interaction. In fact, for most known 3D materials, only a tiny fraction possesses layered structures, and the majority belong to non-layered materials. These non-layered materials possess diverse and tunable electrical, optoelectronic, magnetic, and catalytic properties, thus exhibiting a promising application prospect in 2D scale.

1.2 Importance of studying non-layered 2D metal dichalcogenides

Studying non-layered 2D MDs is important for uncovering their unique properties, exploring new applications, advancing materials innovation, and contributing to fundamental scientific knowledge. It has the potential to drive technological breakthroughs and shape the development of future devices and materials. The study of non-layered 2D metal dichalcogenides (MDs) holds great importance for several reasons:

A. Non-layered 2D MDs exhibit different crystal structures compared to their layered counterparts. Studying these materials helps establish the relationship between their atomic structure and their unique properties. Understanding the structure–property relationship is crucial for tailoring their properties to specific applications.
B. Non-layered 2D MDs often possess distinct properties that differ from their layered counterparts. These materials can exhibit enhanced mechanical strength, improved electrical conductivity, unique optical properties, and exceptional catalytic activity. Investigating these properties opens up new possibilities for developing advanced materials and devices.
C. Non-layered 2D MDs offer a wide range of potential applications. Their unique properties make them promising candidates for electronics, optoelectronics, energy storage and conversion, catalysis, sensors, and biomedical devices. By studying non-layered 2D MDs, researchers

can explore and harness their unique characteristics for various technological advancements.

D. Research in this area fosters materials innovation by expanding the understanding of 2D materials beyond the traditional layered structures. Exploring non-layered 2D MDs contributes to the discovery of new materials and the development of novel synthesis techniques.

E. Studying non-layered 2D MDs deepens our fundamental understanding of materials science and solid-state physics. It provides insights into the behavior of atoms and electrons in low-dimensional systems and helps refine theoretical models. This fundamental knowledge can be applied to other areas of research and enable advancements in diverse scientific disciplines.

2. Structure of different types of non-layered 2D metal dichalcogenides

2.1 Non-layered transition metal dichalcogenides (TMDCs)

Transition metal dichalcogenides (TMDCs) are a prominent class of two-dimensional (2D) materials that have gained significant attention due to their unique electronic and optical properties (Zhang et al., 2017a). While most studies have focused on the layered structures of TMDCs, non-layered TMDCs with different structural types have also emerged as promising materials for various applications.

2.1.1 2D non-layered In$_2$S$_3$

The indium(III) sulfide, or In$_2$S$_3$, may be found in both layered and non-layered forms (Fig. 2). Due to their lower dimensionality, 2D non-layered In$_2$S$_3$ materials have different chemical characteristics from their layered

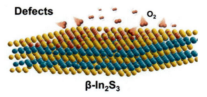

Fig. 2 Schematic diagram of 2D non-layered In2S3 nanosheets. *Reproduced with permission from Huang, W., Song, M., Zhang, Y., Zhao, Y., Hou, H., Hoang, L.H., et al., 2021. Defectsinducedoxidation of two-dimensional β-In2S3 and its optoelectronic properties. Opt.Mater. 119, 111372.*

counterparts (Huang et al., 2021). The chemical composition of 2D non-layered In_2S_3 is the same as that of its bulk equivalent, with a 1:3 ratio of indium (In) to sulfur (S) atoms. By adjusting its size and structure, 2D non-layered In_2S_3's electrical characteristics may be tailored. Wider band gaps in smaller 2D nanostructures relative to bulk In_2S_3 may have an impact on such structures' electrical and optical characteristics. A typical non-layered n-type semiconductor with a moderate bandgap is β-In_2S_3 (Ho and Growth, 2010; Zhao et al., 2019). β-In_2S_3 is a highly photoresponsive material in the visible spectrum due to the several defect states that are brought about by the high-density empty sites in a defect spinel structure. The main photogating effect enables photodetector systems based on heterostructures of β-In_2S_3 to reach ultrahigh responsiveness (Naghavi et al., 2003). The size and shape of 2D non-layered In_2S_3 materials can influence the optical characteristics they display. Compared to bulk In_2S_3, they could have distinct quantum confinement effects and light-absorption properties. 2D non-layered In_2S_3 might have different electrical conductivities. These substances may occasionally behave in a semiconducting manner. Due to its higher surface-to-volume ratio, 2D non-layered In_2S_3 may exhibit different chemical properties from bulk In_2S_3. As a result, 2D In_2S_3 materials may be more vulnerable to surface reactions and chemical interactions. Potential uses for 2D non-layered In_2S_3 materials include catalysts, photodetectors, sensors, and nanoelectronics (Naghavi et al., 2003; Chen et al., 2016).

2.1.2 2D non-layered CdS

The two-dimensional (2D) non-layered CdS materials, like their bulk counterparts, display unique chemical characteristics. The chemical composition of 2D non-layered CdS is the same as that of its bulk equivalent, with a 1:1 ratio of cadmium (Cd) to sulfur (S) atoms (Sun et al., 2018). CdS is a semiconductor material, and its band gap may be modified by changing its size and structure. The band gap of CdS can vary but is generally in the range of 2.4–2.5 (eV). 2D non-layered CdS materials can exhibit unique optical properties, including size-dependent optical absorption and photoluminescence. These properties make them suitable for use in photodetectors, sensors, and optical devices (Yang et al., 2008). The electronic conductivity of 2D non-layered CdS can vary based on its size and structure. Depending on the specific 2D configuration, it may exhibit semiconductor behavior. The chemical reactivity of 2D non-layered CdS can differ from bulk CdS due to its increased surface-to-volume ratio. This can make 2D CdS materials more susceptible to surface reactions and interactions with other chemicals (Huang et al., 2011). 2D non-layered CdS materials have potential applications in nanoelectronics, photodetectors, solar cells, sensors, and catalysis. Their unique size-dependent properties make them promising candidates for various nanotechnology and optoelectronic applications (Jin et al., 2018a).

Successful self-limited epitaxial growth-based CdS and Ge nanoplate synthesis has recently been accomplished (Sun et al., 2018). CdS powder and layered mica were used as the precursor and epitaxy substrate, respectively, to create 2D CdS crystal, and In2S3 powder was used to limit the growth of particular crystal faces, as shown in Fig. 3a. High quality and wurtzite CdS flakes with lateral lengths up to 44 mm and a thickness as low as 6 nm were produced on mica substrates by carefully controlling growth conditions, as illustrated in Fig. 3b (Zhou et al., 2019).

2.1.3 (2D) Non-layered Iron disulfide (FeS$_2$)

Iron disulfide (FeS$_2$), often known as fool's gold or pyrite, can exist in two-dimensional (2D) non-layered forms such as nanosheets or nanoribbons. Due to their decreased dimensionality, these 2D materials have distinctive chemical characteristics. The chemical structure of 2D non-layered FeS$_2$ is the same as that of its bulk equivalent, with a 1:2 ratio of iron (Fe) to sulfur (S) atoms. The crystal structure of 2D non-layered FeS$_2$ may differ from the cubic or pyrite structure of bulk FeS$_2$. Although 2D non-layered FeS$_2$ might have a different band gap depending on its size and structure than bulk FeS$_2$, which is commonly regarded as a semiconductor with a large band gap. FeS$_2$ materials with 2D non-layered structure exhibit distinctive optical properties as size-dependent optical absorbance and photo-luminescence. As a result of these features, they are appropriate for use in photodetectors, sensors, and optical systems.

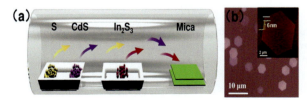

Fig. 3 (A) Schematic illustration of growth of 2D CdS flakes by self-limited epitaxial growth method. Inset: sketch map of 2D CdS flake grown on a mica flake. (B) Optical images of as-synthesized CdS flakes. Inset: AFM image with height profiles. *Reproduced with permission from Zhou, N., Yang, R., Zhai, T., 2019. Two-dimensional non-layered materials. MaterialsToday. Nano 8, 100051.*

The 2D non-layered FeS$_2$'s dimensions and shape can have an impact on its electronic conductivity. According to the unique 2D structure, it may exhibit metallic or even semiconductor behavior. 2D non-layered FeS$_2$ may not be as chemically reactive as bulk FeS$_2$ because of its larger surface-to-volume ratio. As a result, 2D FeS$_2$ materials might be more susceptible to chemical interactions and surface reactions. Among the applications for 2D non-layered FeS$_2$ materials include catalysis, energy storage, photodetectors, sensors, and nanoelectronics.

For non-layered Fe$_{1-x}$S and MoS$_2$, a typical layered TMC, Junyang Tan et al. estimated the defect generation energies as a function of chemical potential of S (ls) (Fig. 4a). For MoS$_2$, they discovered that a S vacancy

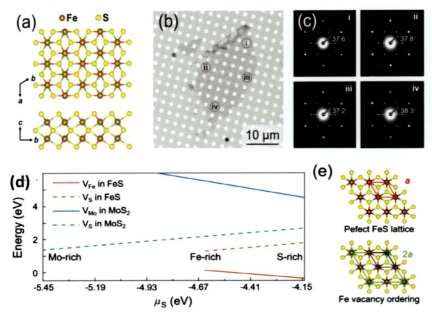

Fig. 4 Structure characterization of non-layered 2D Fe$_{1-x}$S. (A) Top and side views of the atomic structure of hexagonal Fe$_{1-x}$S. Brown: Fe atoms, yellow: S atoms. (B) Low magnification TEM image of a 2D Fe$_{1-x}$S flake. (C) SAED patterns collected at four positions marked (i)–(iv) in (B). (D) Ordered cation Fe vacancies in non-layered 2D Fe$_{1-x}$S. (A) Comparison of the formation energies of a single cation vacancy and a single anion vacancy in non-layered Fe$_{1-x}$S and layered MoS$_2$ as a function of chemical potential of S (ls). (E) Atomic models of perfect hexagonal FeS and ordered Fe vacancies lattice at the perpendicular projection direction (Color online). *Reproduced with permission from Tan, J., Zhang, Z., Zeng, S., Li, S., Wang, J., Zheng, R., et al., 2022a. Dual-metal precursors for the universal growth of non-layered 2D transition metal chalcogenides with ordered cation vacancies. Sci. Bull. 67 (16), 1649–1658.*

(VS) has a significantly lower formation energy than a Mo vacancy (VMo) across the entire range of S chemical potentials. This finding is in line with experimental findings that S vacancies compose the majority of the defects in 2D MoS_2 (Cai et al., 2021; Feng et al., 2020). As opposed to MoS_2, the situation for non-layered $Fe_{1-x}S$ is completely different, and the formation energy of a Fe vacancy (V Fe) is consistently negative, indicating that the production of Fe vacancies is favored. Considering the thickness of the flakes, it is challenging to use HAADF-STEM imaging to directly view each individual point defect in the lattice. The fast Fourier transform (FFT), SAED, EDS, and EELS are employed to identify the existence of Fe vacancies in $Fe_{1-x}S$ flakes. Fig. 4b displays an FFT image from a 2D $Fe_{1-x}S$ atomic-resolution HAADFSTEM image. A second group of superlattice spots with a 2 larger periodicity of 0.58 nm, which is explained by ordered Fe vacancies, can be found in addition to the host lattice spots that show the Fe–S and Fe–Fe spacing in a perfect hexagonal FeS lattice (highlighted by red circles). The framework of hexagonal FeS can be seen in the inverse FFT pictures produced from these two sets of spots (Fig. 4c). The electrical behavior of 2D $Fe_{1-x}S$ was investigated using a two-wire device with a channel thickness of 13.5 nm (Fig. 4d). The temperature-dependent resistance curve (Fig. 4e) exhibits typical semiconductor behavior, with resistance increasing as temperature decreases from 300 to 2 K (Tan et al., 2022a).

2.2 Non-layered metal dichalcogenides beyond TMDCs

Transition metal dichalcogenides (TMDCs) have been extensively studied, but non-layered metal dichalcogenides beyond TMDCs also hold great potential for various applications. In this section, we explore two groups of non-layered metal dichalcogenides: metal selenides (MSe_2) and metal tellurides (MTe_2), as well as mixed metal dichalcogenides that combine different transition metals (Xiong et al., 2023).

2.2.1 Metal selenides (MSe₂)

Metal selenides are non-layered metal dichalcogenides that exhibit intriguing properties (Sun et al., 2022). Similar to their sulfur counterparts, non-layered metal selenides possess a layered structure in their bulk form, but they can be exfoliated into monolayer or few-layer structures (Sun et al., 2022). Non-layered metal selenides have been of particular interest due to their unique electronic, optical, and catalytic properties. Their bandgaps lie in the visible range, enabling efficient light absorption and emission (Huang et al., 2022a). Moreover, non-layered metal selenides demonstrate excellent catalytic activity, which has been explored in applications such as hydrogen evolution reactions and electrocatalysis (Di et al., 2018). The synthesis of non-layered metal selenides follows similar methods used for TMDCs, including chemical vapor deposition (CVD) and exfoliation techniques (Huang et al., 2022b). These methods offer control over the thickness, crystal quality, and doping of the resulting non-layered metal selenides, enabling the exploration of their unique properties. The two-dimensional (2D) non-layered In2Se3 materials display special chemical characteristics similar to their bulk counterparts. Indium (In) to selenium (Se) atoms are distributed in a 1:3 ratio in 2D non-layered In_2Se_3, keeping it chemically identical to its bulk counterpart. The crystal structure of 2D non-layered forms may be different from the layered structure of In_2Se_3. These 2D topologies can take the form of nanosheets, nanoribbons, or other planar arrangements. Depending on the synthesis procedure and environmental factors, the particular 2D structure may change. By adjusting its size and structure, 2D non-layered In_2Se_3's electrical characteristics may be tailored. Compared to bulk In_2Se_3, smaller 2D nanostructures may have different band gaps, which might impact their electrical and optical characteristics. These characteristics make them appropriate for use in photodetectors and other optoelectronic gadgets. Depending on its size and structure, 2D non-layered In_2Se_3 might have different electrical conductivities. It could demonstrate semiconductor behavior, depending on the particular 2D design. This may render 2D In_2Se_3 materials more vulnerable to chemical reactions on the surface and other chemical interactions. Potential uses for 2D non-layered In_2Se_3 materials include nanoelectronics, photodetectors, solar cells, sensors, and catalysis. They are strong prospects for several nanotechnology and optoelectronic applications because to their special size-dependent characteristics. Wenjuan Huang et al. used powders of selenium (Se) and indium iodide (InI_3) in a mixed carrier gas environment of H_2 and Ar, In_2Se_3 nanosheets are epitaxially produced on SiO_2/Si substrates by CVD. A schematic representation of the CVD synthesis of 2D In_2Se_3 nanosheets is shown in Fig. 5a and the crystal structure of $-In_2Se_3$ is was investigated as shown in Fig. 5b (Huang et al., 2022b).

2.2.2 Metal tellurides (MTe₂)

Metal tellurides, such as molybdenum ditelluride ($MoTe_2$), tungsten ditelluride (WTe_2), and others, are another group of non-layered metal dichalcogenides that have attracted significant attention (Cho et al., 2015). Non-layered metal

2D non-layered metal dichalcogenides 73

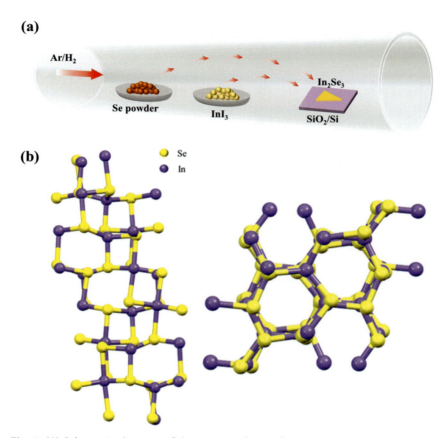

Fig. 5 (A) Schematic diagram of the CVD synthesis of 2D In$_2$Se$_3$ nanosheets. (B) Side and top views of the crystal structure of γ-In$_2$Se$_3$.

tellurides exhibit intriguing electronic and topological properties, making them promising candidates for various applications (Liu et al., 2018). One of the notable properties of non-layered metal tellurides is their high electrical conductivity, which can be attributed to their unique electronic band structures. Non-layered WTe$_2$, for instance, exhibits a type-II Weyl semimetal state, which gives rise to unique electronic transport phenomena (Manzeli et al., 2017). This has led to investigations on their potential use in electronic and quantum devices. Synthesis techniques for non-layered metal tellurides are similar to those used for other non-layered metal dichalcogenides, including CVD and exfoliation methods (Nayak et al., 2016). The controlled synthesis of non-layered metal tellurides enables the exploration of their electronic and topological properties, as well as their potential applications in next-generation electronics.

2.3 Mixed metal dichalcogenides

Mixed metal dichalcogenides refer to non-layered metal dichalcogenides that combine different transition metals in their structures (Ying et al., 2021). These materials offer the possibility of tailoring properties beyond what individual metal dichalcogenides can provide. By varying the composition and arrangement of different transition metals, researchers can tune the electronic, optical, and catalytic properties of mixed metal dichalcogenides. For example, mixed metal dichalcogenides may exhibit enhanced catalytic activity compared to their individual metal dichalcogenide counterparts (Aras et al., 2022). The combination of different transition metals can create synergistic effects, leading to improved catalytic performance in applications such as water splitting, fuel cells, and chemical sensing (Cherusseri et al., 2019). The synthesis of mixed metal dichalcogenides involves strategies such as CVD, alloying, and hybridization techniques (Voiry et al., 2015). By carefully controlling the composition and synthesis parameters, researchers can obtain well-defined mixed metal dichalcogenide structures with tailored properties.

In summary, non-layered metal dichalcogenides beyond TMDCs, including metal selenides, metal tellurides, and mixed metal dichalcogenides, offer unique opportunities for exploring new materials with diverse properties. These materials hold promise for various applications in electronics, optoelectronics, catalysis, and beyond, further expanding the scope of 2D materials research.

3. Structural characteristics of non-layered 2D metal dichalcogenides

Understanding these structural features is crucial for tailoring their properties and exploring their potential applications. In this section, we discuss the structural characteristics of non-layered 2D metal dichalcogenides, including their monolayer and few-layer structures, defects and grain boundaries, and structural phase transitions (Ling et al., 2015).

3.1 Monolayer and few-layer structures

Non-layered 2D metal dichalcogenides can exist in monolayer or few-layer structures, similar to layered materials such as graphene and transition metal dichalcogenides (TMDCs) (Joseph et al., 2023). However, the absence of well-defined layers in non-layered structures introduces distinct structural properties. Monolayer non-layered metal dichalcogenides typically consist of a single atomic layer, where metal atoms are sandwiched between chalcogen atoms. The absence of layer stacking results in reduced interlayer interactions and enhances the intrinsic properties of the materials (Li et al., 2023). Monolayers can be synthesized through various techniques such as chemical vapor deposition (CVD), mechanical exfoliation, and liquid phase exfoliation (Hu et al., 2023). Few-layer structures, on the other hand, consist of a small number of atomic layers, typically up to five layers. The thickness of the few-layer structures can significantly influence the electronic, optical, and mechanical properties of the materials. The controlled synthesis of few-layer structures enables the manipulation of their properties for specific applications.

3.2 Defects and grain boundaries

Non-layered 2D metal dichalcogenides often contain defects and grain boundaries, which can significantly affect their properties (Tantis et al., 2023). Defects can arise from missing atoms, impurities, or structural irregularities within the atomic lattice. Point defects, such as vacancies or substitutional impurities, can modify the electronic and optical properties of the materials. Additionally, line defects, such as dislocations or grain boundaries, can influence the mechanical properties and charge transport in non-layered structures (Nayir et al., 2023). Defect engineering has emerged as an essential strategy for tailoring the properties of non-layered metal dichalcogenides. The controlled introduction of defects can enhance catalytic activity, improve electrical conductivity, and tune the bandgap of these materials (Huang et al., 2018).

3.3 Structural phase transitions

Non-layered 2D metal dichalcogenides can undergo structural phase transitions, where their atomic arrangements and symmetry change at certain temperatures or under external stimuli (Patil et al., 2023). These phase transitions can lead to significant modifications in the material's properties. For instance, some non-layered metal dichalcogenides exhibit reversible structural phase transitions between semiconducting and metallic phases (Kapuria et al., 2023). The transition can be induced by thermal, electrical, or optical stimuli. Understanding the mechanisms behind these phase transitions and their correlation with the electronic and optical properties is essential for the development of next-generation electronic and optoelectronic devices (Irfan et al., 2023). In summary, non-layered 2D metal dichalcogenides possess distinctive structural characteristics, including monolayer and few-layer structures, defects and grain boundaries, and structural phase transitions. These features play a crucial role in determining their properties and offer opportunities for tailoring their functionalities for various applications.

4. Properties of non-layered 2D metal dichalcogenides
4.1 Electronic properties

Non-layered 2D MDs exhibit quantum confinement effects due to their reduced dimensionality. The electronic structure of these materials becomes discretized into energy levels, leading to quantum size effects. The confinement of charge carriers in two dimensions can result in distinct energy bands and density of states compared to bulk materials.

The bandgap of non-layered 2D MDs can be tuned by adjusting the size and shape of the 2D structures. By modifying the dimensions, such as the width or length of the nanostructures, the bandgap can be tailored, allowing for control over their optical and electronic properties. This tunability is particularly useful for optoelectronic applications.

Non-layered 2D MDs can exhibit a strong quantum confinement regime, where the electronic wavefunctions are significantly localized within the nanostructures. This localization leads to discrete energy levels and enhanced quantum effects, such as Coulomb blockade, single-electron charging, and quantum tunneling. These phenomena are of interest for quantum computing and single-electron devices.

Non-layered 2D MDs have large surface-to-volume ratios, resulting in a significant number of surface states. These surface states can affect the electronic properties of the material, leading to enhanced surface reactivity, modified charge transport, and the emergence of unique phenomena, such as topological surface states.

Non-layered 2D MDs can exhibit strong electron–phonon interactions due to their reduced dimensionality and confinement effects. These interactions influence the transport properties, such as electrical conductivity and heat dissipation, and can give rise to novel collective excitations, such as plasmons and excitons.

It's important to note that non-layered 2D MDs represent a diverse class of materials, and their electronic properties can vary depending on the specific composition, size, and structure. The above properties provide a general overview of the characteristics observed in this class of materials, but further research and specific investigation are required to fully understand and exploit their unique electronic properties.

4.2 Mechanical properties

The mechanical properties of non-layered 2D MDs can vary depending on their specific composition, size, and structure. Non-layered 2D MDs can exhibit exceptional mechanical strength. Due to their 2D nature and the absence of weak interlayer bonds, these materials can possess high intrinsic strength and stiffness. Their mechanical properties can rival or even surpass those of traditional engineering materials. The high strength is attributed to strong covalent bonding within the lattice structure.

Non-layered 2D MDs are typically flexible and exhibit good bendability. The absence of interlayer interactions that could restrict movement allows these materials to be highly flexible, enabling them to withstand large deformations without fracture. This property makes them suitable for applications requiring conformability or mechanical flexibility, such as flexible electronics or wearable devices. Non-layered 2D MDs can exhibit high elasticity and have a high elastic modulus. The elastic modulus refers to the resistance of a material to deformation under stress. Due to their strong covalent bonds, non-layered 2D MDs can possess high elastic moduli, providing resistance to deformation and making them suitable for applications that require materials with high mechanical rigidity.

Fracture toughness is a measure of a material's resistance to crack propagation. Non-layered 2D MDs can exhibit varying fracture toughness depending on their specific composition and structure. While the absence of weak interlayer bonding can enhance their intrinsic strength, it can also make them more susceptible to brittle fracture. The fracture toughness of non-layered 2D MDs can be improved through strategies such as defect engineering or hybridization with other materials. Non-layered 2D MDs can display anisotropic mechanical properties. The mechanical behavior of these materials can vary depending on the direction of applied stress due to the anisotropy resulting from their crystal structure. The properties perpendicular to the plane of the 2D MDs may differ from those within the plane.

It is important to note that the mechanical properties of non-layered 2D MDs are influenced by various factors, including the composition, size, defects, and strain. The specific mechanical behavior of a non-layered 2D MD can be further explored and tailored through experimental characterization and computational simulations to understand and optimize their performance for specific applications.

4.3 Chemical properties non-layered 2D metal dichalcogenides

Non-layered 2D MDs can exhibit varying degrees of chemical reactivity depending on their constituent elements. The reactive sites in these materials are primarily located at the edges and defects, where the bonding structure may be altered or exposed to the surrounding environment. The reactivity of non-layered 2D MDs can influence their stability, surface chemistry, and interactions with other substances.

The surface of non-layered 2D MDs can be functionalized through chemical modification. This involves introducing specific chemical groups or molecules onto the surface, allowing for tailored properties or interactions. Functionalization can be achieved through various methods, such as covalent bonding, non-covalent interactions, or electrostatic interactions, enabling the integration of these materials into a wide range of applications.

Non-layered 2D MDs can exhibit catalytic activity due to their unique surface properties and active sites. The chemical composition and structure of these materials can provide sites for catalytic reactions, such as hydrogen evolution, oxygen reduction, or various other redox reactions. The catalytic properties of non-layered 2D MDs make them potential candidates for applications in energy conversion, environmental remediation, and chemical synthesis.

Non-layered 2D MDs can be sensitive to changes in their surrounding chemical environment. Their electronic structure and surface properties can be influenced by the presence of specific molecules or ions, leading to changes in their electrical conductivity, optical properties, or surface charge. This sensitivity to external stimuli makes non-layered 2D MDs promising candidates for chemical sensing and detection applications.

The stability of non-layered 2D MDs can vary depending on their specific composition and the environmental conditions to which they are exposed. Factors such as temperature, humidity, and exposure to reactive gases can influence their stability. It is important to consider the stability of non-layered 2D MDs when designing and implementing them in various applications to ensure their long-term performance.

5. Synthesis of non-layered 2D metal dichalcogenides

To achieve the 2D anisotropic growth of MDs, various strategies have been used including both wet chemistry and dry methods such as chemical vapor deposition (CVD) to demonstrate a variety of applications based on the unique properties of the proposed non-layered materials (Wang et al., 2017b).

5.1 Dry methods

2D non-layered MDs can be directly synthesized through CVD-based methods or by firstly growing 2D layered materials followed by a conversion process to obtain the corresponding 2D non-layered materials (Cheng et al., 2016; Wang et al., 2015a). CVD is a widely used method that has shown superior benefits in preparing 2D nanomaterials with high crystal quality, high yield, and large scale. Multiple stoichiometric ratios of 2D TMDCs with numerous physical and chemical properties have been successfully obtained by controllable CVD system to enable new opportunities in electronics, optoelectonics, valleytronics, spintronics and catalysis area. Chang et al. synthesized 2D InSe nanosheets by a CVD method, the InSe atomic films grown by their method are continuous and uniform over a large area of $1 \, \text{cm} \times 1 \, \text{cm}$ and shows unique optical and electrical properties (Chang et al., 2018). Feng et al. successfully prepared mono-, bi-, tri-layer In_2Se_3 nanosheets through a simple CVD process. By selectively etching the mica substrate by oxygen plasma, they successfully prepared a patterned array of the 2D In_2Se_3 (Feng et al., 2016). Dai et al. have successfully synthesized two types of nickel sulfides nanoplates with different stoichiometries through an atmospheric pressure CVD (APCVD). Tuning the growth temperature and the sulfur partial pressure, 2D non-layered NiS and NiS_2 nanoplates were selectively prepared (Dai et al., 2020). Most non-layered 2D TMDCs possess a sandwiched structure, in which the transition metal atom is between the two chalcogen atoms, giving the stoichiometric MX_2 (Voiry et al., 2015). Nevertheless, when the metallic element has multiple valence states in TMDCs, the controllable synthesis of pure 2D TMDCs remain a challenge due to the lack of a drive force to break the symmetry of bulk crystals and promoting the growth of anisotropy. Ji et al. synthesized vanadium sulfide nanosheets through a simple CVD method and found the component of the resulting nanosheets not only include VS_2 but also consist of V_4S_7 and V_5S_8 (Ji et al., 2017). Therefore, developing the controllable synthesis methods of 2D TMDCs with tunable pure stoichiometry is highly desirable.

Because of the difficulties of anisotropic growth, simple CVD method cannot meet the requirements for synthesizing 2D non-layered crystals. To overcome such an obstacle, a so-called van der Waals epitaxial (vdWE) growth was introduced and has exhibited great ability to realize highly anisotropic growth for 2D non-layered materials. Unlike the conventional epitaxial growth such as metal organic CVD (MOCVD) and molecular beam epitaxy (MBE), vdWE does not require strict requirement of lattice matching between the growing layer and the substrates. That is because the vdWE substrates can act as the thermal mismatch buffer layer to relax the thermal stress. In addition, the migration of the reaction precursors on the substrates can be facilitated as a result of the relatively weak van der Waals interaction at the interfaces during epitaxial growth, thus resulting in highly anisotropic growth with relatively large lateral sizes, high crystal quality, and even defect-free crystals (Wang et al., 2015b, 2015c). Usually, mica, h-BN, graphene and MoS_2 can be used as the vdWE substrates because of their atomically smooth and chemically inert surfaces. Jiang et al. have successfully synthesized non-layered 2D Ni-doped CoO single crystals with an ultrathin thickness via the one-pot vdWE method, which has good homogeneity and reproducibility (Jiang et al., 2023). Making use of 2D layered materials as the vdWE substrates, many 2D non-layered materials were successfully synthesized, such as PdTe, SnTe, CdS, $Pb_{1-x}Sn_xSe$, and PbS nanoplates (Wen et al., 2016; Zhao et al., 2021a). The lateral size, morphology, and layer number of the materials can be adjusted by the growth parameters, such as growth temperature, rate of gas flow, and volume of precursors. The development of 2D material families using this method is clearly described in Fig. 6 (Novoselov, 2020).

A novel synthetic strategy based on space-confined vapor deposition has emerged to obtain ultrathin grains with anisotropic crystal structures, such as ReS_2 and $ReSe_2$ (Wang et al., 2016a; Zhu et al., 2016). These materials are

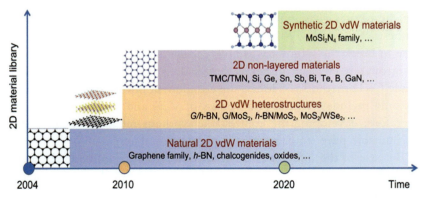

Fig. 6 Development of 2D material families. Natural 2D vdW materials refer to those having known analogous 3D layered allotropes in nature. G represents graphene. *Reproduced with permission from Novoselov, K.S., 2020. Discovery of 2D van der Waals layered $MoSi_2N_4$ family, National Science Review, 7 (12), 1842–1844.*

difficult to obtain as single-layer flakes or even few-layer materials by the traditional vapor deposition process due to their low lattice symmetry and interlayer decoupling characteristics. Moreover, the space-confined growth strategy has also been extended to the fabrication of patterned 2D materials on vdWE substrates (Wang et al., 2015c). Another technique is the self-limited epitaxial growth, Jin et al. artificially controlled the growth kinetics of 2D non-layered CdS crystal, which yielded high-quality single-crystalline wurtzite CdS flakes (as thin as 6 nm) on mica substrate with a large domain size (>40 μm), revealing an anisotropic ratio up to 6×10^3 (Wang et al., 2015b). Hot-pressing method is another conventional technique that has been used to transform a powder to solid body by applying heat and pressure.

Some out-of-plane heterostructures of 2D layered and non-layered MDs such as MoS_2/CdS, $MoS_2/PbSe$, graphene/PbS are fabricated via vdWE (Li et al., 2016; Cui et al., 2017; Wu et al., 2017; Jin et al., 2018b). Another strategy to grow 2D non-layered materials directly on SiO_2/Si wafers are lifting the nanomaterials up of the substrate surfaces. $Pb_{1-x}Sn_xTe$ nanoplates were synthesized on SiO_2/Si substrate. The thickness of the nanoplates mainly distributed in 40 nm with lateral size as large as 20 μm (Wang et al., 2016d).

Tan et al. reported a universal dual-metal precursors method with the use of a mixture of a metal and its chloride as the source to grow various non-layered 2D TMCs (Tan et al., 2022b). Using this method, the evaporation rate to provide a constant metal supply and facilitate the growth of non-layered 2D TMCs can be controlled. Taking $Fe_{1-x}S$ as an example, flakes as thin as 3 nm with a lateral size over 100 mm are grown. The low-melting-point iron chloride ($FeCl_3$, melting point of 315 °C) as an additive to Fe metal (Fe melting point 1535 °C) to control the evaporation rate and promote the growth of Fe-based TMCs. A schematic of the process is illustrated in Fig. 7a. Different metal sources including Fe only, $FeCl_3$ only, and a mixture of Fe and $FeCl_3$ were used. The sulfur powder was used as the sulfur source and the growth was conducted at 500–600 °C in Ar and H_2. They found that when using pure Fe powder as the source, nothing was grown on the substrate. In contrast, the substrate was coated with many small and thick particle-like $Fe_{1-x}S$ flakes when only $FeCl_3$ was used, while 2D $Fe_{1-x}S$ flakes with half-hexagonal-like shape, Fig. 7a, was obtained when a mixture of Fe and $FeCl_3$ was used as the dual metal precursors. At a relatively low temperature, ultrathin $Fe_{1-x}S$ flakes were preferentially obtained, as shown in Fig. 7b which is a representative AFM image of three $Fe_{1-x}S$ flakes with a thickness of 5 nm grown at 500 °C.

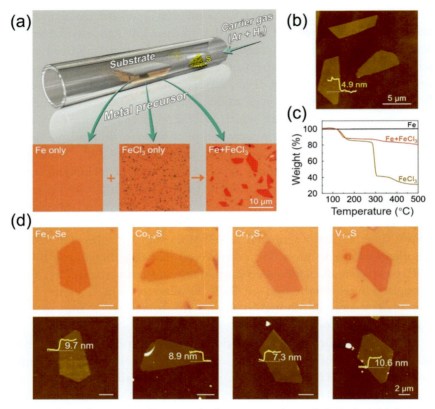

Fig. 7 Dual-metal precursors CVD growth of various non-layered 2D TMCs. (A) Schematic of the growth method. Insets are growth results when using Fe only, FeCl₃ only, or an Fe + FeCl₃ mixture as the metal precursor. (B) AFM image of Fe$_{1-x}$S flakes grown on a SiO₂/Si substrate when using an Fe + FeCl₃ mixture as the dual-metal precursors. (C) TGA curves for Fe, FeCl₃, and a mixture of Fe and FeCl₃. (D) Optical microscopy (upper panel) and the corresponding AFM (lower panel) images of 2D Fe$_{1-x}$Se, Co$_{1-x}$S, Cr$_{1-x}$S, and V1−xS grown by this method. Scale bars: 2 mm. (Color online). *Reproduced with permission from Tan, J., Zhang, Z., Zeng, S., Li, S., Wang, J., Zheng, R., et al., 2022b. Dual-metal precursors for the universal growth of non-layered 2D transition metal chalcogenides with ordered cation vacancies. Sci. Bull. 67 (16), 1649–1658.*

This dual-metal precursors method is universal for growing various non-layered materials in a 2D form, including Fe$_{1-x}$S, Fe$_{1-x}$Se, Co$_{1-x}$S, Cr$_{1-x}$S, and V$_{1-x}$S, paving the way to exploit the properties of these non-layered materials at thickness limit, Fig. 3d. They show the existence of ordered cation vacancies (Fe) in the material, which is distinct from the anion vacancies (S, Se, and Te) in common layered TMCs as MoS₂.

The second approach of synthesizing 2D non-layered MDs is the conversion process of 2D layered materials "precursors". This two-step strategy is relatively facile to control in comparison to the strict requirements of growth parameters of vdWE and vertical growth methods (Zheng et al., 2016; Schornbaum et al., 2014). Nevertheless, it works only on the specific material species, such as organic–inorganic perovskites of RNH_3PbX_3. Xiong et al. demonstrated the synthesis of 2D lead halide perovskite family $CH_3NH_3PbX_3$ (X = Cl, Br, I) by converting 2D layered PbX2 (Wang et al., 2015d). Liu et al. reported the growth of perovskite arrays on graphene, h–BN and MoS_2 substrates (Yakunin et al., 2015). Duan et al. demonstrated the wafer-scale synthesis of perovskite crystals arrays through controlling the periodical hydrophobic/hydrophilic regions by a monolayer hydrophobic (octadecyl) trichlorosilane (OTS) (Liu et al., 2016).

5.2 Wet chemistry methods

Wet-chemical synthesis has been emerging as a very promising alternative to the CVD method towards the high-yield, low-cost, controllability and mass production of non-layer structured ultrathin 2D nanomaterials (Wang et al., 2017b).

Hydrothermal synthesis method is the earliest and most commonly used method for synthesizing non-layered 2D structured nanomaterials using temperature ranges of synthesis from 100 °C to 240 °C. The advantages of the hydrothermal synthesis method are that products have high purity, in general represents a low energy consumption and low environmental impact since the reaction proceed under a controlled atmosphere (Bi et al., 2012). Typically, under the commonly used synthesis process, a sealed vessel usually made of polypropylene or Teflon is used in a steel autoclave, requiring a lower synthesis temperature in comparison to other methods (He et al., 2021). In hydrothermal method, the reactants, water, and surfactants are mixed at room temperature and then placed in a sealed container. The temperature of the reaction system must be higher than the boiling point of the solvent, which can promote the generation of high pressure and further improve the crystallinity of MDs (Khaleque et al., 2020; Zhou et al., 2021). In particular, the hydrothermal method was mainly applied to prepare various non-layered m-TMDCs including $NiSe_2$ nanosheets, CoS nanowires, CoS nanoparticles, and $CoSe_2$ nanosheets, which are hard to be synthesized by the top-down method (Zhou et al., 2019; Wang et al., 2016b; Bao et al., 2008; Li et al., 2015; Peng et al., 2014). Hydrothermal method has the advantages of simplicity, low cost, high yield, and scalability. But the hydrothermal method is very sensitive to experimental conditions such as precursors, solvent systems, surfactants, and temperatures (Fang et al., 2020). Dou et al. developed a generalized solvothermal method for the synthesis of a series of metal oxide nanosheets, including TiO_2, ZnO, Co_3O_4, WO_3, Fe_3O_4 and MnO_2 (Zhao et al., 2021b). The size of the TiO_2 nanosheets is ~200 nm, while the sizes of ZnO, Co_3O_4 and WO_3 are up to 1–10 μm. Similarly, the thickness of these metal oxide nanosheets varied between 1.6 and 5.2 nm, corresponding to 2–7 stacking layers of the monolayer.

There are some wet-chemical methods such as surface-energy-controlled synthesis (SECS), template-directed synthesis (TDS), confined space synthesis (CSS) and colloidal synthesis (CS). Metals such as Ag, Pd, Rh, and Ru, can be synthesized by surface-energy-controlled synthesis (SECS) (Sun et al., 2014; Zhang et al., 2010; Yin et al., 2014; Hou et al., 2013). Q. Zhang et al. prepared Ag nanoplates with high aspect ratio where poly-vinylpyrrolidone (PVP) and citrate ion work as surfactant agents (Yin et al., 2012). Confined space synthesis (CSS) provides a 2D space in the non-layered nanosheets so that the specific reaction is constrained in. Wang et al. fabricated single-crystalline ZnO nanosheets with large lateral sizes for the first time at the water–air interface. To minimize the surface energy of ZnO nanosheets, the reactive precursor ions in the space have to array in a crystallographic face and then for the wanted 2D non-layered nanosheets. By introducing high-concentration sodium dodecyl sulfate, the water–air interface was built up between air and the solution, in which the ZnO growth was only limited. In addition, the sodium dodecyl sulfate also works as surfactant agents to guide the growth of ZnO nanosheets (Wang et al., 2016c). Template-directed synthesis (TDS) offers 2D templates to help the 2D growth. Graphene oxide (GO) has long been used as the 2D template to direct the growth of various non-layered materials such as Au and SnO_2 (Ding et al., 2011; Yan et al., 2018). Compared with dry methods, wet chemical methods occur in solution and can be used for synthesizing more types of 2D non-layered materials, such as metal, metal oxide, metal chalcogenides. Generally, wet chemistry method plays an important role in 2D non-layered material synthesis.

6. Applications of non-layered 2D metal dichalcogenides

Non-layered 2D MDs show superior surface activity and stronger quantum confinement effect than 2D layered MDs. Hybrid heterostructures of layered and non-layered 2D MDs further broadens their applications and might overcome the intrinsic restrictions. In this part, we highlight the latest advances in electronics, optoelectronics, energy conversion, energy storage, catalysis and biomedical applications.

6.1 Electronics

2D field-effect transistors (FETs) represent as the application of 2D materials in microelectronic devices. FETs have attracted tremendous attention due to their excellent electrical characteristics such as ultra-low power consumption, high current switching ratio, and large carrier mobility. Recently, researchers focused on heterojunctions electronic devices s including junction field-effect transistor (JFETs), metal–insulator– semiconductor field-effect transistors (MISFETs) and MESFETs. Yan et al. reported the integration of graphene and n-type β-Ga_2O_3, and a high breakdown electric field in β-Ga_2O_3/graphene vertical barristor heterostructure was obtained (Kim and Kim, 2020). Kim et al. firstly integrated an E-mode with a D-mode MESFET based on graphene/β-Ga_2O_3 heterostructure. Both E-/D-modes β- Ga_2O_3 MESFETs exhibit excellent electrical characteristics, with a high on/off current ratio of 10^7 (Kim et al., 2018). The gain value is high when compared with an AlGaN/GaN HEMTbased direct-coupled FET logic inverter. And an n-channel D-mode WSe_2/β-Ga_2O_3 JFET was also fabricated. The p-WSe_2/β-Ga_2O_3 heterostructure exhibits excellent rectifying behaviors, with a high on/off current ratio of 10^8, a low subthreshold swing of 133 mV/dec and a three-terminal breakdown voltage of +144 V (Kim et al., 2017).

In addition to semiconductor properties, the two-dimensional materials also have insulator properties and are used in electronic devices. For example, the boron nitride can be used as an alternative dielectric material, in which the thermal conductivity of 50 W m^{-1} K^{-1} is higher than the other dielectric materials, such as SiO_2 (1.4 W m^{-1} K^{-1}) and Al_2O_3 (35 W m^{-1} K^{-1}). The BN integrated with β-Ga_2O_3 to fabricate a BN/β-Ga_2O_3 heterostructure MISFET, showed a lower gate leakage current (Yao et al., 2018). Till now, some other heterojunctions devices have also been reported, such as γ-CuI/WS_2(WSe_2), MoS_2/ZnO, ZrO_2/MoS_2 and so on (Su et al., 2019; Weng and Gao, 2019; Ithurria and Tessier, 2011).

6.2 Optoelectronics

Optoelectronic devices, including photodetectors, solar cells and LEDs, etc., are electric devices that can detect, generate, and interact with or control light. Photodetector is mainly used in monitoring, chemical-biological analysis, communication, health care and energy harvesting. The property and performance of an optoelectronic device are strongly dependent on the structures of the functional materials. The specific structure of nanomaterials strongly affects the electronic spectrum of charge carriers then the optoelectronic properties (Cheng et al., 2016). Optoelectronic devices have been one of the most promising applications of nanostructured materials ever since its emergence. This is especially true for nanomaterials with two-dimensional configurations because of their high compatibility with traditional device fabrication process as well as flexible substrates. A growing number of 2D non-layered materials, including metal chalcogenides, like PbS and CdS/Se, metal oxides, like TiO_2 and ZnO, topological crystalline insulators, like $Pb_xSn_{(1-x)}$ Se, and organic–inorganic hybrid perovskites, like CH_3PbI_3 (Wang et al., 2020). Inspired by the pioneer work done, a hybrid Ga_2O_3/graphene solar-blind photodetector was fabricated by sputtering Ga_2O_3 on grapheme. The photodetector shows long-term environmental stability and outstanding mechanical flexibility without any encapsulation (Jiang et al., 2020). The photodetector exhibits a fast response in broad spectra region from ultraviolet (365 nm) to short-wavelength-infrared (1550 nm). A self-powered operation mode photodetector is achieved by the $VO_2/MoTe_2$ heterostructure. The p–n junction exhibits from mid-wavelength infrared to long-wavelength infrared detection with a response time of 17 μs. The photodetectors based on Cd_3As_2/MoS_2 and $VO_2/MoTe_2$ not only exhibit high performance, but also greatly expands the detection range (Zhu et al., 2014). Several studies have led to continuous breakthroughs in the performance parameters of the optoelectronic device. Development of the 2D-LMDs/2D-NLMDs heterojunction device provides numerous platforms for both fundamental research and technological applications.

6.3 Energy conversion and storage

2D non-layered metal oxides and hydroxides have been utilized as the electrodes of supercapacitors. Atomically thin $β$-$Co(OH)_2$ and ultrathin NiO nanosheets with the thickness of 1.16 nm were prepared via 2D oriented attachment and template-directed synthesis method. These atomically thick 2D nanostructures will provide a short ion and electron diffusion path distance during the charge and discharge process. It can also enhance the capacity of the electrodes due to the large amount of electrochemical active sites and electrode–electrolyte interfaces (Mori et al., 2016).

Non-layered structures are used as electrode materials in lithium-ion batteries as well as rechargeable sodium-ion batteries. Non-layered

TMDCs, which are in the category of sulfide minerals, are used as electrode materials in a variety of rechargeable batteries due to their properties such as cost-effectiveness, high capacity, low toxicity, availability, and suitable frequency. One of the most common sulfide minerals in the semiconductor category is pyrite (FeS_2), which has been used both as pure and in composite with carbon fiber, graphene oxide, nanotubes, and other materials in rechargeable lithium-ion and sodium-ion batteries (Liang et al., 2021). CoS_2 has a pyrite structure, and it can play the role of cathode in thermal batteries. It has also been used to replace pyrite in rechargeable batteries as an electrode material. This material is not yet widely used but has good potential for use in rechargeable batteries (Gao et al., 2021).

Non-layered transition metal carbides (NL-TMCs) having diverse morphologies and structures can exhibit considerable prospects toward energy-related applications and triggered worldwide experimental and theoretical research enthusiasm. NL-TMCs generally accommodate a high ratio of sp-electron-rich carbon to form in a low-molecular-weight MC or MC_{1-x} structure; meanwhile, they could also exist as M_2C, M_3C_2, M_8C_7 etc (Chemler and Bovino, 2013; Pajares et al., 2020; Liu et al., 2020). TMCs can effectively improve the cycle stability of the electrodes. Currently, a large amount of NL-TMCs based electrodes, such as Fe_3C, $Ni_3ZnC_{0.7}$, MoC and Mo_2C have been extensively prepared as shown in Fig. 8 (Liu et al., 2019a) and thus displaying high specific capacity and long-life performance in batteries (Song et al., 2018; Yu et al., 2017b; Liu et al., 2019b). Transition metal phosphides (Gupta et al., 2020) and borides (Tang et al., 2022) are also being intensively studied in energy storage devices.

Heterostructures of 2D layered and non-layered MDs have attracted researchers in solar cells due to their unique optoelectronic properties, such as high absorption coefficients and efficient charge separation (Cao et al., 2016). MoS_2–CuS nanocomposite, and MoS_2/FTO as materials with suitable performance in counter electrodes, with ITO/MoS_2/Au, $MoSe_2$/GaN, n-MoS_2/i-SiO_2/p-Si as an active layer, PEDOT:PSS/ MoS_2/ perovskite/PCBM and n-MoS_2/Al2O3/p-S as cavity transfer layer, MoS_2/ P_3HT:PCBM/V_2O_5, and WS_2/P_3HT:PCBM/V_2O_5 as the electron transfer layer (Iqbal et al., 2019).

Despite the challenges facing the perovskite solar cells become a game-changer in the renewable energy sector, researchers are working to develop more stable and environmentally friendly perovskite materials, and to improve the manufacturing process and long-term stability of perovskite solar cells using 2D materials (Krishnamoorthy and Prakasam, 2020).

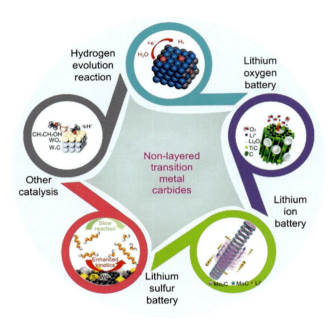

Fig. 8 Schematic illustration of the major applications of NL-TMCs. *Reproduced with permission Gao, Y.H., Nan, X., Yang, Y., Sun, B., Xu, W.L., Dasilva, W.D.L., et al., 2021. Nonlayered transition metal carbides for energy storage and conversion. N. Carbon Mater. 36 (4), 751–778.*

It is demonstrated that high-quality CsPbI$_2$Br perovskite films could be prepared by using 2D non-layered materials as additives, such as In$_2$S$_3$ nanoflakes (Nano-In$_2$S$_3$) with well-matched lattices and unsaturated dangling bonds on the surface (Di Bartolomeo, 2016). In addition, it is found that the introduction of Nano- In$_2$S$_3$ results in not only defect passivation but also remarkable quasi-Fermi level splitting across the perovskite film due to its gradient doping behavior, thereby enhancing the built-in electric field in the inverted perovskite solar cells (PSCs). As a result, the optimal devices based on Nano-In$_2$S$_3$:CsPbI$_2$Br absorber and all-inorganic interfacial layers deliver a champion power conversion efficiency of 15.17% along with excellent ambient and thermal stabilities, superior to those of the pristine devices and comparable to the best organic-free PSCs.

6.4 Catalysis

By using various materials and structures, the photocatalytic reaction triggered by solar energy can be divided into water splitting, CO$_2$ reduction, pollutant removal and sensing applications. Since the photocatalytic efficiency of most photocatalysts is still low, 2D non-layered MDs based heterostructures have been proven to have high-efficiency catalytic capabilities for various catalytic processes (Zhang et al., 2017b).

6.4.1 Catalysis of water splitting

Photocatalytic water splitting gives rise to the green hydrogen hope of powering our planet without any contribution to global warming. In a typical conversion of solar energy to chemical energy, a flux of photon is first absorbed by the designed semiconductor and produce high energy charge carriers. The generated charge carriers then separate from each other and diffuse to catalytically active sites at the semiconductor/liquid interfaces to carry out the water photocatalysis (Jiao et al., 2015). Fujishima and Honda in 1972 demonstrated the electrochemical photolysis of water splitting on TiO_2 semiconductor. Then, various other materials have been developed for photocatalysis of water splitting (Fujishima and Honda, 1972). The unsaturated dangling bonds on the 2D-NLMDs surface provide numerous active surface sites for catalysis, facilitating the activation of the reaction substrates and expanding the scope of application. Recently, the improvement of photo-excited charge separation efficiency and reasonable band offsets design of 2D-LMDs/2D-NLMDs vdW heterojunctions make photocatalytic activity enhance. Yang et al. reported a AlN/BP heterojunction photocatalyst for water splitting (Yang et al., 2016). Xu et al. reported CdS nanosheets with a thickness of 4 nm as visible-light-driven water splitting photocatalysts for hydrogen generation. They used ultra-sonication-induced aqueous exfoliation approach by utilizing inorganic–organic lamellar hybrid of CdS–DETA, water and L-cysteine as starting material, dispersion medium and stabilizing agent, respectively. The obtained ultrathin nanosheets exhibited photocatalytic hydrogen production rate of 41.1 mmol $h^{-1}g^{-1}$ in the presence of sacrificial agent, which is close to 5.5 times and 6.1 times faster than that of CdS–DETA hybrid nanosheets and CdS nanosheet-based aggregates (Xu et al., 2013). The surface modification on such 2D non-layered material enhanced the catalytic capacity toward hydrogen generation as compared to the stoichiometric $NiSe_2$ nanosheet. Experimental and theoretical works made by Xie's group manifested the active role of oxygen vacancy in ultrathin porous In_2O_3 nanosheet that brought a visible light active 2D non-layered photocatalyst for water splitting (Lei et al., 2014). Wu et al. reported a series of Sb_2Se_3/β-In_2Se_3 heterojunctions with different contents of

β-In$_2$Se$_3$. The optimized composites (90% Sb$_2$Se$_3$–10% β-In$_2$Se$_3$) exhibited the highest photocurrent density that was about 283 times higher than that of pure Sb$_2$Se$_3$. Moreover, due to the facilitated separation and transport of photogenerated charge carriers, the photocatalytic activity was enhanced. After 180 min of illumination, 76% of methyl orange dye could be decomposed. And the Sb$_2$Se$_3$/β-In$_2$Se$_3$ composite was stable after the photocatalytic reactions (Wu et al., 2020).

6.4.2 Catalysis of CO conversion

Catalytic oxidation of CO is another catalytic method and is considered to be a promising method to solve the problem of environmental pollution. It is reported that the 2D SnO$_2$ sheet with 40% surface atom occupancy rate shows better CO catalytic performance than bulk SnO$_2$ and SnO$_2$ nanoparticles, which is manifested in the reduced apparent activation energy (59.2 kJ mol^{-1}) and reduced full-conversion temperature. And the catalytic reaction at 250 °C for 54 h did not deactivate, which confirmed the excellent structural stability and stable CO conversion ability (Wu et al., 2020). At present, other 2D-LMDs/2D-NLMDs heterojunction studies have been reported on catalysis applications, such as CdS$_y$Se$_{1-y}$/MoS$_2$, WO$_3$/WS$_2$–MoS$_2$, MoS$_2$/ZnS, and so on (Chen et al., 2021; Mojaddami and Simchi, 2020; Harish et al., 2019). According to these results, it indicates that the unique properties between 2D materials heterojunctions can produce more extraordinary devices in the field.

6.5 Biomedical applications

Recent applications of the biosensor include the detection of CORONA-targeted viruses, cancer cells, DNA hybridization, antibody characterization, and protein conformation. These are extremely sensitive and allow for real-time contaminated analyte analysis in the biosensing sector to identify, assess, and characterize the biomolecule, chemical, environment, and food (Sathya et al., 2022). A biosensor is an analytical system that can identify a particular biological analyte and convert the presented data into analytical information such as electrical, optical, and thermal signals, through a straightforward, inexpensive, and swift process. Due to their inherent enzymatic capabilities, nano enzyme biosensors including transition metal chalcogenides have seen tremendous applications in the biomedical field, particularly diagnostics, since the advent of nanotechnology. Transition metal chalcogenides and nano enzymes have been employed to detect a wide range of biochemical analyzers including macromolecules and small biomolecules like glucose, cholesterol, glutathione (GSH), and cysteine (Yuan et al., 2023).

7. Challenges and future directions
7.1 Current challenges in the research of 2D metal dichalcogenides

The research on 2D metal dichalcogenides (MDCs) has gained significant attention in recent years due to their unique electronic, optical, and mechanical properties. However, several challenges still exist in this field of research. Some of the current challenges in the study of 2D metal dichalcogenides include:

7.1.1 Synthesis and scalability
The synthesis of high-quality, large area 2D Metal Dichalcogenide materials remains a challenge. Many current synthesis methods have limitations in terms of scalability, reproducibility, and control over the desired properties of the materials. Developing scalable synthesis techniques is crucial for the practical application of these materials.

Contamination and defects: During the synthesis process, contamination and defects can be introduced, affecting the properties and performance of 2D metal dichalcogenides. These defects can have a significant impact on the electronic and optical properties, making it challenging to achieve consistent and reliable results.

7.1.2 Stability and degradation
2D metal dichalcogenides are prone to environmental degradation, especially when exposed to moisture, oxygen, and other reactive species. This instability poses challenges for their long-term reliability and practical applications. Developing strategies to enhance the stability and environmental robustness of these materials is a critical area of research.

7.1.3 Integration with other materials
Integration of 2D metal dichalcogenides with other materials is essential for the development of functional devices. However, achieving efficient and reliable interfaces between different materials can be challenging due to the lattice mismatch, chemical incompatibilities, and other factors. Finding suitable methods for the controlled integration of 2D metal dichalcogenides with other materials is a key challenge.

7.1.4 Characterization techniques

Accurately characterizing the properties of 2D metal dichalcogenides at the atomic and nanoscale is crucial for understanding their behavior and optimizing their performance. However, the characterization of these materials presents challenges due to their small size, heterogeneity, and the need for high-resolution imaging and spectroscopy techniques.

7.1.5 Fundamental understanding

Despite significant progress, there are still gaps in our fundamental understanding of the properties and behavior of 2D metal dichalcogenides. Exploring their electronic structure, transport mechanisms, and interaction with light requires further research and theoretical modeling to unravel the underlying physics.

Addressing these challenges requires interdisciplinary research efforts, combining expertise from materials science, chemistry, physics, and engineering. Overcoming these obstacles will pave the way for the development of next-generation electronic and optoelectronic devices based on 2D metal dichalcogenides.

7.2 Future research directions in the field of 2D metal dichalcogenides

The field of 2D metal dichalcogenides (MDCs) continues to be a vibrant area of research, and there are several exciting future directions that researchers are actively pursuing. Some of the key research directions in this field include:

Synthesis and scalability: Developing scalable and reproducible synthesis techniques for large-area 2D Metal Dichalcogenide materials is a crucial research direction. Advances in synthesis methods, such as chemical vapor deposition (CVD), atomic layer deposition (ALD), and solution-based approaches, are being explored to achieve controlled growth and high-quality materials with desired properties.

Tailoring properties: Researchers are investigating ways to tailor the properties of 2D metal dichalcogenides to enhance their performance and enable new functionalities. This includes engineering the bandgap, enhancing carrier mobility, controlling defects, and modulating the electronic and optical properties through strain engineering, doping, and heterostructure engineering.

Integration with other materials: Integration of 2D metal dichalcogenides with other materials is a research direction that enables the development of hybrid structures and devices with enhanced performance. Investigating the interfaces between 2D metal dichalcogenides and different materials, such as metals, semiconductors, and other 2D materials, is important for creating novel devices and exploring new physics at these interfaces.

Device applications: The exploration of 2D metal dichalcogenides for various device applications is an active research area. This includes the development of high-performance transistors, sensors, photodetectors, light-emitting devices, energy storage devices, and flexible electronics. Future research will focus on optimizing device architectures, understanding device physics, and improving device performance and reliability.

Fundamental understanding: Furthering our fundamental understanding of the unique properties and behavior of 2D metal dichalcogenides is an ongoing research direction. This includes investigating the electronic structure, excitonic effects, spin dynamics, and quantum transport properties of these materials. Theoretical modeling and computational simulations are also crucial for predicting and understanding the properties of 2D metal dichalcogenides.

Environmental stability: Enhancing the environmental stability and robustness of 2D metal dichalcogenides is an important research direction. Developing protective coatings, encapsulation techniques, and exploring new material compositions that are less prone to degradation can enable the practical use of these materials in real-world applications.

Emerging phenomena and applications: Exploring emerging phenomena and novel applications in 2D metal dichalcogenides is an exciting research direction. This includes investigating unconventional superconductivity, topological properties, valleytronics, and spintronics in these materials. Exploring the potential of 2D metal dichalcogenides for quantum computing and other quantum technologies is also an active area of research.

Overall, future research in the field of 2D metal dichalcogenides will involve a combination of materials synthesis, advanced characterization techniques, theoretical modeling, and device engineering to unlock the full potential of these materials for a wide range of applications in electronics, optoelectronics, and beyond.

References

Aras, F.G., Yilmaz, A., Tasdelen, H.G., Ozden, A., Ay, F., Perkgoz, N.K., et al., 2022. A review on recent advances of chemical vapor deposition technique for monolayer transition metal dichalcogenides (MX2: Mo, W; S, Se, Te). Mater. Sci. Semicond. Process. 148, 106829.

Balendhran, S., Walia, S., Nili, H., Ou, J.Z., Zhuiykov, S., Kaner, R.B., et al., 2013. Two-dimensional molybdenum trioxide and dichalcogenides. Adv. Funct. Mater. 23 (32), 3952–3970.

Bao, S.J., Li, C.M., Guo, C.X., Qiao, Y., 2008. Biomolecule-assisted synthesis of cobalt sulfide nanowires for application in supercapacitors. J. Power Sources 180 (1), 676–681.

Bi, W., Zhou, M., Ma, Z., Zhang, H., Yu, J., Xie, Y., 2012. CuInSe 2 ultrathin nanoplatelets: novel self-sacrificial template-directed synthesis and application for flexible photodetectors. Chem. Commun. 48 (73), 9162–9164.

Cai, Z.Y., Lai, Y.J., Zhao, S.L., et al., 2021. Dissolution-precipitation growth of uniform and clean two dimensional transition metal dichalcogenides. Natl Sci. Rev. 8, nwaa115.

Cao, X., Tan, C., Zhang, X., Zhao, W., Zhang, H., 2016. Solution-processed two-dimensional metal dichalcogenide-based nanomaterials for energy storage and conversion. Adv. Mater. 28 (29), 6167–6196.

Chang, H.C., Tu, C.L., Lin, K.I., Pu, J., Takenobu, T., Hsiao, C.N., et al., 2018. Synthesis of large-area InSe monolayers by chemical vapor deposition. Small 14 (39), 1802351.

Chemler, S.R., Bovino, M.T., 2013. Catalytic aminohalogenation of alkenes and alkynes. ACS Catal. 3 (6), 1076–1091.

Chen, J., Liu, W.X., Gao, W.W., 2016. Tuning photocatalytic activity of In2S3 broadband spectrum photocatalyst based on morphology. Appl. Surf. Sci. 368, 288–297.

Chen, J., Wu, X.J., Lu, Q., Zhao, M., Yin, P.F., Ma, Q., et al., 2021. Preparation of CdSySe1− γ-MoS2 heterostructures via cation exchange of pre-epitaxially synthesized Cu2− χSySe1− γ-MoS2 for photocatalytic hydrogen evolution. Small 17 (11), 2006135.

Chen, Y.Y., Liu, Z.Y., Li, J.Z., Cheng, X., Ma, J.Q., Wang, H.Z., et al., 2020. Robust interlayer coupling in two-dimensional perovskite/monolayer transition metal dichalcogenide heterostructures. ACS Nano 14, 10258–10264.

Cheng, H.C., Wang, G., Li, D., He, Q., Yin, A., Liu, Y., et al., 2016. van der Waals heterojunction devices based on organohalide perovskites and two-dimensional materials. Nano Lett. 16 (1), 367–373.

Cherusseri, J., Choudhary, N., Kumar, K.S., Jung, Y., Thomas, J., 2019. Recent trends in transition metal dichalcogenide based supercapacitor electrodes. Nanoscale Horiz. 4 (4), 840–858.

Cho, S., et al., 2015. Phase patterning for ohmic homojunction contact in MoTe2. Science 349 (6248), 625–628.

Cui, F., Li, X., Feng, Q., Yin, J., Zhou, L., Liu, D., et al., 2017. Epitaxial growth of large-area and highly crystalline anisotropic ReSe 2 at. layer. Nano Res. 10, 2732–2742.

Dai, C., Li, B., Li, J., Zhao, B., Wu, R., Ma, H., et al., 2020. Controllable synthesis of NiS and NiS$_2$ nanoplates by chemical vapor deposition. Nano Res. 13, 2506–2511.

Di Bartolomeo, A., 2016. Graphene Schottky diodes: an experimental review of the rectifying graphene/semiconductor heterojunction. Phys. Rep. 606, 1–58.

Di, J., Yan, C., Handoko, A.D., Seh, Z.W., Li, H., Liu, Z., 2018. Ultrathin two-dimensional materials for photo-and electrocatalytic hydrogen evolution. Mater. Today 21 (7), 749–770.

Ding, S., Luan, D., Boey, F.Y.C., Chen, J.S., Lou, X.W.D., 2011. SnO2 nanosheets grown on graphene sheets with enhanced lithium storage properties. Chem. Commun. 47 (25), 7155–7157.

Eftekhari, A., 2017. Tungsten dichalcogenides (WS2, WSe$_2$, and WTe$_2$): materials chemistry and applications. J. Mater. Chem. A 5 (35), 18299–18325. https://doi.org/10.1039/C7TA04268J.

Fang, X., Wang, Z., Kang, S., Zhao, L., Jiang, Z., Dong, M., 2020. Hexagonal CoSe2 nanosheets stabilized by nitrogen–doped reduced graphene oxide for efficient hydrogen evolution reaction. Int. J. Hydrog. Energy 45 (3), 1738–1747.

Feng, W., Zheng, W., Gao, F., Chen, X., Liu, G., Hasan, T., et al., 2016. Sensitive electronic-skin strain sensor array based on the patterned two-dimensional α-In$_2$Se$_3$. Chem. Mater. 28 (12), 4278–4283.

Feng, S., Tan, J., Zhao, S., et al., 2020. Synthesis of ultrahigh-quality monolayer molybdenum disulfide through in situ defect healing with thiol molecules. Small 16, 2003357.

Fujishima, A., Honda, K., 1972. Electrochemical photolysis of water at a semiconductor electrode. nature 238 (5358), 37–38.

Gao, Y.H., Nan, X., Yang, Y., Sun, B., Xu, W.L., Dasilva, W.D.L., et al., 2021. Nonlayered transition metal carbides for energy storage and conversion. N. Carbon Mater. 36 (4), 751–778.

Gupta, S., Patel, M.K., Miotello, A., Patel, N., 2020. Metal boride-based catalysts for electrochemical water-splitting: a review. Adv. Funct. Mater. 30 (1), 1906481.

Harish, S., Archana, J., Navaneethan, M., Shimomura, M., Ikeda, H., Hayakawa, Y., 2019. Synergistic interaction of 2D layered MoS2/ZnS nanocomposite for highly efficient photocatalytic activity under visible light irradiation. Appl. Surf. Sci. 488, 36–45.

He, Y., Tang, S., Yin, S., Li, S., 2021. Research progress on green synthesis of various high-purity zeolites from natural material-kaolin. J. Clean. Prod. 306, 127248.

Ho, C.H., 2010. J. Cryst. Growth 312, 2718.

Hou, C., Zhu, J., Liu, C., Wang, X., Kuang, Q., Zheng, L., 2013. Formaldehyde-assisted synthesis of ultrathin Rh nanosheets for applications in CO oxidation. CrystEngComm 15 (31), 6127–6130.

Hu, J., Zhou, F., Wang, J., Cui, F., Quan, W., Zhang, Y., 2023. Chemical vapor deposition syntheses of Wafer-scale 2D transition metal dichalcogenide films toward next-generation integrated circuits related applications. Adv. Funct. Mater. 33, 2303520.

Huang, X., Tang, S., Mu, X., Dai, Y., Chen, G., Zhou, Z., et al., 2011. Nat. Nanotechnol. 6 (1), 28.

Huang, M., Yan, H., Chen, C., Song, A., He, Y., Chen, X., et al., 2018. Direct growth of nonlayered MoS2 on SiO2 substrates. ACS Nano 12 (6), 6371–6379.

Huang, W., Song, M., Zhang, Y., Zhao, Y., Hou, H., Hoang, L.H., et al., 2021. Defects-induced oxidation of two-dimensional β-In2S3 and its optoelectronic properties. Opt. Mater. 119, 111372.

Huang, L., Krasnok, A., Alú, A., Yu, Y., Neshev, D., Miroshnichenko, A.E., 2022a. Enhanced light–matter interaction in two-dimensional transition metal dichalcogenides. Rep. Prog. Phys. 85 (4), 046401.

Huang, W., Song, M., Zhang, Y., Zhao, Y., Hou, H., Hoang, L.H., et al., 2022b. Chemical vapor deposition growth of nonlayered γ-In2Se3 nanosheets on SiO2/Si substrates and its photodetector application. J. Alloy. Compd. 904, 164010.

Iqbal, M.Z., Alam, S., Faisal, M.M., Khan, S., 2019. Recent advancement in the performance of solar cells by incorporating transition metal dichalcogenides as counter electrode and photoabsorber. Int. J. Energy Res. 43 (8), 3058–3079.

Irfan, S., Haleem, Y.A., Irshad, M.I., Saleem, M.F., Arshad, M., Habib, M., 2023. Tunability of the optical properties of transition-metal-based structural phase change materials. Optics 4 (2), 351–363.

Ithurria, S., Tessier, M.D., Mahler, B., Lobo, R.P.S.M., Dubertret, B., Efros, A.L., 2011. Nat. Mater. 10, 936–941.

Ji, Q., Li, C., Wang, J., Niu, J., Gong, Y., Zhang, Z., et al., 2017. Metallic vanadium disulfide nanosheets as a platform material for multifunctional electrode applications. Nano Lett. 17 (8), 4908–4916.

Jiang, W., Zheng, T., Wu, B., Jiao, H., Wang, X., Chen, Y., et al., 2020. A versatile photodetector assisted by photovoltaic and bolometric effects. Light: Sci. Appl. 9 (1), 160.

Jiang, J., Feng, W., Wen, Y., Yin, L., Wang, H., Feng, X., et al., 2023. Tuning 2D magnetism in cobalt monoxide nanosheets via in situ nickel-doping. Adv. Mater. 35, 2301668.

Jiao, Y., Zheng, Y., Jaroniec, M., Qiao, S.Z., 2015. Design of electrocatalysts for oxygen- and hydrogen-involving energy conversion reactions. Chem. Soc. Rev. 44 (8), 2060–2086.

Jin, B., Huang, P., Zhang, Q., Zhou, X., Zhang, X.W., Li, L., et al., 2018a. Adv. Funct. Mater. 28 (20), 1800181.

Jin, B., Huang, P., Zhang, Q., Zhou, X., Zhang, X., Li, L., et al., 2018b. Self-limited epitaxial growth of ultrathin nonlayered CdS flakes for high-performance photodetectors. Adv. Funct. Mater. 28 (20), 1800181.

Joseph, S., Mohan, J., Lakshmy, S., Thomas, S., Chakraborty, B., Thomas, S., et al., 2023. A review of the synthesis, properties, and applications of 2D transition metal dichalcogenides and their heterostructures. Mater. Chem. Phys. 297, 127332.

Kapuria, N., Patil, N.N., Sankaran, A., Laffir, F., Geaney, H., Magner, E., et al., 2023. Engineering polymorphs in colloidal metal dichalcogenides: precursor-mediated phase control, molecular insights into crystallisation kinetics and promising electrochemical activity. J. Mater. Chem. A 11 (21), 11341–11353.

Khaleque, A., Alam, M.M., Hoque, M., Mondal, S., Haider, J.B., Xu, B., et al., 2020. Zeolite synthesis from low-cost materials and environmental applications: a review. Environ. Adv. 2, 100019.

Kim, J., Kim, J., 2020. Monolithically integrated enhancement-mode and depletion-mode β-Ga2O3 MESFETs with graphene-gate architectures and their logic applications. ACS Appl. Mater. interfaces 12 (6), 7310–7316.

Kim, J., Mastro, M.A., Tadjer, M.J., Kim, J., 2017. Quasi-two-dimensional h-BN/β-Ga2O3 heterostructure metal–insulator–semiconductor field-effect transistor. ACS Appl. Mater. Interfaces 9 (25), 21322–21327.

Kim, J., Mastro, M.A., Tadjer, M.J., Kim, J., 2018. Heterostructure WSe2− Ga2O3 junction field-effect transistor for low-dimensional high-power electronics. ACS Appl. Mater. Interfaces 10 (35), 29724–29729.

Krishnamoorthy, D., Prakasam, A., 2020. Preparation of MoS2/graphene nanocomposite-based photoanode for dye-sensitized solar cells (DSSCs). Inorg. Chem. Commun. 118, 108016.

Kumar, R., Goel, N., Hojamberdiev, M., Kumar, M., 2020. Transition metal dichalcogenides-based flexible gas sensors. Sens. Actuators A: Phys. 303, 111875.

Lei, F., Sun, Y., Liu, K., Gao, S., Liang, L., Pan, B., et al., 2014. Oxygen vacancies confined in ultrathin indium oxide porous sheets for promoted visible-light water splitting. J. Am. Chem. Soc. 136 (19), 6826–6829.

Lembke, D., Bertolazzi, S., Kis, A., 2015. Single-layer MoS2 electronics. Acc. Chem. Res. 48, 100–110.

Li, J., Zhou, X., Xia, Z., Zhang, Z., Li, J., Ma, Y., et al., 2015. Facile synthesis of CoX (X = S, P) as an efficient electrocatalyst for hydrogen evolution reaction. J. Mater. Chem. A 3 (24), 13066–13071.

Li, X., Cui, F., Feng, Q., Wang, G., Xu, X., Wu, J., et al., 2016. Controlled growth of large-area anisotropic ReS 2 at. layer and its photodetector application. Nanoscale 8 (45), 18956–18962.

Li, X., Yang, J., Sun, H., Huang, L., Li, H., Shi, J., 2023. Controlled synthesis and accurate doping of Wafer-scale two-dimensional semiconducting transition metal dichalcogenides. Adv. Mater. 2305115.

Liang, Y., Yao, S., Wang, Y., Yu, H., Majeed, A., Shen, X., et al., 2021. Hybrid cathode composed of pyrite-structure CoS2 hollow polyhedron and Ketjen black@ sulfur materials propelling polysulfide conversion in lithium sulfur batteries. Ceram. Int. 47 (19), 27122–27131.

Ling, X., Lee, Y.-H., Lin, Y., Fang, W., Yu, L., Dresselhaus, M.S., 2015. Role of the seeding promoter for MoS2 growth. Nano Lett. 15 (2), 1403–1411.

Liu, X., et al., 2018. Atomic structure of epitaxial MoTe2. Nano Lett. 18 (6), 3850–3855.

Liu, X., Niu, L., Wu, C., Cong, C., Wang, H., Zeng, Q., et al., 2016. Periodic organic–inorganic halide perovskite microplatelet arrays on silicon substrates for room-temperature lasing. Adv. Sci. 3 (11), 1600137.

Liu, X., Li, X., Sun, Y., Zhang, S., Wu, Y., 2019a. Onion-like carbon coated Fe3C nanocapsules embedded in porous carbon for the stable lithium-ion battery anode. Appl. Surf. Sci. 479, 318–325.

Liu, P., Zhang, Z.X., Jun, S.W., Zhu, Y.L., Li, Y.X., 2019b. Controlled synthesis of nickel phosphide nanoparticles with pure-phase Ni 2 P and Ni 12 P 5 for hydrogenation of nitrobenzene. React. Kinet. Mech. Catal. 126, 453–461.

Liu, X., Feng, G., Li, Y., Xu, C., Pan, Q., Wu, Z., et al., 2020. Novel Interlayer on the separator with the Cr3C2 compound as a robust polysulfide anchor for lithium–sulfur batteries. Ind. Eng. Chem. Res. 59 (16), 7538–7545.

Manzeli, S., et al., 2017. 2D transition metal dichalcogenides. Nat. Rev. Mater. 2 (8), 17033.

Mojaddami, M., Simchi, A., 2020. Robust water splitting on staggered gap heterojunctions based on WO3\WS2–MoS2 nanostructures. Renew. Energy 162, 504–512.

Mori, T., Orikasa, Y., Nakanishi, K., Kezheng, C., Hattori, M., Ohta, T., et al., 2016. Discharge/charge reaction mechanisms of FeS2 cathode material for aluminum rechargeable batteries at 55°C. J. Power Sources 313, 9–14.

Mu, X.J., Sun, M.T., 2020. Interfacial charge transfer exciton enhanced by plasmon in 2D in-plane lateral and van der Waals heterostructures. Appl. Phys. Lett. 117, 091601.

Naghavi, N., Spiering, S., Powalla, M., Cavana, B., Lincot, D., 2003. High-efficiency copper indium gallium diselenide (CIGS) solar cells with indium sulfide buffer layers deposited by atomic layer chemical vapor deposition (ALCVD). Prog. Photovolt. Res. Appl. 11, 437–443.

Nayak, A.P., et al., 2016. Pressure-induced semimetal to superconductor transition in the three-dimensional topological Dirac semimetal Cd3As2. Sci. Rep. 6, 31876.

Nayir, N., Mao, Q., Wang, T., Kowalik, M., Zhang, Y., Wang, M., et al., 2023. Modeling and simulations for 2D materials: a ReaxFF perspective. 2D Materials.

Novoselov, K.S., 2020. Discovery of 2D van der Waals layered MoSi2N4 family. Natl Sci. Rev. 7 (12), 1842–1844.

Pajares, A., Prats, H. and Romero, A., 2020. F. Vi nes, PR de la Piscina, R. Sayós, N. Homs and F. Illas. Appl. Catal., B, 267, p.118719.

Patil, S.A., Jagdale, P.B., Singh, A., Singh, R.V., Khan, Z., Samal, A.K., et al., 2023. 2D zinc oxide–synthesis, methodologies, reaction mechanism, and applications. Small 19 (14), 2206063.

Peng, S., Li, L., Mhaisalkar, S.G., Srinivasan, M., Ramakrishna, S., Yan, Q., 2014. Hollow Nanospheres Constructed by CoS2 Nanosheets with a Nitrogen-Doped-Carbon Coating for Energy-Storage and Photocatalysis. ChemSusChem 7 (8), 2212–2220.

Peng, J., Liu, Y., Lv, H., et al., 2021. Stoichiometric two-dimensional non-van der Waals AgCrS2 with superionic behaviour at room temperature. Nat. Chem. 13, 1235–1240.

Ren, B., Wang, Y., Ou, J.Z., 2020. Engineering two-dimensional metal oxides via surface functionalization for biological applications. J. Mater. Chem. B 8 (6), 1108–1127.

Sathya, N., Karki, B., Rane, K.P., Jha, A., Pal, A., 2022. Tuning and sensitivity improvement of Bi-metallic structure-based surface plasmon resonance biosensor with 2-D ε-tin selenide nanosheets. Plasmonics 17 (3), 1001–1008.

Schornbaum, J., Winter, B., Schießl, S.P., Gannott, F., Katsukis, G., Guldi, D.M., et al., 2014. Epitaxial growth of PbSe quantum dots on MoS2 nanosheets and their near-infrared photoresponse. Adv. Funct. Mater. 24 (37), 5798–5806.

Shim, J., Park, H.Y., Kang, D.H., Kim, J.O., Jo, S.H., Park, Y., et al., 2017. Electronic and optoelectronic devices based on twodimensional materials: From fabrication to application. Adv. Electron. Mater. 3, 1600364.

Song, H., Su, J., Wang, C., 2018. Vacancies revitalized Ni3ZnC0. 7 bimetallic carbide hybrid electrodes with multiplied charge-storage capability for high-capacity and stable-cyclability lithium-ion storage. ACS Applied Energy. Materials 1 (9), 5008–5015.

Su, B., He, H., Ye, Z., 2019. Large-area ZnO/MoS2 heterostructure grown by pulsed laser deposition. Mater. Lett. 253, 187–190.

Sun, Q., Wang, X., Li, B., Wu, Y., Zhang, Z., Zhang, X., et al., 2018. Chem. Res. Chin. Universities 34 (3), 344.

Sun, Z., Liao, T., Dou, Y., Hwang, S.M., Park, M.S., Jiang, L., et al., 2014. Generalized self-assembly of scalable two-dimensional transition metal oxide nanosheets. Nat. Commun. 5 (1), 3813.

Sun, J., Zhao, Z., Li, J., Li, Z., Meng, X., 2022. Recent advances in transition metal selenides-based electrocatalysts: Rational design and applications in water splitting. J. Alloy. Compd. 918, 165719.

Susarla, S., Manimunda, P., Jaques, Y.M., Hachtel, J.A., Idrobo, J.C., Amnulla, S.A.S., et al., 2018. Deformation mechanisms of vertically stacked WS2/MoS2 heterostructures: The role of interfaces. ACS Nano 12, 4036–4044.

Tan, J., Zhang, Z., Zeng, S., Li, S., Wang, J., Zheng, R., et al., 2022a. Dual-metal precursors for the universal growth of non-layered 2D transition metal chalcogenides with ordered cation vacancies. Sci. Bull. 67 (16), 1649–1658.

Tan, J., Zhang, Z., Zeng, S., Li, S., Wang, J., Zheng, R., et al., 2022b. Dual-metal precursors for the universal growth of non-layered 2D transition metal chalcogenides with ordered cation vacancies. Sci. Bull. 67 (16), 1649–1658.

Tang, X., Yi, S., Yuan, Q., Shu, Q., Han, D., Feng, L., 2022. 2D Non-Layered In_2S_3 as Multifunctional Additive for Inverted Organic-Free Perovskite Solar Cells with Enhanced Performance. Sol. RRL 6, 2101013.

Tantis, I., Talande, S., Tzitzios, V., Basina, G., Shrivastav, V., Bakandritsos, A., et al., 2023. Non-van der Waals 2D Materials for Electrochemical Energy Storage. Adv. Funct. Mater. 33 (19), 2209360.

Voiry, D., Mohite, A., Chhowalla, M., 2015. Phase engineering of transition metal dichalcogenides. Chem. Soc. Rev. 44 (9), 2702–2712.

Wang, G., Li, D., Cheng, H.C., Li, Y., Chen, C.Y., Yin, A., et al., 2015a. Wafer-scale growth of large arrays of perovskite microplate crystals for functional electronics and optoelectronics. Sci. Adv. 1 (9), e1500613.

Wang, Q., Xu, K., Wang, Z., Wang, F., Huang, Y., Safdar, M., et al., 2015c. Van der Waals epitaxial ultrathin two-dimensional nonlayered semiconductor for highly efficient flexible optoelectronic devices. Nano Lett. 15 (2), 1183–1189.

Wang, Q., Wang, F., Li, J., Wang, Z., Zhan, X., He, J., 2015b. Low-Dimensional Topological Crystalline Insulators. Small 11 (36), 4613–4624.

Wang, Q., Wen, Y., Yao, F., Huang, Y., Wang, Z., Li, M., et al., 2015d. BN-enabled epitaxy of Pb1–xSnxSe nanoplates on SiO2/Si for high-performance mid-infrared detection. Small 11 (40), 5388–5394.

Wang, H., Chen, J., Lin, Y., et al., 2021a. Electronic modulation of non-van der Waals 2D electrocatalysts for efficient energy conversion. Adv. Mater. 33, 2008422.

Wang, Y., Ren, B., Ou, J.Z., Xu, K., Yang, C., Li, Y., et al., 2021b. Engineering two-dimensional metal oxides and chalcogenides for enhanced electro-and photocatalysis. Sci. Bull. 66 (12), 1228–1252.

Wang, Q., Cai, K., Li, J., Huang, Y., Wang, Z., Xu, K., et al., 2016a. Rational design of ultralarge $Pb_{1-x}Sn_xTe$ nanoplates for exploring crystalline symmetry-protected topological transport. Adv. Mater. 28 (4), 617–623.

Wang, F., Li, Y., Shifa, T.A., Liu, K., Wang, F., Wang, Z., et al., 2016b. Selenium-enriched nickel selenide nanosheets as a robust electrocatalyst for hydrogen generation. Angew. Chem. Int. Ed. 55 (24), 6919–6924.

Wang, F., Seo, J.H., Luo, G., Starr, M.B., Li, Z., Geng, D., et al., 2016c. Nanometre-thick single-crystalline nanosheets grown at the water–air interface. Nat. Commun. 7 (1), 10444.

Wang, Q., Wen, Y., He, P., Yin, L., Wang, Z., Wang, F., et al., 2016d. High-performance phototransistor of epitaxial PbS nanoplate-graphene heterostructure with edge contact. Adv. Mater. 28 (30), 6497–6503.

Wang, F., Wang, Z.X., Shifa, T.A., et al., 2017a. Two-dimensional non-layered materials: synthesis, properties and applications. Adv. Funct. Mater. 27, 1603254.

Wang, F., Wang, Z., Shifa, T.A., Wen, Y., Wang, F., Zhan, X., et al., 2017b. Two-dimensional non-layered materials: synthesis, properties and applications. Adv. Funct. Mater. 27 (19), 1603254.

Wang, Y., Yang, Z., Li, H., Li, S., Zhi, Y., Yan, Z., et al., 2020. Ultrasensitive flexible solar-blind photodetectors based on graphene/amorphous Ga2O3 van der Waals heterojunctions. ACS Applied Materials & Interfaces. 12 (42), 47714–47720.

Wen, Y., Wang, Q., Yin, L., Liu, Q., Wang, F., Wang, F., et al., 2016. Epitaxial 2D PbS nanoplates arrays with highly efficient infrared response. Adv. Mater. 28 (36), 8051–8057.

Wen, X.L., Gong, Z.B., Li, D.H., 2019. Nonlinear optics of two dimensional transition metal dichalcogenides. InfoMat 1, 317–337.

Weng, J., Gao, S.P., 2019. Structures and characteristics of atomically thin ZrO2 from monolayer to bilayer and two-dimensional ZrO2–MoS2 heterojunction. RSC Adv. 9 (57), 32984–32994.

Wu, J., Tan, C., Tan, Z., Liu, Y., Yin, J., Dang, W., et al., 2017. Controlled synthesis of high-mobility atomically thin bismuth oxyselenide crystals. Nano Lett. 17 (5), 3021–3026.

Wu, Z., Jie, W., Yang, Z., Hao, J., 2020. Hybrid heterostructures and devices based on two-dimensional layers and wide bandgap materials. Mater. Today Nano 12, 100092.

Xiong, Y., Xu, D., Feng, Y., Zhang, G., Lin, P., Chen, X., 2023. P-Type 2D Semiconductors for Future Electronics. Adv. Mater. 2206939.

Xu, Y., Zhao, W., Xu, R., Shi, Y., Zhang, B., 2013. Synthesis of ultrathin CdS nanosheets as efficient visible-light-driven water splitting photocatalysts for hydrogen evolution. Chem. Commun. 49 (84), 9803–9805.

Yakunin, S., Protesescu, L., Krieg, F., Bodnarchuk, M.I., Nedelcu, G., Humer, M., et al., 2015. Low-threshold amplified spontaneous emission and lasing from colloidal nanocrystals of caesium lead halide perovskites. Nat. Commun. 6 (1), 8056.

Yan, X., Esqueda, I.S., Ma, J., Tice, J., Wang, H., 2018. High breakdown electric field in β-Ga2O3/graphene vertical barristor heterostructure. Appl. Phys. Lett. 112 (3).

Yang, H.G., Sun, C.H., Qiao, S.Z., Zou, J., Liu, G., Smith, S.C., et al., 2008. Nature 453 (7195), 638.

Yang, Q., Tan, C.J., Meng, R.S., Jiang, J.K., Liang, Q.H., Sun, X., et al., 2016. AlN/BP heterostructure photocatalyst for water splitting. IEEE Electron. Device Lett. 38 (1), 145–148.

Yang, Y.S., Liu, S.C., Yang, W., Li, Z.B., Wang, Y., Wang, X., et al., 2018. Air-stable in-plane anisotropic GeSe2 highly polarization-sensitive photodetection short wave region. J. Am. Chem. Soc. 140, 4150–4156.

Yang, Z.B., Hao, J.H., 2019. Recent progress in 2D layered III–VI semiconductors and their heterostructures for optoelectronic device applications. Adv. Mater. Technol. 4, 1900108.

Yao, K., Chen, P., Zhang, Z., Li, J., Ai, R., Ma, H., et al., 2018. Synthesis of ultrathin two-dimensional nanosheets and van der Waals heterostructures from non-layered γ-CuI. npj 2D Mater. Appl. 2 (1), 16.

Ye, T., Li, J.Z., Li, D.H., 2019. Charge-accumulation effect in transition metal dichalcogenide heterobilayers. Small 15, 1902424.

Yeh, C.H., Liang, Z.Y., Lin, Y.C., Wu, T.L., Fan, T., Chu, Y.C., et al., 2017. Scalable van der Waals heterojunctions for high-performance photodetectors. ACS Appl. Mater. Interfaces 9, 36181–36188.

Yin, A.X., Liu, W.C., Ke, J., Zhu, W., Gu, J., Zhang, Y.W., et al., 2012. Ru nanocrystals with shape-dependent surface-enhanced Raman spectra and catalytic properties: controlled synthesis and DFT calculations. J. Am. Chem. Soc. 134 (50), 20479–20489.

Yin, X., Liu, X., Pan, Y.T., Walsh, K.A., Yang, H., 2014. Hanoi tower-like multilayered ultrathin palladium nanosheets. Nano Lett. 14 (12), 7188–7194.

Ying, T., Yu, T., Shiah, Y.S., Li, C., Li, J., Qi, Y., et al., 2021. High-entropy van der Waals materials formed from mixed metal dichalcogenides, halides, and phosphorus trisulfides. J. Am. Chem. Soc. 143 (18), 7042–7049.

Yu, S.L., Wu, X.Q., Wang, Y.P., Guo, X., Tong, L.M., 2017a. 2D materials for optical modulation: challenges and opportunities. Adv. Mater. 29, 1606128.

Yu, H., Fan, H., Wang, J., Zheng, Y., Dai, Z., Lu, Y., et al., 2017b. 3D ordered porous Mo x C (x = 1 or 2) for advanced hydrogen evolution and Li storage. Nanoscale 9 (21), 7260–7267.

Yuan, M., Kermanian, M., Agarwal, T., Yang, Z., Yousefiasl, S., Cheng, Z., et al., 2023. Defect engineering in bomedical sciences. Adv. Mater. 2304176.

Zhang, Q., Hu, Y., Guo, S., Goebl, J., Yin, Y., 2010. Seeded growth of uniform Ag nanoplates with high aspect ratio and widely tunable surface plasmon bands. Nano Lett. 10 (12), 5037–5042.

Zhang, X., et al., 2017a. Non-layered two-dimensional metal chalcogenides. Chem. Soc. Rev. 46 (22), 6777–6803.

Zhang, Z., Zhang, Z., Liang, J., Zhou, Y., Tong, Y., Wang, Y., et al., 2017b. Freestanding single layers of non-layered material γ-Ga2 O3 as an efficient photocatalyst for overall water splitting. J. Mater. Chem. A 5 (20), 9702–9708.

Zhao, X., Yin, Q., Huang, H., Yu, Q., Liu, B., Yang, J., et al., 2021a. Van der Waals epitaxy of ultrathin crystalline PbTe nanosheets with high near-infrared photoelectric response. Nano Res. 14, 1955–1960.

Zhao, B., Shen, D., Zhang, Z., Lu, P., Hossain, M., Li, J., et al., 2021b. 2D metallic transition-metal dichalcogenides: structures, synthesis, properties, and applications. Adv. Funct. Mater. 31 (48), 2105132.

Zhao, Y., Yu, D., Lu, J., Tao, L., Chen, Z., Yang, Y., et al., 2019. Adv. Opt. Mater. 7, 1901085.

Zheng, W., Feng, W., Zhang, X., Chen, X., Liu, G., Qiu, Y., et al., 2016. Anisotropic growth of nonlayered CdS on MoS2 monolayer for functional vertical heterostructures. Adv. Funct. Mater. 26 (16), 2648–2654.

Zhou, N., Yang, R., Zhai, T., 2019. Two-dimensional non-layered materials. Materials Today. Nano 8, 100051.

Zhou, R., Wang, H., Chang, J., Yu, C., Dai, H., Chen, Q., et al., 2021. Ammonium intercalation induced expanded 1T-rich molybdenum diselenides for improved lithium ion storage. ACS Appl. Mater. Interfaces 13 (15), 17459–17466.

Zhu, Y., Cao, C., Tao, S., Chu, W., Wu, Z., Li, Y., 2014. Ultrathin nickel hydroxide and oxide nanosheets: synthesis, characterizations and excellent supercapacitor performances. Sci. Rep. 4 (1), 5787.

Zhu, D.D., Xia, J., Wang, L., Li, X.Z., Tian, L.F., Meng, X.M., 2016. van der Waals epitaxy and photoresponse of two-dimensional CdSe plates. Nanoscale 8 (22), 11375–11379.

CHAPTER FOUR

2D III-V semiconductors

Sattar Mirzakuchaki[a,*] and Atefeh Nazary[b]

[a]School of Electrical Engineering, Iran University of Science and Technology, Tehran, Iran
[b]Department of Electrical, Biomedical and Mechatronics Engineering, Qazvin Branch, Islamic Azad University, Qazvin, Iran
*Corresponding author. e-mail address: m_kuchaki@iust.ac.ir

Contents

1. Introduction	102
2. Non-layered III-V compounds	103
2.1 Boron nitride (BN)	103
2.2 Boron phosphide (BP)	107
2.3 Boron arsenide (BAs)	110
2.4 Aluminum nitride (AlN)	113
2.5 Aluminum antimonide (AlSb)	116
2.6 Indium arsenide (InAs)	119
2.7 Indium phosphide (InP)	123
2.8 Indium antimonide (InSb)	126
2.9 Gallium nitride (GaN)	129
2.10 Gallium arsenide (GaAs)	134
2.11 Gallium arsenide (GaSb)	138
References	141

Abstract

Non-layered 2D III-V compounds represent a distinct class of materials that exhibit notable structural differences compared to their layered counterparts. These compounds, known as III-V semiconductors, are characterized by their non-layered nature. In contrast to the layered materials that are held together by weak van der Waals bonds between interlayers, the non-layered nature of III-V semiconductors arises from the presence of covalent bonds within their layers, which imparts them with unique electronic and optical properties, but preventing their exfoliation and limiting their 2D anisotropic growth. However, when III-V semiconductors are reduced to two dimensions, they exhibit unique properties such as 2D electron and hole gas, blue-shifted band gap, and nonlinear optics, expanding their potential applications especially in catalysis and sensing. This chapter provides a comprehensive examination of important and common structures of two-dimensional non-layered III-V compounds. Additionally, the properties, synthesis methods, and applications of these semiconductors will be explored.

1. Introduction

Semiconductors belonging to the III-V category, known for their distinctive structural characteristics, direct band gaps, and exceptional carrier mobility, exhibit substantial promise in the fabrication of solar cells, lasers, photodetectors, light-emitting diodes, and various other technological devices. This has led to significant interest within the field of 2D materials research (Chen et al., 2016; Cipriano et al., 2020; Del Alamo, 2011; Kobayashi et al., 2012; Wallentin et al., 2013; Zhang et al., 2019). When III-V semiconductors shift into two dimensions, they display unique characteristics, including the formation of a 2D electron and hole gas, a shift towards higher energy in their band gaps (referred to as blue-shifted), and the ability to exhibit nonlinear optical properties. These properties broaden their potential applications (Al Balushi et al., 2016; Ambacher et al., 1999; Chaudhuri et al., 2019; Sanders et al., 2017). It's important to highlight that III-V semiconductors have a notably distinct structural composition compared to 2D layered materials, categorizing them as non-layered compounds. While 2D layered materials are interconnected primarily by robust chemical bonds within the plane and relatively weaker van der Waals forces between layers, non-layered compounds exhibit robust chemical bonds both within individual layers and between them. This structural characteristic renders the process of exfoliating and achieving 2D growth quite challenging (Dou et al., 2017; Wang et al., 2017). III-Vs include elements from groups III (group 3A or IIIA) B, Al, Ga, In, and Tl, and elements from groups V (group 5 A or VA), N, P, As, Sb, Bi, among which metallic Al, Ga, In and nonmetallic N, P, As, Sb have been well manipulated, but Bi and B are still in development (Zhang, 2023). Given the numerous benefits associated with the two-dimensional configuration of materials, it becomes imperative to investigate the ultrathin III-V semiconductor system. Typically, ultrathin III-V semiconductors are characterized by two relatively stable structural forms (Lu et al., 2022). Nitride compounds such as BN, AlN, InN, GaN, and TlN predominantly adopt the wurtzite crystal structure within the P63mc space group, while other materials like BP, BAs, AlAs, AlP, AlSb, InAs, InP, InSb, GaP, GaAs, and GaSb exhibit the zinc blende structure within the F43m space group (Kuech, 2016).

Within a two-dimensional context, non-layered substances experience pronounced lattice distortion, resulting in the emergence of numerous exposed, reactive bonds on ultrathin III-V semiconductors. This, in turn, augments their capacity to capture gas molecules, rendering them auspicious for applications in catalysis and sensing. Furthermore, it is observed that monolayer nitrides and AlP commonly take on a planar structure, whereas AlSb, GaP, GaSb, InP, and InSb tend to adopt a buckled configuration (Wang et al., 2017; Zhuang et al., 2013). We can understand the growing popularity of these structures by considering two key factors; the bonding energy associated with the hybridization of group III and V ions in both sp^2 and sp^3 configurations, and also the electrostatic potential energy. On one side, the layered hexagonal materials exhibit a staggered or buckled structure that results in a dipole moment passing through each layer. The electrostatic energy associated with this dipole moment is directly corresponded to the square of the ionic charge (Q) and inversely corresponded to the square of the lattice parameter (a). The significant variation in electronegativity between the elements in nitrides and AlP results in a substantial ionic charge. Simultaneously, the small lattice parameters result in a significant electrostatic energy. Consequently, the buckled structure in nitrides and AlP becomes less favorable. However, explaining the structural trend solely through electrostatic interactions is insufficient. We also need to consider the sp^2 and sp^3 hybridization of ions in III-V semiconductors. III-group elements preferentially form a planar triangular structure with D_{3h} symmetry through sp^2 hybridization, while V-group elements tend to adopt a pyramidal trigonal configuration via sp^3 hybridization. Nitrogen atoms can participate in aromatic bonding through p orbitals, forming a triangular plane configuration. Conversely, phosphides and heavier group V compounds favor sp^3 hybridization, leading to a pyramidal trigonal configuration with C_{3V} symmetry. Combining these factors, we can conclude that the planar hexagonal configuration is unique to nitrides and AlP (Zhuang et al., 2013; Şahin et al., 2009). Therefore, in this chapter, we aim to explore the properties, applications, and synthesis strategies of common non-layered III-V compound semiconductors. We also anticipate that it will serve as a foundation for exploring 2D non-layered III-V semiconductors characteristics and potential uses in the near future.

2. Non-layered III-V compounds
2.1 Boron nitride (BN)
2.1.1 BN properties
BN is a compound made up of boron and nitrogen atoms linked together alternately. It comes in various crystalline forms: cubic BN (c-BN), which resembles diamond; wurtzite BN (w-BN), similar to lonsdaleite; and two

layered structures with sp^2 bonding – hexagonal BN (h-BN) with AB stacking, akin to h-graphite, and rhombohedral BN (r-BN) with ABC stacking, resembling r-graphite (Jiang et al., 2015). As we, discussed on non-layered III-V Compounds in this chapter, so we focused on cubic BN (c-BN), which known as cubic zinc-blend structure. As described in the Fig. 1, it is consisted of boron and nitrogen atoms bonded in a sp^3 arrangement, with {111} planes organized in a three-layer ABCABC stacking pattern. As mentioned before, Cubic BN, known as c-BN, shares a crystal structure similar to diamond. However, while diamond comprises two face-centered cubic (FCC) carbon cells, c-BN features one boron FCC cell and one nitrogen FCC cell. The lattice constant for c-BN is approximately 0.3615 nm, very close to diamond's 0.3567 nm. In terms of atomic bonding, diamond is entirely covalent, whereas c-BN exhibits ionic characteristics, ranging from 0.26 to 0.48 on the ionicity scale. Like diamond, c-BN forms under high-temperature and high-pressure conditions and possesses wide indirect bandgaps, measuring 5.5 and 6.4 eV, respectively. Due to their robust atomic bonds, both c-BN and diamond are

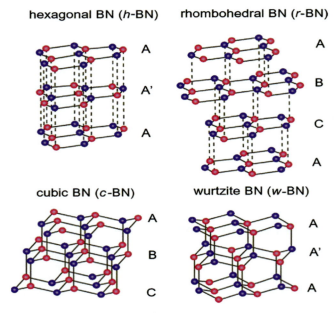

Fig. 1 The optimized structure for the four primary crystalline BN phases: h-BN, r-BN, c-BN, and w-BN (Zhang and Meng, 2019).

exceptionally hard and efficient thermal conductors. These properties explain why c-BN is currently used as coatings on cutting tools and is sought after for wear- and corrosion-resistant coatings, heat dissipation solutions, and high-temperature electronic devices (Zhang and Meng, 2019). Cubic BN (c-BN) stands out as the lightest member of the III-V family of compounds. Therefore, what makes it particularly remarkable are its exceptional physical traits, including extraordinary hardness, a high melting point, excellent thermal conductivity, a substantial band-gap energy, low dielectric constant, and remarkable chemical stability. If we can overcome the challenges related to growing crystals of this material, it holds the potential to become a valuable resource in various industrial applications (Adachi, 2013; Gao, 2012). Moreover, the most important physical and chemical properties of crystalline Cubic BN is listed in Table 1 (Zhang and Meng, 2019). In the next section, the band structure of various BN phases is shown in Fig. 3.

2.1.2 Synthesis and development of BN

Cubic Boron Nitride (c-BN) has been obtainable in bulk form, consisting of small crystallites, since the 1960s. This is achieved through the High-Temperature High-Pressure (HTHP) synthesis technique pioneered by Wentorf (Wentorf, 2004). In this method, mixtures containing boron and nitrogen are exposed to extreme HTHP conditions inside a metal container, with typical settings reaching temperatures of 1800 °C and pressures of 85,000 atmospheres (8.6 GPa). Ongoing efforts in this field are directed at reducing the required pressure and temperature for this process while also exploring novel catalysts to enable the growth of larger c-BN crystals. The stability of c-BN as a phase gives hope for achieving even lower pressures and temperatures during synthesis. However, the equipment required for high-temperature and high-pressure (HTHP) synthesis is complex, and the resulting crystals are typically just a few millimeters in size, which isn't suitable for most industrial applications. To produce larger, high-quality c-BN crystals, it's crucial to control the level of BN saturation in the solvent. Ideally, pure c-BN crystals should be colorless, but those synthesized using the HTHP method often exhibit amber to yellow hues, suggesting impurities. The crystal color depends on the solvent; for example, Li_3BN_2 produces amber crystals, while a $Ba_3B_2N_4$ catalyst system yields colorless ones (Taniguchi and Yamaoka, 2001).

Since Sokolowski first reported the synthesis of c-BN thin films under low pressure in 1979, substantial efforts have been made to grow these films. From the 1990s onward, various ion-assisted physical vapor deposition (PVD) and plasma-assisted chemical vapor deposition (CVD) methods have successfully

Table 1 Physical and chemical features of c-BN and diamond (Zhang and Meng, 2019).

Characteristics	c-BN	Diamond
Structure	Zinc blende	Diamond
Lattice constant	0.3615	0.3567
Cleavage	(011)	(111)
Bonding length (nm)	0.157	0.154
Density (g/cm^3)	3.48	3.52
Bulk modulus (GPa)	367	435
Microhardness (GPa)	75–90	80–120
Melting point (K)	3500 (10.5 MPa)	4000 (13 MPa)
Expansion coefficient ($10^{-6}/K$)	4.7	3.1
Thermal conductivity (kW/mK)	1.3	2.0
Thermal stability (K)	1573–1673	833–973
Graphitize (K)	> 1773	1673–2073
Chemical stability (with metal)	Better	Bad
Resistivity (Ω-cm)	10^2–10^{10}	10^{10}–10^{16}
Dopable	type p- and n-type	p-type
Dielectric constant	4.5	5.58
Refraction index	2.117 (583 nm)	2.417 (589.3 nm)
Bandgap (eV)	6.4	5.5

produced c-BN films. These methods include pulsed laser deposition (PLD), radio frequency (RF) sputtering, ion beam-assisted deposition (IBAD), mass-selected ion beam deposition, molecular beam epitaxy (MBE), direct-current (DC) jet-plasma CVD, radio-frequency plasma CVD (RF-CVD), inductively coupled plasma CVD (ICP-CVD), and electron-cyclotron resonance microwave plasma CVD (ECR-MPCVD) Regardless of the methods of film synthesis, it is essential and a common feature to conduct energetic-ion bombardment during film growth by irradiation with an ion beam or substrate biasing (Zhang and Meng, 2019).

2.1.3 Applications of BN

Given c-BN's outstanding physical and chemical attributes, it is regarded as a potential wide bandgap semiconductor for robust electronic devices operating at high power and elevated temperatures in challenging conditions. Nevertheless, the development of device fabrication techniques for c-BN remains relatively unexplored, and it is constrained to the construction p-n junctions using both intrinsic and doped c-BN (Zhang and Meng, 2019). In fact, pure (Native defects) or accidentally doped (Impurity doping) c-BN can demonstrate either n-type or p-type semiconductor properties, depending on how it is prepared (Zhang and Meng, 2019).

2.2 Boron phosphide (BP)

2.2.1 BP properties

In the category of III-V binary compounds, boron phosphide (BP) has garnered significant interest. It is distinguished by its exceptional blend of mechanical, thermal, and electrical characteristics, as well as impressive thermoelectric capabilities and remarkable chemical and high-temperature resilience, this material holds great promise for a broad spectrum of engineering applications. While BP is predominantly found in its zincblende form in nature, it can also occur in other phases like the known cubic polymorph (c-BP), stable crystal structures of wurtzite (w-BP) and recently discovered rhombohedral (rh-BP) (Solozhenko and Matar, 2022; Hu et al., 2021). In Fig. 2, different phases of boron phosphide are described in details. Cubic sphalerite BP, often referred to as c-BP, is the naturally stable form of boron phosphide under standard conditions. Furthermore, it maintains its stability even under pressures as high as 110 GPa. The other form of BP is known as the hexagonal wurtzite polymorph (w-BP), but there is a notable absence of information in the literature regarding its structure, stability, and properties. More recently, a third polymorph, rhombohedral BP (rh-BP), was identified using transmission electron microscopy [TEM] (Solozhenko and Matar, 2022). In Solozhenko and Matar (2022), it is investigated and concluded that all three phases of BP display remarkably similar thermodynamic and mechanical characteristics. When it is examined their electronic band structures using comprehensive calculations, it becomes evident that they possess band gaps close to 2.0 eV, suggesting a stronger covalent nature compared to the polymorphs of BN. Therefore, in this chapter, we focused on cubic boron phosphide (c-BP) as a stable structure. Thanks to its robust covalent bonds, cubic BP boasts high decomposition and oxidation resistance, making it chemically stable against concentrated mineral acids and

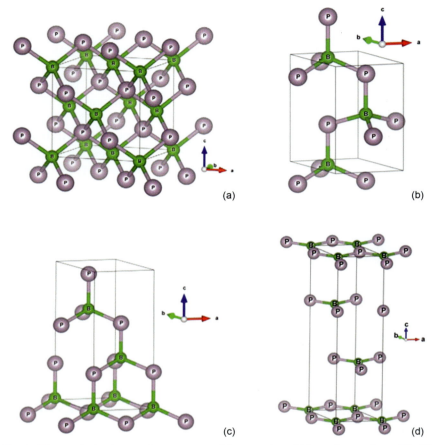

Fig. 2 The optimized crystal structures of boron phosphide polymorphs: c-BP (A); w-BP (B); rh-BP (hexagonal axes) – 3D (C) and 2D (D) (Solozhenko and Matar, 2022).

aqueous alkali solutions. These characteristics have positioned cubic BP as a promising material for electronic devices designed to operate under extreme conditions. Cubic BP functions as an indirect band gap semiconductor with a band gap of approximately 2.0 eV. This energy relationship alongside the relatively small band gap of cubic BP, suggests its potential for photocatalytic water splitting when exposed to visible light. Shi et al. recently demonstrated that n–type cubic BP particles, created through a solid-state reaction, can catalyze the production of hydrogen (H_2) from water without the need for metal cocatalysts. This marks a significant advancement in enabling water splitting under visible light conditions (Sugimoto et al., 2019). The band structures of various structures of BP compared to BN is depicted in Fig. 3.

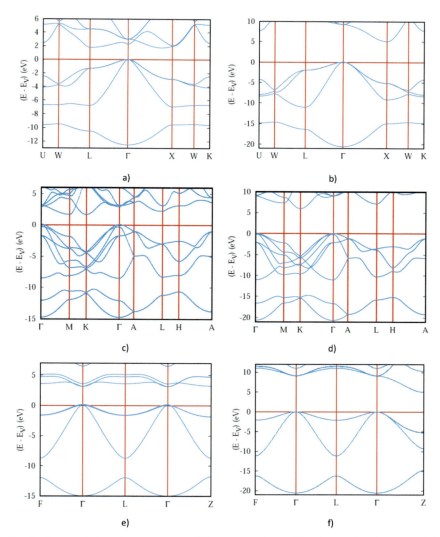

Fig. 3 Electronic band structures of BP compared to BN polymorphs: (A) c-BP; (B) c-BN; (C) w-BP; (D) w-BN; (E) rh-BP; (F) rh-BN (Solozhenko and Matar, 2022).

2.2.2 Synthesis and development of BP

Despite having numerous advantageous technological properties, cubic BP is known for its difficult synthesis, thus preventing it from being a widely used material. The three different methods of synthesis of cubic boron phosphide, BP, is developed and compared including; a high-temperature reaction of elements, Sn flux–assisted synthesis, and a solid-state metathesis reaction (Woo et al., 2016). The structural and optical properties of the three methods' products were thoroughly characterized. In terms of reaction temperature and time, solid-state metathesis is the cleanest and most efficient method. Sn flux synthesis produced a novel Sn-doped BP compound. According to DFT calculations, BP monolayer has a hexagonal planar structure and is expected to behave as a semiconductor with a band gap of approximately 0.9 eV. However, there is currently no experimental data available on the structural and electronic properties of boron phosphide monolayer, as it has not yet been successfully synthesized through experiments (Vu et al., 2021).

2.2.3 Applications of BP

Boron phosphide with high thermal conductivity Λ of 540 W/m-K have been synthesized, making it suitable candidate for applications in thermal management (Mahat et al., 2021) in micro/nano electronics (Ullah et al., 2019). BP also has been studied as a potential anode nominee for alkali-based batteries (Cakmak et al., 2020). It is can be functionalized via atomic adsorption and can be suitable for various applications in optoelectronics and spintronics (Du et al., 2018). Furthermore, BP-based heterostructures, such as BP/Blue-p and F4TCNQ/Blue-p heterostructures, have superior electronic, optical, and transport properties, making them promising candidates for electronic and optoelectronic device applications (Li et al., 2018a). BP-based van der Waals heterostructures have been investigated for their photocatalytic performance (Alrebdi and Amin, 2020).

2.3 Boron arsenide (BAs)

2.3.1 BAs properties

The compound Boron Arsenide (BAs) exhibits exceptional physical properties, such as low density, low dielectric constant, and remarkably high thermal conductivity. It also displays a unique characteristic where the anions and cations exhibit an inverse contribution, with boron acting as an anion due to its small size and lack of p-electrons in its core. Recent research has shown that BAs is a promising candidate for use in photovoltaic and photo-electrochemical applications, as demonstrated in an experimental study (Rastogi et al., 2019). Boron arsenide (BAs) has the stable cubic, zinc-blende structure with lattice constant of 4.777 A (Adachi, 2013). Cubic boron arsenide (BAs) is an emerging semiconductor material, drawing significant research attention due to its remarkable thermal conductivity, which makes it promising for applications in managing electronic device heat (Kang et al., 2019). However, our understanding of BAs is limited because high-quality single crystals have only recently become available. In Kang et al. (2019), it is conducted systematic experiments to measure crucial properties of BAs, including its bandgap, optical refractive index, elasticity, thermal expansion, and heat capacity. It is found an optical bandgap of 1.82 eV and a refractive index of 3.29 at room temperature. The elastic modulus was notably high at 326 GPa, twice that of silicon. Therefore, a summary of experimentally measured physical properties of BAs is mentioned in Table 2 (Kang et al., 2019). On the other hand, the schematic crystallin structure and band structure of zincblende BAs or c–BAs described in Fig. 4.

2.3.2 Synthesis and development of BAs

Effective heat control becomes more critical as device dimensions are shrunk and enhance computing capabilities. Engineering materials like boron arsenide (BAs) with excellent thermal conductivity poses challenges since it's crucial to prevent defects and impurities during production, as these would impede the flow of heat. Researcher used some new methods for c–BAs synthesizing. Three separate research teams have managed to create BAs with an impressive thermal conductivity of approximately 1000 watts per meter-kelvin. Kang et al. (2018), Li et al. (2018b), and Tian et al. (2018), in their respective studies, successfully produced high-purity BAs with thermal conductivities surpassing those of typical metals by more than double, although still half that of diamond. This achievement is noteworthy, as discussed in the Perspective by Dames (2018). In Li et al. (2018b), a single crystal of cubic BAs is synthesized by a modified chemical vapor transport method using high–purity source materials (B and As powders) and transport agents. Among the different vapor transport agents, it is investigated (I_2, H_2, Br_2, and NH_4I), and found that NH_4I yielded the highest–thermal conductivity crystals. Moreover, it is optimized the ratio of starting materials and temperature profiles for their synthesis. The thermal conductivity of BAs, measuring at 1000 ± 90 watts per meter-kelvin, more than of silicon carbide by a factor of 3 and exceeded only by diamond and the basal-plane value of graphite. This study demonstrates that BAs belongs to a category of materials with exceptionally high thermal conductivity, as foreseen by a recent theory. It also suggests that BAs has the potential to serve as an effective material for managing heat in high-power electronic devices (Li et al., 2018b).

Table 2 Recently short list of experimentally measured physical properties of BAs (Kang et al., 2019).

Physical properties	Experimental values
Crystal structure	Zinc-blende cubic ($\overline{F4}$ 3 m)
Lattice constant (A°)	4.78
Bandgap (eV)	1.82
Refractive index	3.29 (657 nm)
	3.04 (908 nm)
Mass density (g/cm^3)	5.22
Stiffness C_{11} C_{12}, C_{44} (GPa)	285, 79.5, 149
Compliance S_{11} S_{12}, S_{44} ($\times 10^{-12}$ Pa^{-1})	3.99, −0.87, 6.71
Averaged elastic modulus (GPa)	326
Averaged shear modulus (GPa)	128
Bulk modulus (GPa)	148
Poisson's ratio	0.22 ($\langle 100 \rangle$ on (100))
Longitudinal sound velocity (m/s)	7390 ($\langle 100 \rangle$)
	8150 ($\langle 111 \rangle$)
Transverse sound velocity (m/s)	5340 ($\langle 100 \rangle$)
Thermal conductivity (W/m·K)	1300
Volumetric heat capacity (J/cm^3.K)	2.09
Thermal expansion coefficient (10^{-6} K^{-1})	3.85 (linear)
	11.55 (volume)
Grüneisen parameter	0.82

2.3.3 Applications of BAs

According to the calculations, the Seebeck coefficient and power factor values indicate that the BAs sheet is suitable for both thermal management and thermoelectric applications. Additionally, the thermodynamic properties of the BAs nanosheets, based on phonon frequencies, show

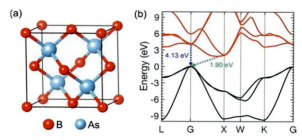

Fig. 4 (A) An illustration of the unit cell structure of Cubic BAs crystals, (B) Calculated band structure of cubic or zincblende BAs using the HSE hybrid functional. The calculated indirect gap is 1.90 eV and the minimum direct gap at Γ is 4.13 eV (Chae et al., 2018).

promising potential for applications in nanoelectronics and photovoltaics (Manoharan and Subramanian, 2018). Furthermore, in Shin et al. (2022), it is employed an optical transient grating (TG) technique to measure both electrical mobility and thermal conductivity at the same location on c-BAs single crystals. Therefore, it is validated that c-BAs exhibits notably high thermal conductivity as well as strong mobility for both electrons and holes. It is illustrated that ionized impurities significantly scatter charge carriers, while neutral impurities primarily contribute to the reduction in thermal conductivity through our ab initio calculations. These discoveries put c-BAs as the sole semiconductor known to possess this exceptional combination of advantageous characteristics, making it a promising material for advanced microelectronics applications in the future.

2.4 Aluminum nitride (AlN)
2.4.1 AlN properties

Aluminum nitride (AlN) is a highly interesting optoelectronic material with a wide energy bandgap. It possesses several unique characteristics, including relatively high hardness and excellent thermal conductivity. AlN thin films have been successfully grown on various substrates for applications like piezoelectric materials. There are various phases for AlN under different conditions. Under normal temperature and pressure conditions, AlN adopts a hexagonal wurtzite structure, referred to as the B4 type. It can take on a cubic zincblende structure (known as zb-AlN) as a metastable state or a cubic rock-salt variant (rs-AlN) under high pressure, as predicted through density functional theory (DFT) calculations. There has been recently an increasing fascination with the metastable cubic AlN films. This interest stems from the pursuit of innovative and improved mechanical and functional characteristics that are not typically found in the hexagonal structure. A recent comprehensive analysis from scratch reveals that under normal conditions of temperature and pressure, the most secure form of AlN is the hexagonal wurtzite structure. However, under elevated pressures and temperatures, the cubic rock–salt phase becomes the more stable configuration. Interestingly, the zincblende phase was not observed to be stable under any combination of pressure and temperature. This finding eliminates the possibility that either temperature or pressure influences the formation of zincblende AlN (Yadav et al., 2016). It is discussed that a consistent lattice parameter, 'a_i,' for various crystal phases is established. 'a_i' is defined as the distance between the nearest aluminum (Al) atoms. It varies for different phases, such as wurtzite, wz–AlN, zb–AlN, and rs–AlN, based on the crystal structure. The introduction of biaxial strain allows the lattice to relax, reducing stress. As the in-plane lattice parameter decreases for wz–AlN and zb–AlN, their relative energy decreases, but zb–AlN never becomes more stable than wz–AlN. Additionally, rs–AlN becomes the most stable phase at an 'a_i' of 2.9 Å. In Table 3 (Yadav et al., 2016), a comparison of DFT-calculated values and actual experimental data for lattice parameters, bulk modulus, and AlN in different phases (wz–AlN, zb–AlN, and rs–AlN) is listed, as well as TiN in the rock–salt crystal structure. It's noteworthy that there is an outstanding level of agreement between the DFT calculations and the experimental measurements. Moreover, different phases of AlN are depicted in Fig. 5.

2.4.2 Synthesis and development of AlN

In Yadav et al. (2016), it is mentioned that by customizing substrates based on their crystal structure, orientation, and elasticity, it's possible to grow cubic AlN layers in superlattice systems. The stability of these cubic structures is highly reliant on their layer thickness. For instance, when annealed at 600 °C, it is managed to synthesize epitaxial metastable zb–AlN through a solid-state reaction between single crystal Al(001) and TiN(001). In magnetron-sputtered AlN/TiN(001) epitaxial superlattices, it is observed the stabilization of the high-pressure rs–AlN when the AlN layer thickness was less than or equal to 2.0 nm. Using reactive sputtering, it is also witnessed the epitaxial stabilization of rs–AlN in AlN/VN(001) and AlN/TiN(001) superlattices, but this occurred with significant layer thicknesses of AlN at 3 to > 4 nm and 2–2.5 nm, respectively. Moreover, the stabilization of zb–AlN was noted in AlN/W(001) superlattices when the AlN thickness was less than or equal to 1.5 nm. It's worth noting that interfaces play a substantial role in controlling the properties of these structures. Additionally, AlN can

Table 3 Contrast between computed and existing experimental data of lattice parameters, bulk modulus, of Al, TiN, and AlN in wurtzite (wz), zincblende (zb), and rock-salt (rs) phases (Yadav et al., 2016).

Structures	Al		wz-AlN		zb-AlN	rs-AlN	TiN	
Data source	DFT	Exp.	DFT	Exp.	DFT	DFT	DFT	Exp.
Lattice parameter (Å)	4.04	4.05	3.12 (a) 5.01 (c)	3.11 (a) 4.98 (c)	4.40	4.06	4.24	4.24
Bulk modulus (GPa)	76	79	202	211 208 220	224	298	306	318
Relative energy (meV/f.u.)			0.0		43	345		

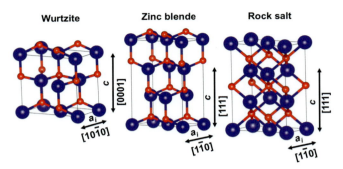

Fig. 5 Supercells of various AlN crystal phases and definition of lattice parameters ai and c (Yadav et al., 2016).

undergo different forms of stress during bulk AlN formation and when subjected to pressure from indentations. It is taken into account three main types of stress that effectively capture the structural limitations for the specific interfaces AlN with TiN and Al within the (111) plane: (1) hydrostatic stress, (2) biaxial stress within the (111) plane, and (3) uniaxial stress applied along a single direction in the [111] direction. In Yadav et al. (2016), it is determined that the development of zinc-blende AlN thin films is primarily driven by kinetics rather than thermodynamics.

2.4.3 Applications of AlN

As mentioned earlier, Aluminum nitride (AlN) is a wide-bandgap material with significant technological significance. AlN thin films have been grown on diverse substrates for applications such as in piezoelectric materials. Moreover, it possesses a range of distinctive attributes, including notable hardness and high thermal conductivity which it valuable for a wide range of device applications (Adachi, 2013; Yadav et al., 2016).

2.5 Aluminum antimonide (AlSb)

2.5.1 AlSb properties

Aluminum antimonide (AlSb) is a semiconductor with zinc-blende crystal structure in nature and an indirect gap of 1.6 eV at room temperature (Adachi, 2013). The size of atoms and the distance between them in a substance called aluminum antimonide is measured in (Singh and Gupta, 2016). It is found that the equilibrium lattice constant was 4.46 $\mathring{A}$, and the bond length between Al and Sb was 2.57 $\mathring{A}$. These values were very similar to those of low-buckled aluminum antimonide (AlSb), which had a lattice constant of 4.33 $\mathring{A}$, and the same bond length. The schematic low-buckled

AlSb is described in Fig. 6A. It is also calculated the cohesive energy of the substance, which was 6.85 eV, compared to 8.04 eV in low-buckled AlSb and similar to what other research studies have found. Furthermore, the electronic band structure of two-dimensional planar AlSb is analyzed and found that it behaved like a semiconductor, showing a bandgap. The band structure revealed that there were multiple ways for electrons to move from the valence band maximum (VBM) to the conduction band minimum (CBM), including direct and indirect transitions. At the K-point, we found three possibilities for electron transitions, with direct band gap energy of 1.62 eV, indirect band gap energies of 1.31 eV and 0.57 eV as shown in Fig. 6B. These findings provide important numerical information about the properties of aluminum antimonide and its electronic behavior.

Therefore, the single-layered 2D AlSb material is anticipated to undergo a transformation into a semiconductor with a direct bandgap (1.62 eV) at room temperature, which mobility of both electrons and holes

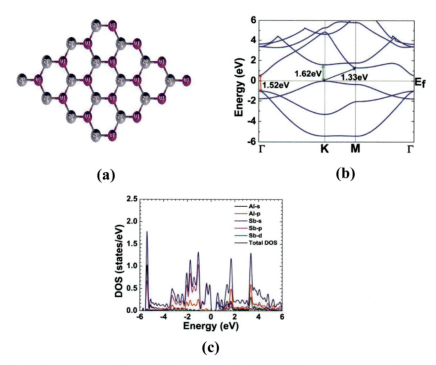

Fig. 6 (A) Schematic diagram depicting low-buckled AlSb (B) Electronic band structure and (C) Total and partial density of states of 2D planar AlSb monolayer (Singh and Gupta, 2016).

limited by phonons expected to be around 1700 cm^2/Vs Qin et al., (2021). To understand the role of different orbitals in the electronic states of 2D planar AlSb, the partial density of states (PDOS) is presented in Fig. 6C. The PDOS graph clearly shows that the electronic states near the Fermi level are mainly contributed by the p-orbitals of both aluminum (Al) and antimony (Sb). However, the p-orbital of Sb has a much greater contribution than that of Al, and it dominates the valence band maximum (VBM). At the conduction band minimum (CBM), the contribution of s-orbitals of both Al and Sb in total DOS is higher near the Fermi level (Qin et al., 2021).

2.5.2 Synthesis and development of AlSb

There are two methods for obtaining monolayer van der Waals (vdW) materials. The first method involves top-down exfoliation techniques, where 2D sheets are mechanically removed from the bulk 3D network (zinc-blend or cubic). However, this method is inefficient and only works for parent crystals with weak interlayer interactions. The second method involves bottom-up thin film growth techniques, such as chemical vapor deposition (CVD) or molecular beam epitaxy (MBE). These methods can synthesize large-scale films with controlled thickness, and the resulting crystal structures may be completely different from their bulk counterparts. For example, some materials like $1TNbSe_2$ and $1T' - WSe_2$ only be present in thin films whose unit layer in the bulk is the 1H structure. This opens up the possibility of growing novel 2D phases of materials that have strong 3D covalent structures, such as technologically mature zinc-blende (ZB) or wurtzite semiconductors. The ability to grow 2D vdW phases of materials that are typically associated with covalent 3D bulk would expand the range of 2D materials available for exploring new physics and functionalities in device applications (Dong and Li, 2021). In theory, it is anticipated that III-V semiconductors will achieve stability in an ultra-thin, bi-layer honeycomb formation. Recently, two-dimensional AlSb was grown in the double-layer honeycomb structure through molecular beam epitaxy (MBE) on graphene-covered SiC(0001), and a fundamental gap of 0.93 eV was found, similar to Si. Furthermore, using first-principles calculations and the Bethe Salpeter equation (BSE), the ground-state geometric and electronic structure of monolayer AlSb was investigated, revealing an excitonic instability because the ground-state exciton has a binding energy that is approximately 0.1 eV greater than the corresponding one-electron gap. This suggests that monolayer AlSb could be an intrinsic excitonic insulator. Spin-orbit coupling (SOC) was found to be necessary for predicting the ground state of monolayer AlSb accurately. The intrinsic carrier mobility of monolayer AlSb was calculated using deformation potential theory since the many-body state in monolayer AlSb is likely to become unstable above the phase transition temperature. It is also reported the successful fabrication of InSb on a Si (100) substrate utilizing GaSb and AlSb buffers with high responsivity (Qin et al., 2021).

2.5.3 Applications of AlSb

AlSb, a relatively understudied III-V compound semiconductor, has received limited attention because of its challenging chemical properties and problems related to bulk crystal growth. Nevertheless, in recent years, there has been a growing interest in this material due to its significant relevance in the realm of GaSb/AlSb superlattices in next-generation electronic devices, both from a physical and technological perspective (Adachi, 2013). Its related alloys, such as $Al_x Ga_{1-x}Sb$ and $Al_x Ga_{1-x}$ As_y Sb_{1-y} are also suitable materials for many optoelectronic devices in the near-IR spectral region (Cipriano et al., 2020; Yadav et al., 2016). The potential applications of the wurtzite structure of AlSb in semiconductor engineering is included the fabrication of transistors, photo-detectors, and solar cells (Singh and Gupta, 2016).

2.6 Indium arsenide (InAs)

2.6.1 InAs properties

InAs, a semiconductor with a direct-band-gap crystalline structure in zincblende, has drawn attention due to its high photoresponse, small direct band gap $(-0.36\,eV$ at $300\,K)$ and resulting high electron mobility (approximately $10^4\,cm^2/V\cdot s$ at $300\,K$). These attributes have made it a compelling choice for Hall-effect devices. InAs can serve as both a substrate and an active layer in combination with various semiconductors like AIGaSb or InAsSb, presenting intriguing possibilities for heterojunctions in modulation-doped field-effect transistors, lasers, and quantum-well structures (Adachi, 2013). Moreover, InAs very useful for making high-speed and low power consumption devices, as well as highly sensitive photo-detectors that can detect light from ultraviolet to infrared wavelengths. Researchers have been exploring different ways to use InAs in various applications. For example, researchers are integrated ultrathin InAs film onto a SiO_2 substrate to create high-performance nanoscale transistors. While 2D InAs is easy to integrate and fabricate at a nanoscale, its large surface area makes it particularly beneficial for photodetectors and sensors. However, research on low-dimensional InAs has primarily focused on quantum dots and nanowires, and high-quality 2D InAs crystals smaller than 10 nanometers have not yet been reported (Dai et al., 2022; Mendoza-Estrada et al., 2017). Fig. 7 shows the atomic structure of bulk InAs, where each In anion has hexagonal coordination with four neighboring As cations, while each As anion coordinates with four In cations to create a tetrahedral bonding environment. However, the non-layered structure of InAs makes it difficult to synthesize ultra-thin InAs films. Recent studies on low-dimensional semiconductor materials with honeycomb structures, such as graphene and hexagonal boron nitride, have revealed exciting physical and chemical properties. Based on these works, researchers examined the possibility of a two-dimensional (2D) hexagonal monolayer InAs using DFT within first-principles calculations. They found that monolayer InAs tends to adopt a low-buckled (BK) hexagonal structure rather than a hexagonal planar (PL) structure like graphene. Non-planar structures are due to the formation of sp^3-like orbitals, which weaken the π-bonds. Further theoretical studies proposed additional potential stable 2D monolayer InAs structures, including 2D tetragonal, hexagonal zigzag, and hexagonal armchair. It has been predicted that InAs can exist in a stable layered form, with the double layer honeycomb (DLHC) InAs being more energetically favorable than the single layer

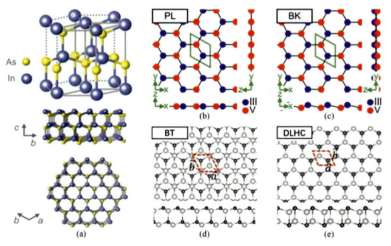

Fig. 7 (A) Atomic structure of bulk InAs. Top and side view of optimized structures for (B, C) monolayer BK structure and PL structure and (D, E) single layer honeycomb or truncated bulk and double layer honeycomb InAs with highlighted primitive unit cells (Kun et al., 2023).

honeycomb (SLHC) or truncated bulk (BT) InAs associated with the three-dimensional bulk phase. All cations in DLHC InAs are structura14lly bound to three anions from the same layer and one anion from another layer, forming a distorted tetrahedral coordination. Within the thickness range of 3–12 monolayers, the truncated haeckelite structure is energetically favorable, characterized by the presence of alternating octagonal and square rings. However, so far, it has been impossible to fabricate pure InAs van der Waals (vdW) materials in experiments, regardless of geometry. Despite this, researchers remain optimistic and are working to make advances in experimental preparation technology, as monolayer and bilayer InAs as fundamental building blocks can open new avenues for both perfect and hybrid vdW integrations (Ahmed et al., 2021; Kun et al., 2023).

Moreover, to determine the energetic stability of a single-layer two-dimensional material, it is often used its formation energy (E_f) compared to its bulk form. This is calculated using the equation $E_f = E_{2D}/N_{2D} - E_{3D}/N_{3D}$, where E_{2D} and E_{3D} are the energy of the 2D and bulk systems, respectively, and N_{2D} and N_{3D} represent the number of atoms in the crystal structures (Zhuang et al., 2013; Liu and Ling, 2023). The buckled structures of monolayer InAs, its band structure and (projected density of states) PDOS also are described in Fig. 8.

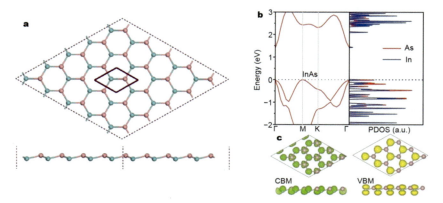

Fig. 8 Top and side views of the geometric structure of monolayer InAs. (A) Pink and cyan balls represent In and As atoms, respectively. (B) Calculated band structures and PDOS of monolayer InAs using HSE06 functional. (C) Calculated charge density corresponding to conduction-band minimum (CBM) and valence-band maximum (VBM) of monolayer InAs (isovalue = 1×10^{-3} $eÅ^{-3}$) (Liu and Ling, 2023).

It is reported that a 2D single crystals of InAs with thicknesses as low as 4.8 nm and lateral sizes up to approximately 37 μm is generated. The resulting InAs flakes are of high crystalline quality and are uniform in nature (Dai et al., 2022). The thickness can be adjusted by altering the growth time and temperature. Additionally, the optical properties of these InAs flakes are studied, noting that the properties vary depending on the thickness. Results show that 2D InAs has high conductivity and carrier mobility. This breakthrough introduces InAs to the 2D materials family and opens up new possibilities for using 2D InAs in high-performance electronics and optoelectronics (Dai et al., 2022).

2.6.2 Synthesis and development of InAs

The synthesis of 2D InAs single crystals is challenging because of the non-layered structure. InAs flakes and hexagonal wurtzite InAs are different versions of InAs. Indium Arsenide flakes are very thin, almost like a flat sheet, and can be made using different methods like mechanical exfoliation or chemical vapor deposition (CVD). They are only a few atoms thick and have unique properties due to their flat, two-dimensional structure. According to Dai et al. (2022), researchers reported that van der Waals epitaxy was utilized to generate 2D single crystals of InAs with thicknesses as low as 4.8 nm and lateral sizes up to approximately 37 μm. The InAs flakes had high quality crystal structures and were uniform in thickness. Moreover, it was demonstrated that wurtzite InAs nanowires could be coated with aluminum using an epitaxial growth method, which resulted in a strong and adaptable growth process. This process enables the creation of a close connection between the aluminum superconductor and the 2D InAs nanowire (Liu and Ling, 2023; Kang et al., 2017).

2.6.3 Applications of InAs

(2D) indium arsenide (InAs) is promising for future electronic and optoelectronic applications such as high-performance nanoscale transistors, flexible and wearable devices, and high-sensitivity broadband photodetectors, and is advantageous for its heterogeneous integration with Si-based electronics. Furthermore, as InAs flakes' optical properties vary with thickness, high conductivity and carrier mobility, it expands 2D materials applications in high-performance electronics and optoelectronics (Dai et al., 2022).

2.7 Indium phosphide (InP)

2.7.1 InP properties

InP is an important material in modern technology because it is a direct–gap semiconductor and zinc blende crystal structure in nature that is commonly used as the substrate for optoelectronic devices that operate at a communications wavelength of 1.55 mm. Many studies have been conducted to better understand the band structure parameters of InP and its alloys. For example, it is found that the direct band gap at low temperatures is 1.423 eV, which has been corroborated by other studies with slight variations. When observing absorption measurements, exciton transitions are usually seen rather than interband transitions, so a binding energy of 5 meV is typically added to the spectral position of the resonance. The temperature dependence of the direct band gap has also been studied and reported in Vurgaftman et al. (2001). Furthermore, InP is typically found as very thin films or layers that are only a few atoms thick. Despite their small size, they possess distinctive electronic and optical properties that make them valuable for a range of electronic and optoelectronic applications. Researchers have conducted theoretical studies on 2D hexagonal indium phosphide and indium arsenide compounds and found that they have a low buckled geometry, with buckling displacements of 0.52 and 0.67 $A^°$, respectively. Using the HSE06 hybrid functional, the lattice constant of monolayer InP is reported to be 4.25 $A^°$, while that of monolayer InAs is 4.38 $A^°$. The band gap of monolayer InP is larger than that of monolayer InAs, at 1.80 eV and 1.41 eV, respectively, and both materials have an indirect band gap, similar to h-InN. In comparison to various bulk phases of InP and InAs, the zinc blende atomic arrangement is found to be the most stable. The lattice constant value of the InP unit cell is reported to be 5.87 $A^°$, while that of the InAs unit cell is 6.06 $A^°$. Both InP and InAs have direct band gaps of 1.34 eV and 0.35 eV, respectively, at room temperature (Shah and Roy, 2023). It is employed ab initio density functional theory calculations to found that a slightly buckled hexagonal InP sheet (HInPS) remains stable in its pristine form and when doped with Zn atoms. This same stability was also observed in hydrogen-passivated zigzag InP nanoribbons (ZInPNRs), which are quasi-one-dimensional versions of the quasi-two-dimensional material (Longo et al., 2013). Moreover, the electrical and thermoelectric properties of monolayer In-VA (where VA is an element from group VA of the periodic table) is examined in Bi et al. (2018). The researchers found that the ultralow intrinsic thermal conductivities of InP at room temperature are similar to those of good thermoelectric (TE) materials. The thermoelectric properties of materials are determined by a figure of merit called ZT, which is calculated using the equation $ZT = \sigma S^2 T / \kappa_e + \kappa_l$. In this equation, σ represents the electrical conductivity, S is the Seebeck coefficient, T is the absolute temperature, and κ_e and κ_l are the electronic and lattice thermal conductivity, respectively. The power factor (PF) is another important parameter, which is defined as $PF = \sigma S^2$ and characterizes the electric power output. Large ZT values are particularly important for practical applications of thermoelectric materials. Therefore, the study calculated the maximal ZT values of p-type InP at 900 K, indicating that In-VA has the potential to be an effective TE material that could work at medium to high temperatures (Bi et al., 2018). The hexagonal buckled structures of monolayer InP, its band structure and PDOS are described in Fig. 9.

2.7.2 Synthesis and development of InP

Various techniques are used to synthesize or grow III-V semiconductors, including InP. These techniques include molecular beam epitaxy (MBE), metal-organic chemical vapor deposition (MOCVD), and liquid phase epitaxy (LPE). MBE is a precise growth technique that deposits atom-by-atom under controlled growth dynamics, making it ideal for electronic and optical devices. However, it is expensive and slow, limiting its applicability. In contrast, MOCVD is a chemical vapor deposition technique that offers

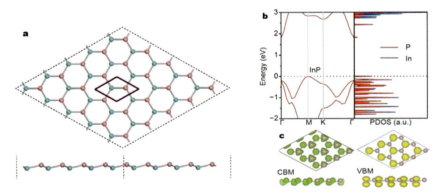

Fig. 9 (A) Top and side views of the geometric structure of monolayer InP. Pink and cyan balls represent In and P atoms, respectively. (B) Calculated band structures and PDOS of monolayer InP using HSE06 functional. (C) Calculated charge density corresponding to conduction-band minimum (CBM) and valence-band maximum (VBM) of monolayer InP (isovalue = $1 \times 10^{-3}\ e\text{Å}^{-3}$) (Liu and Ling, 2023).

faster growth rates and lower costs, making it widely used for integrated circuit applications. LPE is a low-cost deposition technique that deposits material from a supersaturated solution, but it cannot form sharp junctions between two material layers. Other cost-effective and faster growth techniques such as hybrid vapor phase epitaxy (HVPE), close-spaced vapor transport (CSVT), and thin-film vapor-liquid solid (TF-VLS) growth are being explored for III-V semiconductors (Vyas et al., 2022). For instance, it is reported that the 2D single-crystalline InP was successfully fabricated by low-pressure chemical vapor deposition in (Chen et al., 2016; Lu et al., 2022). These techniques provide a pathway towards the commercialization of a cost-effective approach for III-V semiconductors for photonics. To fabricate integrated optical devices like waveguides, photonic crystals, ring resonators, modulators, and interferometers, a standard top-down approach is followed. This involves depositing a mask, writing the pattern on the mask, and transferring the pattern to the underlying material layers. The process is followed by post-processing if required. Before starting the top-down processes, the substrate with the epitaxial layers undergoes cleaning using several chemical cleaning steps that usually involve acetone, isopropyl alcohol (IPA), and de-ionized water (DI water). After cleaning, the material is dried either by passing nitrogen over the surface or heating it on a hot plate. Some of the key steps in the top-down fabrication process are resist deposition, lithography, and etching (Vyas et al., 2022).

2.7.3 Applications of InP

Bulk InP and InAs are commonly used in electronics and optoelectronics devices as substrate material for a variety of applications such as photodiodes, detectors, sensors, LEDs, and switches (Shah and Roy, 2023). The InP substrate is part of a system that includes quaternary alloys of AlGaInAs, InGaAsP, and AlInAsP along three composition lines, as well as two ternary alloys of InAlAs and InGaAs. All of these compositions have a direct bandgap ranging from 0.74 eV to 1.5 eV, offering a wide range of potential applications (Zhang, 2023). InP is becoming an increasingly significant platform for photonic integrated circuits due to its ability to enable the monolithic integration of several active components, including lasers, semiconductor optical amplifiers, photodetectors, and modulators, with passive devices. Additionally, InP-based membranes have been created with high index contrasts, which has resulted in the production of micron-scale devices with a full range of photonic functions (Vyas et al., 2022). Researcher found that properties of monolayer hexagonal InXs including InP are so unique which made them good candidate for photocatalysts in water splitting. Firstly, the intrinsic dipole in monolayer InP produces an electric field that can enhance the energy conversion efficiency during the photocatalytic reaction, thereby reducing the recombination of photogenerated electron-hole pairs. Moreover, strain engineering can effectively tune the band gap and light absorption of monolayer InP, which can significantly improve the solar-to-hydrogen efficiency for water splitting (Liu and Ling, 2023).

2.8 Indium antimonide (InSb)

2.8.1 InSb properties

The Indium Antimonide (InSb) is a semiconductor material with a direct narrow bandgap of 0.17 eV at 300 K. InSb is also known for its very high electron mobility and high electron saturation velocity of $5 \times 10^7 \, cm \times s^{-1}$, which is an attractive material for applications involving galvanomagnetic and Seebeck devices. It has a ballistic length of up to 0.7 μm, which is the highest of any known semiconductor, except possibly for carbon nanotubes. Notably, InSb boasts the lowest melting point (T_m = 800 K) and the largest lattice constant 6.47937 Å at 298.15 K among III-V semiconductors. The narrow energy bandgap of InSb is equivalent to the midwavelength infrared transmission window of 3–5 μm (Adachi, 2013; Penchev, 2012). It is determined that there are some different phases of InSb including zinc blende (ZB), wurtzite (WZ), and 4 H polytypes, using X-ray diffraction (XRD) and transmission electron microscopy (TEM), which can be used in nanowires (Kriegner et al., 2011). In recent decades, antimony-based monolayer materials have become increasingly popular due to their unique properties. In particular, InSb has been found to possess unusual characteristics, such as a small electron effective mass, high charge density, robust spin-orbit interactions, and outstanding carrier mobility. It has been predicted that free-standing InSb is a stable semiconductor with a bandgap of 0.69 eV and ZB single crystals (Liu and Ling, 2023; Bafekry et al., 2021). The high electronic mobility of InSb (78,000 cm^2/Vs) suggests that it has the potential to be used as an optoelectronic material. In bulk form, InSb has a narrow bandgap of 0.18 eV at room temperature, while InSb nanosheets and nanowires exhibit high electron mobility (2.5 $\times$ 10^4 cm^2/Vs) and ambipolar behavior. Monolayer InSb has a low effective mass and other interesting electronic properties that make it a fascinating candidate for future electronic applications, especially when strain is applied to improve its electronic properties. On the other hand, the 2D buckled structure of InSb structure has a direct bandgap of 1.29 eV, high electron mobility, and an intrinsic electric polarization of approximately 12.6 pC/m. Additionally, InSb contains two heavy atoms (In and Sb), leading to strong spin-orbit coupling that can significantly affect its electronic properties (Jalil et al., 2019). By using density functional theory calculations along with Perdew-Burke-Ernzerhof (PBE) functional with spin-orbit coupling (SOC), the InSb monolayer exhibits a band structure with metallic characteristics. However, when using the Heyd-Scuseria-Ernzerhof (HSE) functional, InSb becomes a semiconductor with a very small bandgap of 0.06 eV. The optical properties of InSb monolayers suggest that they could be used as active materials in the ultraviolet region due to the prominent absorption peak at around 5 eV. Therefore, electronic properties of InSb depend on the method used to calculate its band structure. In Fig. 10, schematics of the unit cells of the different polytypes of InSb is shown, where the theoretically derived from deformation compared to the geometrically converted unit cell dimensions (dashed lines). Also, the buckled structures of monolayer InSb, its band structure and PDOS are described in Fig. 11. All of the studies suggest that InSb will be an essential 2D semiconductor material in optoelectronics, low-power electronics, photonic devices and future applications (Bafekry et al., 2021).

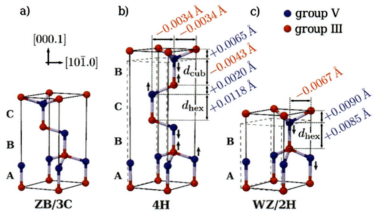

Fig. 10 The sketch showing unit cell structures for different phases of ZB/3C, 4H, and WZ/2H of InSb, highlighting how the unit cell distorts compared to an ideal cubic lattice (dashed lines). It provides numerical values for distortions in materials like InSb, especially of the bond lengths dhex and dcub, and arrows is used to indicate the direction of atomic displacements due to internal parameters (Kriegner et al., 2011).

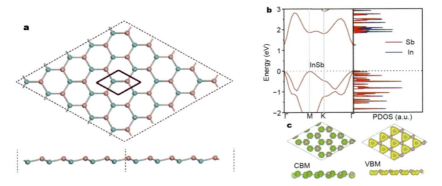

Fig. 11 (A) Top and side views of the geometric structure of monolayer buckled InSb. Pink and cyan balls represent In and Sb atoms, respectively. (B) Calculated band structures and PDOS of monolayer InSb using HSE06 functional. (C) Calculated charge density corresponding to conduction-band minimum (CBM) and valence-band maximum (VBM) of monolayer InSb (isovalue = $1 \times 10^{-3}\ e\text{Å}^{-3}$) (Liu and Ling, 2023).

2.8.2 Synthesis and development of InSb

In a recent work, InAs and InSb nanowire samples, grown using metal organic vapor phase epitaxy (MOVPE), and compared them to InAs nanowires grown through chemical beam epitaxy (CBE). The samples were carefully chosen to minimize polytypism and defects (like stacking faults and twin planes). All were grown on InAs (111) B substrates. All efforts aimed to obtain pure InAs and InSb nanowires (Kriegner et al., 2011). The synthesis of InSb on Si (100) is reported using AlSb and GaSb buffers and observed that the InSb on AlSb exhibit a higher crystal quality than the InSb on GaSb. Moreover, InSb (1 1 0), InSb (1 1 0), InSb (111), and InSb ($\bar{1}\,\bar{1}\,\bar{1}$) surfaces are investigated in very low electron energy diffraction (VLEED) and low electron diffraction (LEED). Furthermore, an InSb (001) surface is synthesized on sulfur-treated substrate with a remarkable structural quality. Finally, the successful growth of 2D nanostructures is reported by using a molecular beam epitaxy (MBE) technique. The produced InSb nanosheets had a high electron mobility and an ambipolar behavior (Touski, 2021).

2.8.3 Applications of InSb

As mentioned previously, InSb has been employed in a wide variety of applications, such as infrared sensors and emitters, magnetic field sensors, toxic gas sensors, and most recently in low-power high speed electronic devices due to its unique electronic and optoelectronic properties

(Penchev, 2012). The indium-based compounds demonstrate materials with small and direct bandgap. Due to their narrow bandgap, they proposed suitable applications for free-space communications, infrared imaging systems, infrared detectors, Schottky diodes, thermal imaging, and gas phase-detection systems (Jalil et al., 2019). Moreover, nanotubes made from 2D InSb monolayers due to the flexibility of its band structure have great potential for use as one-dimensional structures (Touski, 2021).

2.9 Gallium nitride (GaN)

2.9.1 GaN properties

GaN-based semiconductors are renowned for their exceptional characteristics, including a tunable and direct band gap, high thermal conductivity, and robust chemical stability (Jia et al., 2020). Gallium nitride (GaN) usually crystallizes in the hexagonal wurtzite structure at normal environmental conditions characterized by the space group P63 mc, where the lattice constants a and b are equal, as is c, and the lattice angles α and β are 90°, while γ is 120°. Conversely, zinc-blende GaN can be defined as a cubic structure with equal lattice constants a, b, and c, and equal angles α, β, and γ, all at 90°, belonging to the space group F-43 m. Fig. 12 illustrates both these structures for GaN crystals (Adachi, 2013; Qin et al., 2017a). Cubic phase gallium nitride (c-GaN) offers several advantages over the hexagonal phase. These advantages include higher mobility, no spontaneous polarization in the growth direction, ease of p-type doping, and faster electron drift. However, there are limited options for substrates for homoepitaxial growth, which is important for reducing dislocations and lattice mismatch. Notably, it is achieved high-quality crystalline c-GaN films using GaP substrates, which is a less commonly reported choice compared to GaAs (Santis et al., 2023). Theoretical studies have shown that the 2D semiconductor material GaN exhibits two types of hexagonal structures, namely a buckled structure and a planar structure, upon relaxation. The buckled structure, which is properly passivated using

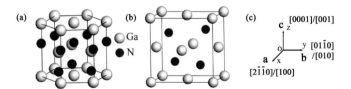

Fig. 12 Crystal structures of: Wurtzite GaN (A); and zinc-blende GaN (B); and the relationship between a, b and c axes and crystallographic directions (C) (Qin et al., 2017a).

partially charged pseudo-hydrogen, exhibits a direct band gap of 5.28 eV (Guo et al., 2018). On the other hand, the hexagonal geometry and the band structure and related projected density of state (PDOS) of planar structure are calculated and illustrated in Fig. 13. The findings revealed that 2D GaN exhibited a direct band gap at the Γ-point, with a value of 4.12 eV. This value significantly differed from the indirect band gap of 3.42 eV previously determined for 2D planar GaN using the HSE06 method. The increase in the band gap can be attributed primarily to the presence of N-H bonding and H-Ga bonding, which were demonstrated to lower the energy of the valence band maximum (VBM) and enhance the chemical stability of planar GaN (Chen et al., 2019).

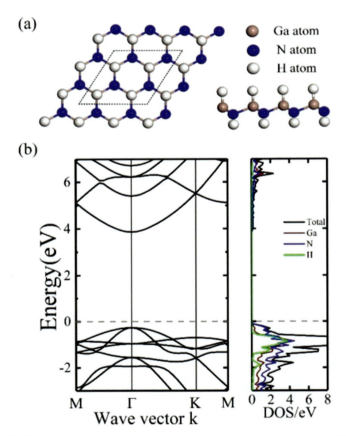

Fig. 13 (A) 2D GaN structure shown from top and side. Rhombus plotted in dashed line represents 2 × 2 supercell used in calculations. (B) Band structure and corresponding projected density of state (PDOS) of 2D GaN (Chen et al., 2019).

The mechanical behavior of ML GaN under uniaxial tension using classical molecular dynamics has been studied. Experimentally, a monolayer GaN with honeycomb structure has been synthesized by reconstructing wurtzite GaN into a 2D graphitic structure. Moreover, the phonon dispersion of 2D GaN exhibits a unique Raman longitudinal optical (LO) phonon mode different from that of bulk GaN, and is anticipated to present different polariton behavior in nano-photonic applications. Overall, the physical and chemical properties of 2D GaN may show diversity due to its low dimensionality, which can further affect the applications of relevant electronic devices. It is important to note that the various applications of monolayer GaN in nano-electronics are closely related to its thermal transport properties, which calls for systematic investigation of the phonon transport properties of this emerging 2D GaN material (Qin et al., 2017b).

2.9.2 Synthesis and development of GaN

Researchers have used various methods like molecular beam epitaxy (MBE), hydride vapor phase epitaxy (HVPE), and metal–organic chemical vapor deposition (MOCVD) to grow c-GaN. In Qin et al. (2017a), the growth of thin films of cubic GaN (c-GaN) on both GaP and GaAs substrates is reported, employing a low-pressure metal-organic chemical vapor deposition (MOCVD) technique. The research investigates how these substrates impact the material properties of GaN. To analyze the structural, morphological, and compositional characteristics of the epitaxial GaN thin films, various analytical methods such as X-ray diffraction (XRD), Raman spectroscopy, and high-resolution scanning electron microscopy (HRSEM) are employed. Additionally, the study explores the nitridation process and the mechanisms involved in epitaxial growth (Santis et al., 2023). The high cost and lack of control over sample quality obtained by molecular beam epitaxy (MBE) have significantly impeded the widespread commercial application of 2D III-nitride materials. Over the past three years, researchers from Pennsylvania State University and South China University of Technology have successfully synthesized 2D GaN using metal organic chemical vapor deposition, aided by graphene encapsulation (Jia et al., 2020; Guo et al., 2018). These remarkable achievements have greatly accelerated the industrialization process of 2D GaN-based semiconductors. However, the progress achieved in lighting devices has been primarily impeded by the growth of polar materials along the c-axis [0001]. GaN-based semiconductors typically crystallize in the wurtzite structure, where the c-axis to a-axis ratio is lower than that of perfectly hexagonally close-packed atoms. This, coupled with the absence of inversion symmetry and the strong ionic nature of III-N bonds, gives rise to a spontaneous dipole moment along the [0001] direction. The dipole moment causes opposing bound polarization charges on the metal-polar [0001] and N-polar [0001] surfaces, generating a robust internal electrostatic field (IEF) along the [0001] direction. This IEF spatially separates electrons and holes on either side and induces an energy band slope, leading to a decrease in emission efficiency and resulting in a redshift in lighting devices (known as the quantum confined Stark effect) (Jia et al., 2020). This suggests that the internal electric field (IEF) can be reduced or even eliminated in these configurations. Consequently, it is highly anticipated that the quantum efficiency of lighting devices based on GaN-based semiconductors will be increased. Hence, the removal of the internal electric field (IEF) in GaN-based materials is anticipated to significantly enhance the efficiency of lighting devices (Jia et al., 2020).

Considerable efforts have been dedicated to addressing the challenges posed by the internal electrostatic field (IEF) in GaN-based materials. Early attempts, such as doping, failed to effectively screen the IEF due to the exceedingly high doping concentrations required (around $10^{19} \ cm^{-3}$). To overcome these fundamental obstacles, researchers have explored the feasibility of utilizing semi- or nonpolar materials that can significantly reduce or eliminate the IEF (Jalil et al., 2019). One approach involved the deposition technique of molecular beam epitaxy (MBE) using $LiAlO_2$ as a substrate to produce nonpolar GaN. However, the presence of a high density of stacking faults in the resulting material compromised its quality and rendered it unsuitable for device structures. Another intuitive option was to replace the wurtzite phase with the nonpolar zinc-blende phase of nitride semiconductors. However, the inherent thermodynamic metastability of this phase and the absence of ideal substrates hindered the realization of device-quality materials, despite extensive efforts. The fundamental solution to this predicament lies in the search for a stable alternative material that lacks an IEF while maintaining comparable optoelectronic performance to that of wurtzite structures (Jia et al., 2020).

Furthermore, the researchers have investigated the direct growth of wurtzite GaN on SiC(0001) and found that it typically results in the formation of three-dimensional (3D) islands due to surface energy constraints and lattice mismatch. However, they have developed a process called "migration-enhanced encapsulated growth" (MEEG) that enables the growth of two-dimensional (2D) GaN. The process involves utilizing an epitaxial graphene layer on SiC(0001) as a starting substrate. The graphene layer is hydrogenated to passivate the dangling bonds, converting it into quasi freestanding epitaxial graphene (QFEG). The QFEG/SiC(0001) substrate is then exposed to cycles of trimethylgallium, which decomposes into gallium adatoms that diffuse on the graphene surface and intercalate between the QFEG and SiC(0001). The intercalated gallium is then transformed into 2D GaN through ammonolysis. In contrast, without the graphene capping layer, the ammonolysis of gallium on SiC(0001) leads to the formation of 3D doughnut-shaped GaN structures. The regions of 2D GaN are found near 3D islands of GaN and graphene wrinkles, indicating that defects in graphene facilitate the intercalation of gallium. The intercalated gallium nitride exhibits unique properties, and aberration-corrected scanning transmission electron microscopy confirms the presence of 2D GaN layers at the graphene/SiC(0001) interface. The researchers also observe that even thicker layers of GaN grown through MEEG retain the same surface termination as 2D GaN, highlighting the role of graphene in stabilizing the 2D buckled quintuple R3m structure. The polarity of the nitrogen-gallium termination in 2D GaN is inverted compared to thick GaN, potentially due to nitrogen replacing hydrogen during ammonolysis. The study demonstrates that MEEG with QFEG/SiC(0001) enables the formation of 2D GaN and thicker GaN layers, both exhibiting unique properties. The graphene layer plays a crucial role in stabilizing the 2D GaN structure and passivating high-energy sites, leading to the stability and charge neutrality of 2D GaN. The findings provide insights into the synthesis and properties of 2D nitrides using MEEG (Al Balushi et al., 2016).

2.9.3 Applications of GaN

Gallium nitride (GaN) has been extensively studied for its use in opto-electronic devices operating in the visible and ultraviolet spectrum, including photo-detectors, light-emitting diodes, lasers, and solar cells. It has been discovered that monolayer (ML) GaN exhibits a low thermal conductivity (κ) driven by its orbital properties. This finding highlights the potential of 2D GaN for energy conversion applications, particularly in thermoelectrics. By investigating the phonon transport in ML GaN using the Boltzmann transport equation (BTE) and exploring its electronic structure, it is provided a foundational understanding of these phenomena. This knowledge contributes to the field of nanoscale phonon transport in 2D materials, and it also illuminates new directions for future research endeavors (Qin et al., 2017b).

2.10 Gallium arsenide (GaAs)

2.10.1 GaAs properties

In the realm of device applications, GaAs is presently among the most adaptable semiconductors in practical use. From a solid-state physics standpoint, GaAs is highly captivating as an exemplary direct-band-gap semiconductor (Adachi, 2013). The most common crystal structures of GsAs are zinc blende (ZB) and wurtzite (WZ) as shown in Fig. 14 (Du et al., 2013). Both the cubic zinc-blende and the hexagonal wurtzite crystal structures remain stable because of the nanowire's specific geometry and the resulting significant surface-to-volume ratio in III-V semiconductor nanowires (Hubmann et al., 2016).

In the 2D III-V material, monolayer hexagonal gallium arsenide has a buckled structure with a buckling distance of 0.59Å. The lattice constant of monolayer GaAs is greater than that of monolayer GaP and GaN, with numerical values of 4.06Å, 3.91Å, and 3.25Å, respectively. The entire GaAs material has an indirect band gap, and among the three, monolayer GaAs has the lowest band gap, with monolayer GaN, and monolayer GaP having higher values. The semiconducting band gap value of monolayerGaAs is 1.83 eV (Shah and Roy, 2023).

The small effective mass of electrons in GaAs makes it a suitable material for the production of ultra-fast transistors. Despite extensive experimental and theoretical studies, there is still a need for a fundamental understanding of the physical properties of this novel material, given its crucial role in industrial applications. GaAs exists in the zinc-blende phase at ambient

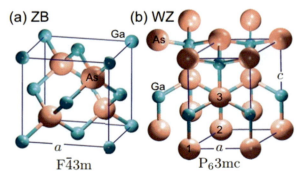

Fig. 14 (Color online) The structures of ZB and WZ GaAs crystals. The conventional unit cells for a ZB and a WZ GaAs crystal are depicted in (A) and (B), respectively. The As and Ga atoms are represented by large pink and small green spheres, respectively (Du et al., 2013).

temperature and pressure, with an experimental lattice constant of 5.65$\overset{\circ}{A}$ (Anua et al., 2012). However, to the best of our knowledge, there have been no previous investigations into the two-dimensional hydrogenated buckled GaAs system (González-García et al., 2020; Mishra and Bhattacharya, 2020). The hexagonal crystalline structure of monolayer GaAs is depicted in Fig. 15 (Mishra and Bhattacharya, 2020). The different configurations of the 2D-GaAs hexagonal unitcell, and hydrogenated buckled GaAs system primitive cell, which consists of one Ga atom and one As atom that were constructed from the zinc-blende structure are illustrated in Fig. 16 (González-García et al., 2020).

Theoretical predictions suggest that 2D GaAs is a mechanically and dynamically stable semiconductor with a buckled geometry, both in its pristine hexagonal unit cell and in forming a graphene-GaAs van der Waals heterostructure. The electronic properties of GaAs and graphene can be modulated in a graphene-GaAs heterostructure, making it an exciting prospect for future practical applications in plasmonic and photonic technology. However, there have been no previous studies on the two-dimensional hydrogenated buckled GaAs system, and despite its theoretical prediction, this material has never been synthesized (González-García et al., 2020).

In González-García et al. (2020), it is used first-principles calculations to investigate the stability, structural, and electronic properties of hydrogenated two-dimensional (2D) GaAs with chair, zigzag-line, and boat

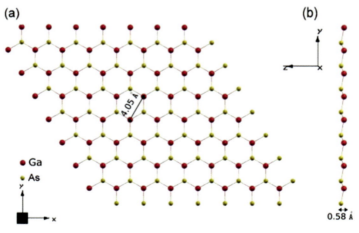

Fig. 15 (A) Crystalline structure of monolayer GaAs exhibiting a hexagonal C3v3m point group symmetry. An in-plane lattice constant a = 4.05 $\overset{\circ}{A}$ and (B) a buckling height along the out of plane (z-direction) is detected to be 0.58 $\overset{\circ}{A}$ after performing a ground state energy relaxation calculation (Mishra and Bhattacharya, 2020).

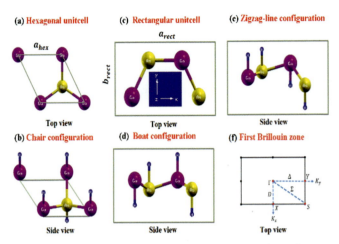

Fig. 16 The schematic representation shown in color online illustrates (A) the 2D-GaAs hexagonal unit cell, as well as the different configurations of the hydrogenated buckled GaAs system, including (B) the 2D H-GaAs chair configuration, (C) the 2D H-GaAs rectangular unit cell, (D) the 2D H-GaAs boat configuration, and (E) the 2D H-GaAs zigzag-line configuration. (F) high symmetry points in the Brillouin zone for the rectangular unit cell of 2D H-GaAs (González-García et al., 2020).

configurations. All configurations were found to be energetically and dynamically stable and exhibited a semiconducting character. The chair and boat configurations displayed a direct bandgap nature, while pristine 2D-GaAs and zigzag-line were indirect semiconductors. The bandgap sizes of all configurations were hydrogen-dependent, and wider than that of pristine 2D-GaAs. Two-dimensional buckled GaAs is a good candidate for hydrogen surface passivation, as the hydrogenation of 2D-GaAs tunes its bandgap, making it a potential candidate for optoelectronic applications.

Furthermore, the electronic band structures of pristine 2D-GaAs and hydrogenated 2D-GaAs with chair, boat, and zigzag-line configurations with the PBE functional are analyzed in Fig. 17. The red lines represent the GaAs-4pz orbitals, the black lines represent the GaAs-4pxy mixed orbitals, and the yellow lines represent the GaAs-4s orbitals for pristine 2D-GaAs or GaAs-4s and H-1s mixed orbitals for 2D H-GaAs. The valence bands of the chair, boat, and zigzag configurations showed an increased overlapping between pz and both planar pxy and s orbitals compared to the pristine 2D-GaAs, resulting in weaker sp^2 hybridization and stronger sp^3 hybridization. This led to a larger average bond length between Ga and As atoms for all three configurations compared to pristine 2D-GaAs. The conduction bands of pristine 2D-GaAs contained 4s and

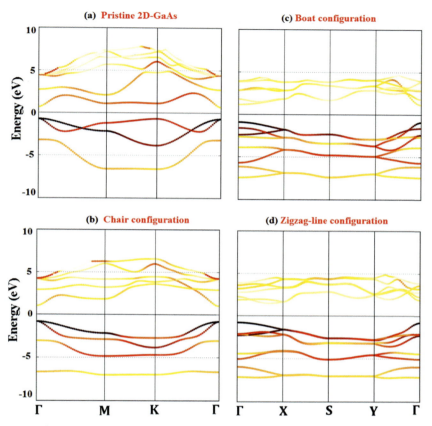

Fig. 17 The electronic band structure for (A) pristine 2D-GaAs sheet and (B) chair, (C) boat, and (D) zigzag-line configurations for 2D H-GaAs are shown in the color online figure, using the DFT-PBE functional. The contributions of GaAs-4pz orbitals are represented by red lines, the contributions of GaAs-4pxy mixed orbitals are represented by black lines, and the yellow lines represent the contributions of either the GaAs-4s orbitals (pristine 2D-GaAs) or GaAs-4s and H-1s mixed orbitals (2D H-GaAs) (González-García et al., 2020).

4p mixed unoccupied orbitals, with a 4s orbital at the bottom of the conduction band. The average bond length between Ga and As atoms for the chair, boat, and zigzag-line configurations of 2D hydrogenated GaAs was found to be significantly larger than that of pristine 2D-GaAs. Hydrogenation of the 2D-GaAs layer was found to increase the bandgap, which makes it a potential material for optoelectronic applications in the visible range, similar to the effect seen in graphene. In contrast, hydrogenation of the 2D-BN sheet resulted in a smaller bandgap compared to the pristine material (González-García et al., 2020).

2.10.2 Synthesis and development of GaAs

Experimental findings of the thermal stability of hydrogenated boron nitride adsorbed on Ni (111) and the synthesis of buckled gallium nitride by hydrogen surface passivation and graphene encapsulation on SiC (0001) suggest a possible route towards synthesizing other group III-V materials in 2D form, including buckled 2D-GaAs. It is investigated the synthesis of 2D-GaAs by exploring the dynamic stability of the material when it is functionalized with hydrogen. It is assessed the structural, mechanical, and electronic properties of both pristine 2D-GaAs and hydrogenated 2D-GaAs to determine whether hydrogen passivation could improve these properties. Moreover, it is noted that the synthesis of these various structures of pristine 2D-GaAs and hydrogenated 2D-GaAs with chair, boat, and zigzag phases could be feasible using appropriate growth conditions or a specific substrate (González-García et al., 2020).

2.10.3 Applications of GaAs

GaAs one of III-V compound semiconductor that exhibit direct band gap has attracted much attention of researchers due its potential application in the field of optoelectronic and microelectronic devices like lasers, photovoltaic cells, photo detectors, modulators, filters, integrated circuits and light emitting diodes (Anua et al., 2012). Furthermore, it is suggested that the presence of hydrogen on 2D-GaAs can adjust the bandgap, making it a promising material for optoelectronic applications in the blue and violet ranges of the visible electromagnetic spectrum. Moreover, it is also pointed out that 2D buckled-GaAs may be a suitable candidate for hydrogen surface passivation (González-García et al., 2020).

2.11 Gallium arsenide (GaSb)

2.11.1 GaSb properties

Gallium antimonide (GaSb) is a semiconductor belonging to the III-V compound group, and it has a crystal structure known as zinc-blende. GaSb has a melting point of 985 K. Moreover, GaSb, along with its alloy systems AIGaAsSb and InGaAsSb, which share the same lattice structure, have found applications in various optoelectronic and electronic devices like lasers, photodetectors, and heterojunction bipolar transistors. To create these GaSb-based devices, it is essential to study low carrier concentration GaSb layers. Recent advancements in molecular beam epitaxy (MBE) and metal-organic chemical vapor deposition (MOCVD) have enabled the growth of high-quality, undoped GaSb with an incredibly low residual carrier density (Adachi, 2013). Gallium antimonide has a band gap of 0.72 eV at room temperature (Zhang et al., 2022). GaSb monolayers exhibit honeycomb-like atomic structures similar to graphene, with low buckling degree. The in-plane lattice constant (a), bond lengths (d), bond angles (β) and buckling heights (δ) of isolated GaSb monolayers are a = 4.310 Å, d = 2.595 Å, β = 112.254° and δ = 0.738 Å, respectively, which are in good agreement with other results (a = 4.38 Å, d = 2.62 Å, β = 113.14° and δ = 0.70 Å for GaSb monolayer in Ref. Bahuguna et al. (2018). Fig. 18A, is exhibited GaSb monolayer honeycomb-like atomic structures similar to graphene, with low buckling degree. Then, the band structures of isolated GaSb monolayers, is calculated and analyzed, as demonstrated in Fig. 18B. The results reveals that the isolated GaSb monolayer has a direct bandgap of 1.097 eV, and the conduction band minimum (CBM) and the valence band maximum (VBM) are both located at Γ-point. The calculated bandgap values of single-layer structures are within the range of previous data (0.789 eV in Bahuguna et al. (2018) and 1.43 eV in Zhuang et al. (2013). As the thickness of GaSb decreases towards the 2D limit, it can adopt a double-layer honeycomb (DLHC) structure (Akiyma et al., 2019). Stacking layers of different materials to form a heterojunction is a promising strategy to enhance the electronic and optical properties of two-dimensional materials. The structure, electronic, and optical properties of GaSb/InAs heterostructure were investigated through first-principles calculations based on density functional

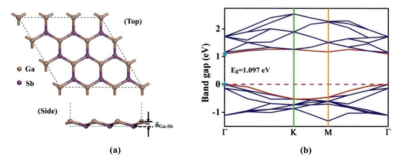

Fig. 18 (A) Top and lateral views of the optimized structures of GaSb monolayers. The brown and purple atoms refer to Ga, Sb, respectively. δGa-Sb represent the vertical distance between the upper and lower atoms. (B) Electronic band structures along the highly symmetric path in Brillouin region (Γ→K→M→Γ) of GaSb monolayer. The Fermi level is determined as 0 eV and indicated by the horizontal magenta dashed lines. The conduction band minimum (CBM) and valence band maximum (VBM) are marked with cyan solid pentagons (Zhang et al., 2022).

theory in Zhang et al. (2022). The band gap of the heterostructure was found to be adjustable by changing the interlayer spacing, external strain, and electric field. Under compression strain, the maximum band gap was found to be 0.611 eV, and under the influence of an electric field, the maximum band gaps were 0.637 eV and 0.129 eV for different models. The heterostructure exhibited enhanced ultraviolet absorption ability compared to GaSb and InAs monolayers, making it a promising material for use in two-dimensional optoelectronic devices.

2.11.2 Synthesis and development of GaSb

Several methods for synthesizing GaSb structure are utilized in some studies. Firstly, at 350 °C, faceted GaSb islands with a density of 10^{11} cm^{-2} are formed on Ga/Si (111), while hexagonal- and pyramid-shaped islands are formed at 400 °C. Only at 450 °C are dome-shaped islands, similar to those grown on clean Si (111), formed. The morphology of GaSb islands grown on Ga/Si (111) is affected by the atomic species terminated on Si (111) and the growth temperature (Hara et al., 2013). Secondly, it is demonstrated the growth of atomically smooth, exfoliatable GaSb films on a graphene-terminated GaSb (001) substrate through pinhole-seeded lateral epitaxy.

The pinholes are created by native oxide desorption from the substrate, and GaSb nucleates directly from the underlying GaSb substrate in the pinholes, followed by lateral overgrowth and coalescence of a continuous film. The resulting films have similar structural quality as previous reports and can be exfoliated to produce free-standing membranes. It is suggested that the seeded lateral epitaxy mechanism may explain the growth of other materials on transferred graphene. The evolution of GaSb films grown on graphene/GaSb (001) was tracked through a combination of molecular beam epitaxy (MBE) synthesis, in-situ electron diffraction and photo-emission spectroscopy (XPS), and ex-situ atomic force microscopy (AFM) and scanning electron microscopy (SEM) (Manzo et al., 2022).

2.11.3 Applications of GaSb

GaSb-based materials have great potential applications in optical fiber communication and infrared detectors (Zhang et al., 2022). Furthermore, due to the weak Van der Waals interactions and small pinholes (10–300 nm diameter) graphene masks permit the etch-free exfoliation of large-scale single-crystalline GaSb membranes for applications in flexible electronics (Hara et al., 2013). GaSb and its derivatives are used in spintronic applications due to their magnetic properties (Doukkali et al., 1992).

References

Adachi, S., 2013. Optical Constants of Crystalline and Amorphous Semiconductors: Numerical Data and Graphical Information. Springer Science & Business Media.

Ahmed, S., et al., 2021. The first-principles prediction of two-dimensional indium-arsenide bilayers. Mater. Sci. Semiconductor Process. 134, 106041.

Akiyma, T., et al., 2019. Realization of honeycomb structures in octet ANB8−N binary compounds under two-dimensional limit. Appl. Phys. Express 12.

Al Balushi, Z.Y., et al., 2016. Two-dimensional gallium nitride realized via graphene encapsulation. Nat. Mater. 15 (11), 1166−1171.

Alrebdi, T.A., Amin, B., 2020. Optoelectronic and photocatalytic applications of hBP-XMY (M = Mo, W; (X ‡ Y) = S, Se, Te) van der Waals heterostructures. Phys. Chem. Chem. Phys. 22 (40), 23028−23037.

Ambacher, O., et al., 1999. Two-dimensional electron gases induced by spontaneous and piezoelectric polarization charges in N- and Ga-face AlGaN/GaN heterostructures. J. Appl. Phys. 85 (6), 3222−3233.

Anua, N.N., et al., 2012. DFT investigations of structural and electronic properties of gallium arsenide (GaAs). In: AIP Conference Proceedings. American Institute of Physics.

Bafekry, A., et al., 2021. Novel two-dimensional AlSb and InSb monolayers with a double-layer honeycomb structure: a first-principles study. Phys. Chem. Chem. Phys. 23 (34), 18752−18759.

Bahuguna, B.P., et al., 2018. Strain and electric field induced metallization in the GaX (X = N, P, As & Sb) monolayer. Phys. E: Low-Dimensional Syst. Nanostructures 99, 236−243.

Bi, J., et al., 2018. Thermoelectric properties of two-dimensional hexagonal indium-VA. Chin. Phys. B 27, 026802.

Cakmak, N., et al., 2020. Functionalisation of hexagonal boron phosphide (h-BP) monolayer via atomic adsorption. Philos. Mag. Lett. 100, 116−127.

Chae, S., et al., 2018. Point defects and dopants of boron arsenide from first-principles calculations: donor compensation and doping asymmetry. Appl. Phys. Lett.

Chaudhuri, R., et al., 2019. A polarization−induced 2D hole gas in undoped gallium nitride quantum wells. Science 365 (6460), 1454−1457.

Chen, J., et al., 2019. Strong selective oxidization on two-dimensional GaN: a first principles study. Phys. Chem. Chem. Phys. 21 (11), 6224−6228.

Chen, K., et al., 2016. Direct growth of single-crystalline III−V semiconductors on amorphous substrates. Nat. Commun. 7 (1), 10502.

Cipriano, L.A., et al., 2020. Quantum confinement in group III−V semiconductor 2D nanostructures. Nanoscale 12 (33), 17494−17501.

Dai, J., et al., 2022. Controlled growth of two-dimensional InAs single crystals via van der Waals epitaxy. Nano Res. 15 (11), 9954−9959.

Dames, C., 2018. Ultrahigh thermal conductivity confirmed in boron arsenide. Science 361 (6402), 549−550.

Del Alamo, J.A., 2011. Nanometre-scale electronics with III−V compound semiconductors. Nature 479 (7373), 317−323.

Dong, S., Li, Y., 2021. Excitonic instability and electronic properties of AlSb in the two-dimensional limit. Phys. Rev. B 104 (8), 085133.

Dou, Y., et al., 2017. Atomically thin non-layered nanomaterials for energy storage and conversion. Chem. Soc. Rev. 46 (23), 7338−7373.

Doukkali, A., et al., 1992. Morphology of the Ag GaSb (110) interface: a study by quantitative AES. J. De. Phys. Iii 2, 275−285.

Du, L., et al., 2018. Novel electronic structures and enhanced optical properties of boron phosphide/blue phosphorene and F4TCNQ/blue phosphorene heterostructures: a DFT + NEGF study. Phys. Chem. Chem. Phys. 20 (45), 28777–28785.

Du, Y.A., Sakong, S., Kratzer, P., 2013. As vacancies, Ga antisites, and Au impurities in zinc blende and wurtzite GaAs nanowire segments from first principles. Phys. Rev. B 87 (7), 075308.

Gao, S.-P., 2012. Cubic, wurtzite, and 4H-BN band structures calculated using GW methods and maximally localized Wannier functions interpolation Comput. Mater. Sci 61, 266–269.

González-García, A., et al., 2020. Two-dimensional hydrogenated buckled gallium arsenide: an ab initio study. J. Phys. Condens. Matter 32 (14), 145502.

Guo, Y., et al., 2018. n- and p-type ohmic contacts at monolayer gallium nitride–metal interfaces. Phys. Chem. Chem. Phys. 20 (37), 24239–24249.

Hara, S., et al., 2013. Growth process and morphology of three-dimensional GaSb islands on Ga/Si(111). Phys. Status Solidi (c) 10, 865–868.

Hu, R., et al., 2021. Recent advances of monoelemental 2D materials for photocatalytic applications. J. Hazard. Mater. 405, 124179.

Hubmann, J., 2016. GaAs nanowires: epitaxy, crystal structure-related properties and magnetic heterostructures.

Jalil, A., et al., 2019. New physical insight in structural and electronic properties of InSb nano-sheet being rolled up into single-wall nanotubes. Appl. Surf. Sci. 487, 550–557.

Jia, Y., et al., 2020. Elimination of the internal electrostatic field in two-dimensional GaN-based semiconductors. npj 2D Mater. Appl. 4 (1), 31.

Jiang, X.-F., et al., 2015. Recent progress on fabrications and applications of boron nitride nanomaterials: A review. J. Mater. Sci. Technol. 31 (6), 589–598.

Kang, J.H., et al., 2017. Robust epitaxial Al coating of reclined InAs nanowires. Nano Lett. 17 (12), 7520–7527.

Kang, J.S., et al., 2018. Experimental observation of high thermal conductivity in boron arsenide. Science 361 (6402), 575–578.

Kang, J.S., et al., 2019. Basic physical properties of cubic boron arsenide. Appl. Phys. Lett. 115 (12).

Kobayashi, Y., et al., 2012. Layered boron nitride as a release layer for mechanical transfer of GaN-based devices. Nature 484 (7393), 223–227.

Kriegner, D., et al., 2011. Unit cell structure of crystal polytypes in InAs and InSb nano-wires. Nano Lett. 11 (4), 1483–1489.

Kuech, T.F., 2016. III-V compound semiconductors: growth and structures. Prog. Cryst. Growth Charact. Mater. 62 (2), 352–370.

Kun, T.-T., et al., 2023. Research progress on first-principles calculations of interfacial charge transfer characteristics in InAs-based van der Waals heterojunctions. J. Infrared Millimeter Waves, 42 (5), 666–680.

Li, J., et al., 2018a. Strain-induced band structure modulation in hexagonal boron phosphide/blue phosphorene vdW heterostructure. J. Phys. Chem. C.

Li, S., et al., 2018b. High thermal conductivity in cubic boron arsenide crystals. Science 361 (6402), 579–581.

Liu, X., Ling, F., 2023. 2D indium-VA semiconductors: promising photocatalysts with intrinsic electric fields for water splitting. Sci. China Mater. 1–9.

Longo, R.C., et al., 2013. Electronic properties of pure and p-type doped hexagonal sheets and zigzag nanoribbons of InP. J. Phys. Condens. Matter 25 (8), 085506.

Lu, F., et al., 2022. Infinite possibilities of ultrathin III-V semiconductors: starting from synthesis. Iscience 25 (3).

Mahat, S., et al., 2021. Elastic constants of cubic boron phosphide and boron arsenide. Phys. Rev. Mater. 5 (3), 033606.

Manoharan, K., Subramanian, V., 2018. Exploring multifunctional applications of hexagonal boron arsenide sheet: a DFT study. ACS Omega 3 (8), 9533–9543.

Manzo, S., et al., 2022. Pinhole-seeded lateral epitaxy and exfoliation of GaSb films on graphene-terminated surfaces. Nat. Commun. 13 (1), 4014.

Mendoza-Estrada, V., et al., 2017. Structural, elastic, electronic and thermal properties of InAs: a study of functional density. Rev. Facultad de. Ingeniería 26 (46), 81–91.

Mishra, H., Bhattacharya, S., 2020. Exciton-driven giant nonlinear overtone signals from buckled hexagonal monolayer GaAs. Phys. Rev. B 101 (15), 155132.

Penchev, M.V., 2012. Indium Antimonide Nanowires: Synthesis, Characterization, and Applications. University of California, Riverside.

Qin, H., et al., 2017a. Mechanical, thermodynamic and electronic properties of wurtzite and zinc-blende GaN crystals. Materials 10 (12), 1419.

Qin, L., et al., 2021. Realization of AlSb in the double-layer honeycomb structure: a robust class of two-dimensional material. ACS Nano 15 (5), 8184–8191.

Qin, Z., et al., 2017b. Orbitally driven low thermal conductivity of monolayer gallium nitride (GaN) with planar honeycomb structure: a comparative study. Nanoscale 9 (12), 4295–4309.

Rastogi, A., Rajpoot, P., Verma, U., 2019. Properties of group III–V semiconductor: BAs. Bull. Mater. Sci. 42, 1–11.

Şahin, H., et al., 2009. Monolayer honeycomb structures of group-IV elements and III-V binary compounds: first-principles calculations. Phys. Rev. B 80 (15), 155453.

Sanders, N., et al., 2017. Electronic and optical properties of two-dimensional GaN from first-principles. Nano Lett. 17 (12), 7345–7349.

Santis, J.A., Marín-García, C.A., Sánchez-R, V.M., 2023. Effect of different substrates on material properties of cubic GaN thin films grown by LP-MOCVD method. J. Cryst. Growth 601, 126944.

Shah, E.V., Roy, D.R., 2023. Group III–V hexagonal pnictide clusters and their promise for graphene-like materials. Atomic Clusters with Unusual Structure, Bonding and Reactivity. Elsevier, pp. 139–155.

Shin, J., et al., 2022. High ambipolar mobility in cubic boron arsenide. Science 377 (6604), 437–440.

Singh, D., Gupta, S.K., Sonvane, Y. 2016. Structural and opto-electronic properties of 2D AlSb monolayer. In: AIP Conference Proceedings. AIP Publishing, 1731 (1).

Solozhenko, V.L., Matar, S.F., 2022. Polymorphism of boron phosphide: theoretical investigation and experimental assessment. J. Mater. Chem. C 10 (10), 3937–3943.

Sugimoto, H., et al., 2019. Size-dependent photocatalytic activity of cubic boron phosphide nanocrystals in the quantum confinement regime. J. Phys. Chem. C 123 (37), 23226–23235.

Taniguchi, T., Yamaoka, S., 2001. Spontaneous nucleation of cubic boron nitride single crystal by temperature gradient method under high pressure. J. Cryst. Growth 222 (3), 549–557.

Tian, F., et al., 2018. Unusual high thermal conductivity in boron arsenide bulk crystals. Science 361 (6402), 582–585.

Touski, S.B., 2021. Strain induced modification in electronic properties of monolayer InSb. Superlattices Microstructures 156, 106979.

Ullah, S., Denis, P.A., Sato, F., 2019. Hexagonal boron phosphide as a potential anode nominee for alkali-based batteries: a multi-flavor DFT study. Appl. Surf. Sci. 471, 134–141.

Vu, T.V., et al., 2021. Structural, elastic, and electronic properties of chemically functionalized boron phosphide monolayer. RSC Adv. 11 (15), 8552–8558.

Vurgaftman, I., Meyer, J.A., Ram-Mohan, L.R., 2001. Band parameters for III–V compound semiconductors and their alloys. J. Appl. Phys. 89 (11), 5815–5875.

Vyas, K., et al., 2022. Group III-V semiconductors as promising nonlinear integrated photonic platforms. Adv. Phys. X 7 (1), 2097020.

Wallentin, J., et al., 2013. InP nanowire array solar cells achieving 13.8% efficiency by exceeding the ray optics limit. Science 339 (6123), 1057–1060.

Wang, F., et al., 2017. Two-dimensional non-layered materials: synthesis, properties and applications. Adv. Funct. Mater. 27 (19), 1603254.

Wentorf, R.H.Jr, 2004. Synthesis of the cubic form of boron nitride. J. Chem. Phys. 34 (3), 809–812.

Woo, K., Lee, K., Kovnir, K., 2016. BP: synthesis and properties of boron phosphide. Mater. Res. Express 3 (7), 074003.

Yadav, S.K., Wang, J., Liu, X.-Y., 2016. First-principles modeling of zincblende AlN layer in Al-AlN-TiN multilayers. arXiv preprint arXiv:1604.02590.

Zhang, G., et al., 2019. Telecom-band lasing in single InP/InAs heterostructure nanowires at room temperature. Sci. Adv. 5 (2), 88–96.

Zhang, X., et al., 2022. DFT study on the controllable electronic and optical properties of GaSb/InAs heterostructure. J. Mater. Res. 37 (2), 479–489.

Zhang, X., Meng, J., 2019. Chapter 4 – Recent progress of boron nitrides. In: Liao, M., Shen, B., Wang, Z. (Eds.), Ultra-Wide Bandgap Semiconductor Materials. Elsevier, pp. 347–419.

Zhang, Y.-G., et al., 2023. The Magic of III-Vs. in Earth and Space: From Infrared to Terahertz, Proc. SPIE 12505, (ESIT 2022), 1250501 (31 January 2023); https://doi.org/10.1117/12.2664543.

Zhuang, H.L., Singh, A.K., Hennig, R.G., 2013. Computational discovery of single-layer III-V materials. Phys. Rev. B 87 (16), 165–415.

CHAPTER FIVE

Two dimensional perovskites

Memoona Qammar[a,b,*] and Faiza[b,c]

[a]Department of Chemistry, The Hong Kong University of Science and Technology (HKUST), Kowloon, Hong Kong (SAR), P.R. China
[b]School of Natural Sciences, National University of Science and Technology, Islamabad, Pakistan
[c]Department of Macromolecular Engineering, Case Western Reserve University, Cleveland, OH, United States
*Corresponding author. e-mail address: mqammar@connect.ust.hk

Contents

1. Introduction	145
2. Synthesis of 2D perovskites	149
2.1 Hot injection (HI)	150
2.2 Ligand assisted reprecipitation (LARP)	151
2.3 Other methods	154
3. Structure and physical properties	155
4. Applications	158
5. Conclusions and perspective	163
References	164

Abstract

Perovskite materials have attracted tremendous attention in the field of material science and technology owing to their unique properties like tunable band gap, high carrier mobility, long diffusion length and high molar absorptivity coefficient. Among 2D perovskites the layered materials have been extensively studied in comparison to non-layered 2D perovskites. But non layered perovskites have emerged as a class of potential materials possessing unique structure and optoelectronic properties which make them suitable for many applications. This chapter will focus on the comprehensive review of synthesis, structure, properties and applications of non-layered 2D perovskites. We will discuss the crystal structure of these materials in comparison to their layered counterparts, followed by an insight of their physical properties and finally their applications. Overall, this chapter will provide insights into the potential of non-layered 2D perovskites as an emerging class of materials.

1. Introduction

History of perovskites can be traced back to their discovery in 1839 where they were discovered by a German scientist Gustav Rose and were named after a Russian minerologist Lev Von Perovski (De Graef and McHenry, 2012). Since the recent decade perovskites are emerging as a rock star class of semiconducting materials. Recently there is an aroused interest in these materials on account of their astonishing and distinctive optoelectronic properties (Dong et al., 2015). Their outstanding physical properties like, long carrier diffusion lengths, enhanced defect tolerance, high absorption coefficient, tunable band gap, elevated charge carrier mobility, better charge carrier lifetime, facile and economic solution casting fabrication have raised their demand in research and development (Xing et al., 2013; Chen et al., 2021; Zhao et al., 2022). These unique properties extended their deployment in almost every optoelectronic field including photovoltaics (Zhao et al., 2022; Kim et al., 2020), photodetectors (Wu et al., 2023), X-ray scintillators and imaging (Xu et al., 2023; Wibowo et al., 2023), field effect transistors (FETs) (Zhang et al., 2023a; Ji et al., 2023), lasing, photocatalysis (Wu et al., 2018a), light emitting diodes (LEDs) (Bai et al., 2023), thermoelectric applications (Kattan et al., 2023), batteries (Zhang et al., 2023b) etc.

Most research in the field of perovskites has been carried out on three dimensional (3D) perovskites. Generally, 3D perovskites are represented by ABX_3 formula and their 3D structure is presented in Fig. 1A (Green et al., 2014). Here, A represents a monovalent cation which can be cesium (Cs^+), formadinium (FA^+), methylammonium (MA^+), B is a divalent cation, which is lead Pb^{2+} in most cases but can be Ge^{2+}, Sn^{2+} in lead free cases and X represents halide anion (usually, I^- but can be Cl^-, Br^-). In a typical 3D perovskite framework, perovskite adopts a regular cubic structure where, A^+ is surrounded by $[BX_6]^{4-}$ octahedra which ultimately shares its each six sides. This makes a favorable electronic structure facilitating the charge carrier movement and improved conduction making it viable for optoelectronic and photovoltaic applications (Green et al., 2014). It is important to note that A^+ must be small enough to set in octahedral voids to sustain the regular structure of 3D perovskite. The size of this cation is restricted by Goldschmidt tolerance factor and it limits the radius to 2.6 Å. Thus, here are limited number of cations that fulfills this requirement (Kieslich et al., 2014).

Despite incredible properties and development 3D perovskites are facing hindrance in commercialization for various applications. One of the major challenges they are facing is the instability of these materials against thermal and environmental ingress. Many solutions like encapsulation, compositional engineering, altering device structures etc. have been deployed (Berhe et al., 2016; You et al., 2016). Another approach that has

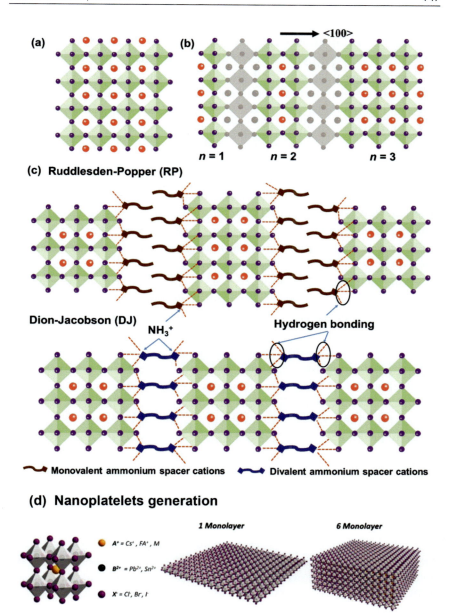

Fig. 1 (A) Crystallographic representation of 3D perovskites (B) incorporation of spacer cations and cleaving the 3D perovskite along the ⟨100⟩ plane (C) RP phase and DJ phase perovskites (Huang et al., 2019a). (D) Generation of monolayer and six layered nanoplatelets from 3D bulk crystal. *(A–C) Reprodutd with permission Huang, P., Kazim, S., Wang, M., Ahmad, S., 2019a. Toward phase stability: Dion–Jacobson layered*

opened a new direction of research is dimensional tailoring to improve photophysical properties and environmental stability of perovskites.

Morphological control and dimensional tailoring are important in tuning physical properties of semiconducting nanomaterials. Final properties of material under test can strongly be standardized by final dimension. For an instance density of states (DOS) changes from discrete points to saw tooth type and finally to steps for zero dimensional (0D), one dimensional (1D) and two dimensional (2D) respectively (Wang et al., 2017). Generally, when we talk about 2D materials, ultrathin nanoplates with atomic layer thickness are considered here. Certain unusual phenomena like 2D electron gas (2DEG), quantum hall effect, spintronics can be observed in 2D materials owing to the confinement of charge carriers. Thus 2D materials are being deployed in sensing, bio imaging, energy storage, catalysis, electronic, optoelectronic, energy conversion (batteries, supercapacitors), etc (Ling et al., 2015; Qammar et al., 2023). Research on layered 2D materials like graphene, silicene, gemanene, boron nitride, MXenes, metal chalcogenides etc. has already been very advanced (Qammar et al., 2023; Li et al., 2016).

Layered materials suffer some limitations due to the restricted amount of functional groups available and their mode of synthesis has limited their variety. Secondly, some 2D materials don't offer superior attributes needed for certain applications. Without any doubt 2D layered materials have offered promising traits in various fields, it is still very important to expand research beyond them to explore and develop new materials with 3D crystal structures and are different from layered materials hence they will be called as 2D non layered materials. These materials possess some distinctive properties like dangling bonds on their surface which improve their chemical activity and make them prone to be deployed in multiple applications like sensing, catalysis and charge carrier transfers (Wang et al., 2017).

In order to resolve the biggest issues like stability and degradation persisting in 3D perovskites, the search for 2D perovskites has emerged as a potential solution (Ahmad et al., 2019). Recently, Ruddlesden popper (RP) and Dion Jacobson (DJ) phase 2D perovskites have achieved highest

Fig. 1—Cont'd *perovskite for solar cells. ACS Energy Lett. 4(12), 2960–2974. Copyright 2019, American Chemical Society. (D) Reproduced with permission from Otero-Martínez, C., Ye, J., Sung, J., Pastoriza-Santos, I., Pérez-Juste, J., Xia, Z., et al., 2022. Colloidal metal-halide perovskite nanoplatelets: thickness-controlled synthesis, properties, and application in light-emitting diodes. Adv. Mater. 34(10), 2107105. Copyright 2021, WILEY.*

attention in field of organic inorganic layered perovskites. Apart from these another alternating cation among interlayer (ACI) phase is also seeking interest. The general formulas of RP phase and DJ phase are slightly different depending upon the bulky cation present in them. RP phase is represented by $R'_2A_{n-1}B_nX_{3n+1}$ and DJ phase perovskites are represented by $R''A_{n-1}B_nX_{3n+1}$ generally. Where R' and R'' represent monoammonium and diammonium bulky cations which act as spacer cations between inorganic layers, A stands for monovalent organic cation, B is metallic cation (divalent) X defines the halogen anion and n represents number or inorganic layers held together and the number of these layers can tuned by controlling stoichiometry of precursors (Ahmad et al., 2019; Grancini and Nazeeruddin, 2019). On the basis of value of n, DJ phase perovskites can be categorized as 2D, quasi2D and 2D/3D heterojunction perovskites. General structural representation of 3D, RP and DJ phase perovskites can be observed in Fig. 1. Apart from these layered various other quantum confined nanosheets (NSs) and nanoplatelets (NPls) has been reported (Weidman et al., 2017; Kostopoulou et al., 2023). Generation of NPls from 3D parent source are depicted in Fig. 1.

Here this chapter is organized in a way that, it will give an overview of 2D non layered perovskites including inorganic NPls and organic inorganic hybrid 2D structures and mono layered structures. Then different synthetic routes used for synthesis of 2D NPls and NSs are summarized followed by photophysical properties are explained. In the last section highlights of several applications are mentioned followed by the current challenges and perspective in the last section.

2. Synthesis of 2D perovskites

Till now various synthetic methodologies for the fabrication of 2D non layered materials (monolayers, nanoplatelets, and nanosheets) have been discovered and deployed. They comprise of wet chemical approaches (hot injection (HI) and ligand assisted reprecipitation (LARP)), vapor phase deposition and solid–state methodologies like mechanical exfoliation and melt quenching. In the following sections we will provide an overview of different synthetic routes along with their pros and cons with some examples of 2D NPls and NSs.

2.1 Hot injection (HI)

Primarily, we will consider a colloidal methodology called hot injection for synthesis these nanomaterials. In 2016, Vybornyi et al. introduced a synthetic method without using polar solvents (Vybornyi et al., 2016). In this reaction lead salt (BX_2) and methylamine cation are dissolved in a non–polar solvent i.e., 1-octadecene (ODE) at a high temperature. The synthesis was assisted by the addition of organic ligands named oleic acid (OA) and oleyl amine (OLA) and to improve colloidal stability. As synthesized $MAPbX_3$ perovskite nano-crystals demonstrated photoluminescence quantum yield efficiency (PLQY) of 10–15%. These nanocrystals were accompanied by some other nanostructures including nano cubes, nanowires and NPls (Vybornyi et al., 2016). After one year another report with the same methodology was reported by Protesescu et al. for highly luminescent $FA_{0.1}Cs_{0.9}PbI_3$ and $FAPbI_3$ perovskite nano-crystals. The injection temperature was set to 50 °C to obtain $FAPbI_3$ NSs. In this report they deployed two precursor approach and three precursor approach was deployed but almost similar results were produced by both ways (Vybornyi et al., 2016).

General representation of HI methodology is shown in Fig. 2A. In essence, PbX_2 salt is dissolved in 1-octadecene at an elevated temperature.

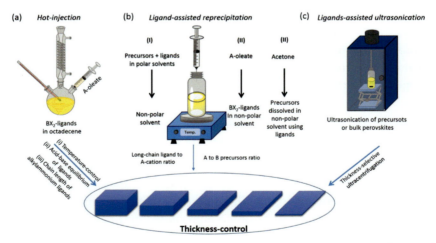

Fig. 2 Different methods for synthesis of NPls. (A) Hot injection. (B) Ligand assisted reprecipitation method. (C) Ligand assisted ultrasonication. *Reproduced with permission from Otero-Martínez, C., Ye, J., Sung, J., Pastoriza-Santos, I., Pérez-Juste, J., Xia, Z., et al., 2022. Colloidal metal-halide perovskite nanoplatelets: thickness-controlled synthesis, properties, and application in light-emitting diodes. Adv. Mater. 34(10), 2107105. Copyright 2021, WILEY.*

In the same mixture organic ligand OLA and OA are added for colloidal stability. In another flask cationic precursor is prepared and added to the above solution at high temperature to generate NPls. Sometimes this step is followed by the addition of an antisolvent like acetone to generate precipitates. At the end the product is purified and the final product is dispersed in some suitable organic solvents. The size of NPls can be tailored bay varying sizes of ligands used during synthesis. The major product of this synthetic route is NPls ranging from monolayers to multiple layers, with the injection temperature less than or equal to 130 °C (Akkerman et al., 2016). Otero–Martínez et al. reported a reduction in NPls thickness with a decrement in the reaction temperature while studying $CsPbBr_3$ NCs (Otero-Martínez et al., 2021).

The role of ligand concentration and acidity on the shape and size of $CsPbBr_3$ nanomaterials was studied by Almeida et al. (2018). According to their work $CsPbBr_3$ NPls can be generated by increasing the amine (RNH_3^+) to monovalent metallic ion (Cs^+) ratio at high temperature (190 °C) (Almeida et al., 2018). The synthesis and tuning of optoelectronic properties of NPls have strong dependence on different factors like temperature, concentration, reaction duration, ratio of precursors and metallic ions etc. The role of an additional halide on tailoring the properties and structure of $CsPbBr_3$ was reported by Sheng et al. (2018). The effect of ratio of $PbBr_2$ to the additional metal source $(CuCl_2)$ on the structural evolution of the product is presented in Fig. 3A–D. Moreover, the replacement of typical long chain ligand i.e., OLA to short chain ligand like hexylphosphonate can lead to generation of very thin NPls (Shamsi et al., 2020).

2.2 Ligand assisted reprecipitation (LARP)

It is noteworthy that HI methodology requires inert atmosphere and high temperature that is not industrial friendly but ligand assisted reprecipitation (LARP) offers a cost effective and facile methodology being able to be executable at room temperature and under ambient atmosphere. LARP is a widely utilized synthetic route and generally in the first step perovskite precursors are dissolved in a good solvent (Dimethyl sulfoxide (DMSO) and N, N–dimethylformamide (DMF)). In the second step a poor solvent accompanied with ligands is added to the above solution in order to induce supersaturation which eventually leads to precipitation (Otero-Martínez et al., 2022; Schmidt et al., 2014; Weidman et al., 2016). Many pure inorganics as well as hybrid (organic inorganic halide) perovskites nano

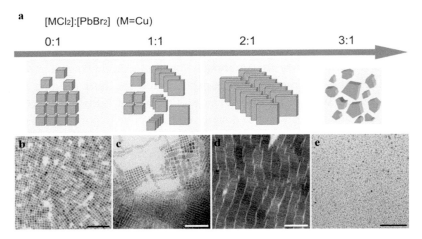

Fig. 3 Morphology of CsPbBr$_3$. (A) The change in morphology with the variation in [MCl$_2$]: [PbBr$_2$]. (B–E) Corresponding TEM images. *Reproduced with permission from Sheng, X., Chen, G., Wang, C., Wang, W., Hui, J., Zhang, Q., et al., 2018. Polarized optoelectronics of CsPbX$_3$ (X = Cl, Br, I) perovskite nanoplates with tunable size and thickness. Adv. Funct. Mater. 28(19), 1800283. Copyright, 2018 WILEY.*

cubes and NPls have been synthesized via similar approach (Akkerman et al., 2016; Bohn et al., 2018; Huang et al., 2019b).

In the case of organic inorganic NPls the ratio of short chain and long chain ligand effects the thickness of NPls and longer chain ligands favor the formation of thinner NPls. This opens the way to tune the layer number of NPls similarly this method is applicable to all inorganic perovskites as well. Where BX$_2$-ligand solution is prepared in non-polar organic solvent and A-oleate is added to it to develop NPls (Huang et al., 2019b). Moreover, a polar solvent like acetone can be added to a solution containing precursors and ligands to induce the nucleation and eventually generation of NPls (Bohn et al., 2018). All of these three variants are schematically summarized in Fig. 2B. Here we will consider the progress in synthesis MAPbBr$_3$ NPls by using LARP methodology. The quantum size effect was studied on the optical properties of MAPbBr$_3$ NPls by Sichert et al. in 2015, the emission of these NPls ranged from blue to green (Sichert et al., 2015). Where they observed a thickness variation upon varying the ratio of organic cations. Their TEM, emission and absorption spectra are depicted in Fig. 4A. Later on, MAPbBr$_3$ NPls emitting from deep blue to intense green were fabricated by Yuan et al. They enacted the variation of ligand chain length and ratio to control the thickness of NPls and synthesized quasi 2D NPls

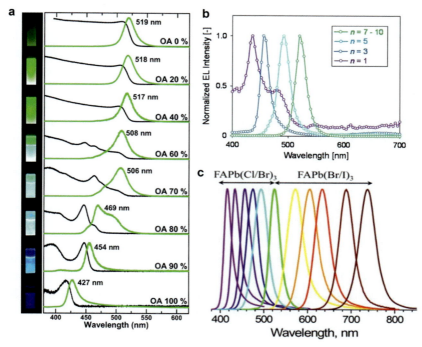

Fig. 4 (A) Optical photographs of suspensions under UV illumination, PL emission and UV absorption of suspensions prepared with variable OA/ MA ratio for MAPbBr$_3$ NPls. (B) EL spectra of MAPbBr$_3$ NPls LEDs fabricated by using suspensions with different n values. (C) PL spectra of FAPbX$_3$ (X = Cl, Br, I, mixed) NPls. *(A) Reproduced with permission from Sichert, J.A., Tong, Y., Mutz, N., Vollmer, M., Fischer, S., Milowska, K.Z., et al., 2015. Quantum size effect in organometal halide perovskite nanoplatelets. Nano Lett. 15(10), 6521–6527. Copyright 2015, American Chemical Society. (B, C) Reproduced with permission from Kumar, S., Jagielski, J., Yakunin, S., Rice, P., Chiu, Y.-C., Wang, M., et al., 2016. Efficient blue electroluminescence using quantum-confined two-dimensional perovskites. ACS Nano 10(10), 9720–9729. Copyright 2016, American Chemical Society.*

(Yuan et al., 2016). Thickness and optical properties of MAPbBr$_3$ NPls can be tailored by varying metal, organic cation and halide ratios, ligand chain lengths and ligand ratios (Kostopoulou et al., 2023; Levchuk et al., 2017a). Apart from these strategies, additive addition like pyridine can also play role in tuning the thickness and optical properties of MAPbBr$_3$ NPls by tailoring 3D structure to 2D structures (Ahmed et al., 2017). Ultrathin MAPbBr$_3$ NPls (down to n = 1) with high electroluminescence (EL) were fabricated by Kumar et al. The EL spectra for various n values of NPls is shown in Fig. 4B. In this study they deployed low k-organic hosts and studied the alteration in their ratios (Kumar et al., 2016).

As extensive work has been presented here for MAPbBr$_3$, there are several reports covering the synthesis of MAPbI$_3$ NPls as well. Researchers have tailored their emissive properties by manipulating the composition, precursor ratio and by changing chemical pressure inside the crystal. They have deployed LARP under different circumstances to generate MAPbBr$_3$ NPls with emission ranging from 512 to 677 nm (Weidman et al., 2016; Stoumpos et al., 2016; Blancon et al., 2017; Hautzinger et al., 2020). Similarly, NPls with long alkyl chain like FA are also under attention of research community. A facile and ambient temperature synthesis of FAPbX$_3$, with X = Cl, Br, I or mixture was reported by Levchuk et al. (2017b). The tunable broad PL emission spectra is shown in Fig. 4C. Apart from the lead-based researchers are exploring synthetic ways to fabricate lead free perovskite NPls. For example synthesis of FASnI$_3$ with tunable optical properties was reported by Weidman et al. (2016).

Apart from organic inorganic perovskites, pure inorganic perovskites, especially CsPbBr$_3$ NPls have also been synthesized via LARP. A monolayer controlled CsPbBr$_3$ NPls were synthesized by colloidal method at low temperature by Akkerman et al. and their properties were compared to thin films and nano cubes (Akkerman et al., 2016). The thickness control can be attributed to the low temperature of the reaction. They used acetone as an antisolvent and were able to vary the thickness of NPls by altering the amount of HBr in the reaction. Generation of layered structure can be attributed to the protonation of OLA due to increased HBr, which results in the competition for lattice sites among OLA cations and Cs$^+$. Later on, CsPbBr$_3$ NPls were synthesized by modifying LARP by Tong et al. They used OLA and OA ligands and acetone as an antisolvent but made this method DMF free (Bohn et al., 2018). Acetone being widely used as an antisolvent can cause surface defects which can add an extra post synthesis step for surface passivation. Later it was found that acetone can be avoided and still NPls can crystallized just by varying the Cs$^+$ to Pb^{2+} precursor ratio (Huang et al., 2019b).

2.3 Other methods

LARP and HI are the widely and extensively studied methods for the synthesis of NPls providing control over thickness, sized and photophysical and optoelectronic properties. Apart from this some other methods have also been tried and are getting limelight to be extensively studied. One of these methods in ligand assisted ultrasonication, its schematic diagram is presented in Fig. 2C. Generally, the perovskite precursor along with organic ligands and a non–polar solvent are ultrasonicated (by tip sonicator) in this route. It is a top-down approach where the bulk perovskite crystals are transformed into NPLs and at the end centrifugation can be applied for purification (Hintermayr et al., 2018). Another approach called pre dissolution assisted solvothermal method has also been used for synthesis of $CsPbBr_3$ NPls. Precisely, PbX_2 was dissolved completely in ODE along with added ligand (OA and OLA) at high temperature (120 °C). Simultaneously, Cs–oleate was also prepared and injected into the above solution after being cooled down to room temperature. Whole mixture was transferred to a Teflon flask and kept in an oven at high temperature (Zhai et al., 2018; Chen et al., 2018). The lower reaction temperature (100 °C) resulted in homogenous and sharp edged NPls as compared to the NPls prepared at higher temperature. Microwave assisted synthetic route has also been tried for the synthesis of 2D perovskites. Shortly, the reactants are encapsulated in a microwave tube and placed into a microwave at 80 °C for only 5 min (Pan et al., 2017). Researchers were able to synthesize nano cubes and nano rods along with NPls by using this fast and efficient methodology.

3. Structure and physical properties

Quantum confinement effect can be approached by reducing the size below Bohr's diameter of respective material. Physical properties especially, optoelectronic properties can be tuned by controlling quantum confinement effect. Spatial confinement of charge carriers along one axis is prominent in 2D and quasi 2D materials and their properties can be tuned by controlling their thickness (number of layers) (Akkerman et al., 2016; Bohn et al., 2018). In 2D materials a range from strongly confined (monolayered) to weakly quantum confined NPls can be fabricated. The regular 3D ABX_3 structure (Fig. 1A) doesn't exist anymore when dimensionality and thickness of perovskite is reduced. At this stage the composition is represented by using $L_2[ABX_3]_{n-1}BX_4$ general formula. Here, L represents ligand, n is used to describe number of octahedral mono layers and A is the spacer cation in layered structures but absent in monolayer structure. Organic ligands (OA and OLA) not only enhance the stability but also play role in limiting the growth in certain directions. As depicted in Fig. 1C monolayer consists of $[PbX_6]^{4-}$ octahedra arranged in a 2D array (Otero-Martínez et al., 2022). Weidman et al. synthesized NPls with $n = 1$

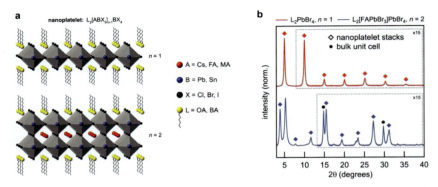

Fig. 5 (A) Perovskite crystal structure for n = 1 and n = 2, there is no A cation in the structure with n = 1. (B) XRD reflection pattern for n = 1 (L$_2$PbBr$_4$) and n = 2 (L$_2$[FAPbBr$_3$] PbBr$_4$). Reproduced with permission from Levchuk, I., Osvet, A., Tang, X., Brandl, M., Perea, J.D., Hoegl, F., et al., 2017. Brightly luminescent and color-tunable formamidinium lead halide perovskite FAPbX$_3$ (X = Cl, Br, I) colloidal nanocrystals. Nano Lett. 17(5), 2765–2770. Copyright 2016, American Chemical Society.

and 2, and studied the effect of halide, metal and cation composition and type on the optoelectronic properties (Weidman et al., 2016). The crystal structure of NPls i.e., n = 1 (L$_2$PbBr$_4$) and n = 2 (L$_2$[FAPbBr$_3$]PbBr$_4$) and their respective XRD reflection patterns are depicted in Fig. 5A and B. The reflections from NPls stacking are represented by diamonds and the strong reflections were attributed to the lateral dimensions. Typical perovskite peaks at 14.9° and 29.9° corresponding to (100) and (200) planes were observed for n = 2 variant. The average spacing of 1.7 and 2.3 nm was observed for n = 1 and n = 2 respectively.

Colloidal NPls are appealing for research community due to a strong quantum confinement effect, and its influence on the intrinsic optical properties. The variation in optical properties upon thickness variation is presented in Fig. 6D and E. According to a classical model the exciton binding energy (E_b) increases with a decrement in the dimension of NPls, and their relationship can be represented by following Eq. (1).

$$E_b = \left(\frac{2}{\alpha - 1}\right)^2 E_e \text{ with } \alpha = 3 - \gamma e^{-\frac{L_W}{2\alpha_0}} \tag{1}$$

Where α, α_0, L_w, γ, and E_e represent dimensional parameter, exciton's Bohr radius, width of quantum well, empirical correction factor and effective Rydberg constant respectively (Blancon et al., 2018). NPls

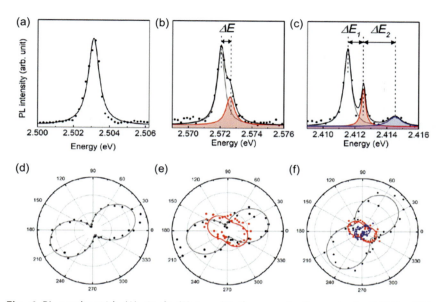

Fig. 6 PL results with (A) single (B) two (C) three peaks for a single CsPbBr$_3$ NPls. Linear polarization properties correspond to (D) single (E) double (F) triple peaks. The splittings are calculated by fitting ΔE (0.71 meV), ΔE$_1$ (0.57 meV) and ΔE$_2$ (1.84 meV). *Reproduced with permission from Huo, C., Fong, C.F., Amara, M.-R., Huang, Y., Chen, B., Zhang, H., et al., 2020. Optical spectroscopy of single colloidal CsPbBr$_3$ perovskite nanoplatelets. Nano Lett. 20(5), 3673–3680. Copyright 2020, American Chemical Society.*

are also called as dielectric confined 2D quantum wells because they show dielectric confinement along with quantum confinement. In bulk perovskite dielectric screening is very prominent as compared to NPls thus enhancing the exciton binding energy. Owing to these confinements a blue shift in band gap, high photoluminescence quantum yield (PLQY) and faster radiative decay is observed in NPls (Wang et al., 2018).

So far, few studies have been attempted to learn the linear optical properties of NPls. Huo et al. studied CsPbBr$_3$ at 5 K and observed single photon emission with no blinking effect. They extended their studies on multiple single NPls and observed multiple PL peaks. Their findings are depicted Fig. 6A–C where they showed multiple PL peaks. The effect of polarization intensity on respective peaks are shown in Fig. 6D–F (Huo et al., 2020). Dey et al. studied the angle dependent polarization for NPls (Dey et al., 2021). All these findings lead to an opportunity for tunning the optoelectronic properties of 2D NPls.

The basics of nonlinear optics can be well understood by Eq. (2).

$$p(t) = \varepsilon_0 (\chi^{(1)} E(t) + \chi^{(2)} E^2(t) + \chi^{(3)} E^3(t) + \dots$$

$$\chi^{(n)} E^n(t)) \tag{2}$$

In this equation $p(t)$, ε_0 and $E(t)$ stands for macroscopic polarizability and vacuum permittivity respectively. The first term tells about linear optical properties and higher terms tells about the nonlinear optic response but they become insignificant only when the strength electric field ($E(t)$) isn't high enough. Higher electrical field strength, quantum and dielectric quantum confinement leads to nonlinear optical processes in NPls (Ferrando et al., 2018).

Multiphoton absorption process has a deeper penetration depth as compared to single photon due to longer wavelength of incident radiation and the materials with multiphoton cross section can find their utilities in imaging and single molecule detection. Usually, the photon absorption cross section is presented in Goeppert-Mayer (GM) units. Nonlinear optical properties can be directed by dimensionality, composition, size and morphological aspects of materials. Zhao et al. carried out a study about single and double photon absorption cross section for $CsPbBr_{2.7}I_{0.3}$ and $CsPbI_3$. They found that $CsPbBr_{2.7}I_{0.3}$ NPls exhibited higher double photon absorption cross section (346 GM nm^{-3}) than respective nano cubes (Zhao et al., 2019). Later on another study observed opposite trend for this dimensional conversion for nonlinear refraction coefficients for $CsPbBr_3$ and $CsPbBr_{1.5}I_{1.5}$ nanocrystals (Roy et al., 2020). As carrier dynamics plays a crucial role for the majority of optoelectronic applications, researchers are trying efficiently to explore these properties for perovskite NPls in comparison to their bulk counterparts. Various studies can be found which have discussed the carrier dynamics of perovskite NPls in detail (Xing et al., 2013; Tan et al., 2014).

4. Applications

Owing to unique physical properties, 2D NPls have been deployed in various optoelectronic applications including light emitting diodes (LEDs). photovoltaics (PV), photodetectors (PDs) and even photocatalysis. In this section we will cover a short glimpse of every application with some examples of NPls. First of all, we will consider utilization of NPls in the field of LEDs.

The first green LED possessing narrow emission based on MAPbBr$_3$ NPls was reported back in 2015 by Ling et al. They synthesized MAPbBr$_3$ colloidal NPls by using solvothermal method. The ligand capping strategy improved the stability of NPls in an ambient atmosphere. Later, Kumar et al. studied the effect of number of layers on electroluminescence (EL) of MAPbBr$_3$ devices (Kumar et al., 2016). A drastic blue shift was observed when the number of layers (n) was reduced to 3 and the EQE obtained for blue LED (n = 3) was 0.23%. The key finding of the study was to develop a room temperature system with EL in deep blue region (470 nm). The TEM images of NPls, the variation of PL along various n values and device parameters are presented in Fig. 7.

Later, in the following years a device made with ultrathin pure metallic perovskite (CsPbBr$_3$) NPls was reported by Yang et al. (2018). Despite very low external quantum efficiency (EQE) this study provided a new direction for the utilization of ultrathin NPls to exploit their applications. Bohn et at. also fabricated an LED by using PbBr$_2$-ligands passivated CsPbBr$_3$ NPls and obtained 0.01% EQE only (Bohn et al., 2018). Another report highlighting the importance of in situ HBr passivation on CsPbBr$_3$ NPls with EQE 0.12% was reported by Wu et al. (2018b).

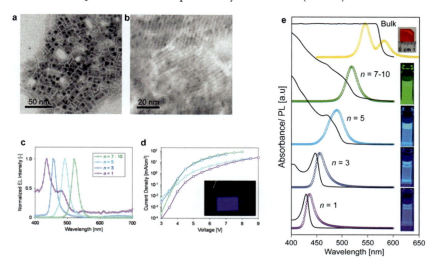

Fig. 7 TEM micrographs of MAPbBr$_3$ NPls (A) n = 7–10 (B) n = 3. Device parameters (C) EL spectrum for different monolayers (D) JV curves for devices. (E) PL and UV Vis spectrum of NPls with different monolayers. *Reproduced with permission from Kumar, S., Jagielski, J., Yakunin, S., Rice, P., Chiu, Y.-C., Wang, M., et al., 2016. Efficient blue electroluminescence using quantum-confined two-dimensional perovskites. ACS Nano 10(10), 9720–9729. Copyright 2016, American Chemical Society.*

One reason for poor EQE of $CsPbBr_3$ was the radiative recombination loses, Hoye et al. used polymeric layer insertion into the structure of NPls to reduce these loses and by deploying poly(triarylamine) they got EQE of 0.3% and 0.55% for blue and sky-blue LEDs respectively (Hoye et al., 2019). Formadinium perovskites have better thermal stability as compared to their methylammonium counterparts and Fang et al. used theses NCs in green LEDs. They prepared few layer (n = 1–4) $FAPbBr_3$ NPls which outperformed its bulk counterpart and achieved 3.53% EQE (Fang et al., 2019). Peng et al. also reported a deep blue LED based on the similar organic inorganic halide perovskite NPls, where they used trioctylphosphine oxide (TOPO) to treat the surface of NPls to reduce surface defects, recombination loses and improve stability (Peng et al., 2020). Effect of surface treatment and size of NPls on $CsPbBr_3$ NPls based LED was studied by Yin et al. In this device they used polyethylenimine (PEI) to treat NPls and enhanced the size of NPls. These modifications led to reduce the number of trap states and coalescence feature which eventually improved the PLQY (Yin et al., 2021). Device architecture, energy band diagram, EL/PL spectrum and device operation are depicted in Fig. 8.

Apart from LEDs, 2D perovskites have also gained attention and are being deployed in different kinds of photodetectors as well. Liu et al. used typical 2D $MAPbI_3$ in a rigid photodetector by synthesizing ultrathin unit cell (Liu et al., 2016). They deployed a hybrid of solution processing and vacuum phase synthesis methodology to fabricate 2D $MAPbI_3$ nanosheets, the schematic representation of synthetic protocol and photodetector are depicted in Fig. 9A and B. The photoresponsivity of the photodetector under 405 nm and 532 nm laser conditions was 22 and 12 AW^{-1} respectively. The graphical results for device tests are shown in Fig. 9C–F. The thickness of active material used in PDs plays a crucial role in determining the performance of the device. Niu et al. highlighted the importance of this parameter and obtained best results with 30–40 nm thick NSs of $MAPbI_3$ (Niu et al., 2016). The improvement of device performance owing to dimensional alteration was reported by Qin et al. Where low dimensional PDs outperformed bulk counterparts and achieve ON/OFF ratio of $> 10^2$ (Qin et al., 2016).

Solution processed scattered $CsPbI_3$ NPls were deployed in PD made via soldering route was reported by Liu et al. They compared the PD performance of single scattered NPls and densely packed NPls. The later one resulted in better performance owing to higher absorption and better carrier transport (Liu et al., 2017). An in-depth study related to the effect of

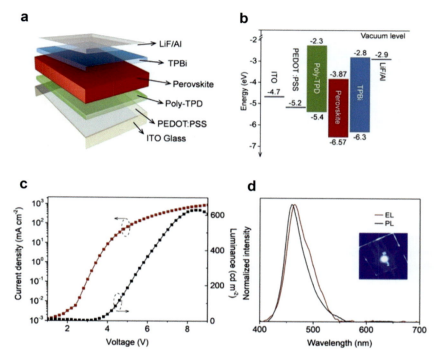

Fig. 8 PEI-CsPbBr$_3$ NPls based (A) complete device architecture (B) band alignment diagram. (C) J-V curve and brightness of LED. (D) Normalized PL (black) and EL (red) spectrum with an LED in inset. *Reproduced with Permission from Yin, W., Li, M., Dong, W., Luo, Z., Li, Y., Qian, J., et al., 2021. Multidentate ligand polyethylenimine enables bright color-saturated blue light-emitting diodes based on CsPbBr$_3$ nanoplatelets. ACS Energy Lett. 6(2), 477–484. Copyright 2021, American Chemical Society.*

thickness tunability on mixed halide 2D perovskites was carried out by Mandal et al. in 2021 (Mandal et al., 2021). In this study ~4.9 nm thick CsPbBr$_{1.5}$I$_{1.5}$ NSs demonstrated better carrier mobility and stability. Moreover, various flexible PDs based on 2D perovskites have also been reported (Song et al., 2016; Li et al., 2017). The first report regarding flexible PD based on CsPbBr$_3$ can date back to 2016 where Song et al. casted the active material on ITO patterned polyethylene terephthalate (PET). This PD achieved photoresponsivity (0.25 AW^{-1}) near to commercial Si based and they endured almost 10,000 bending cycles with less than 3% current loses (Song et al., 2016).

Photocatalysis is a procedure where toxic organic compounds are broken down into useful organic compounds in presence of light is

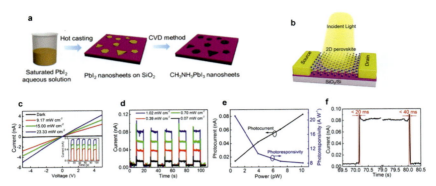

Fig. 9 (A) Schematic representation of solution and vacuum deposition for development of 2D MAPbI$_3$ NSs. (B) Photodetector architecture. (C) I-V response of PD under natural light with variable powers. Under 405 nm laser with 1 V bias voltage. (D) Photocurrent Vs time measurements. (E) Photocurrent and photoresponsivity. (F) Temporal photocurrent response. *Reproduced with Permission from Liu, J., Xue, Y., Wang, Z., Xu, Z.-Q., Zheng, C., Weber, B., et al., 2016. Two-dimensional CH$_3$NH$_3$PbI$_3$ perovskite: synthesis and optoelectronic application. ACS Nano 10(3), 3536–3542. Copyright 2016, American Chemical Society.*

described as photocatalysis for organic chemistry. Researchers have found that 2D perovskites have potential to be used in this field too. Dai et al. reported a lead free Cs$_3$Bi$_2$Br$_9$ NPls for oxidation of toluene into benzaldehyde selectively (Dai and Tüysüz, 2021). They synthesized 100–500 nm thick NPls by using H$_2$SO$_4$ and ethyl acetoacetate (EA), the schematic representation can be found in Fig. 10A. In photocatalytic experimentation 232 μmol of toluene was converted into benzaldehyde (selectively > 88%) with benzyl alcohol as a major byproduct. A constant and stable conversion rate for 28 h was achieved for NPls after irradiation of 8 h. The photooxidation of toluene with thinnest NPls is shown in Fig. 10B.

2D NPls can also be used in CO$_2$ reduction, recently in 2021 Liu et al. reported the utilization of Cs$_2$AgBiBr$_6$ NPls for this application and compared their performance with their nano cubes (Liu et al., 2021). They showed > 99% selectivity for CO$_2$ reduction and their reaction rates were higher as compared to nano cubes. The representation of CO$_2$ reduction reaction, evolution of CO$_2$ and CH$_4$ over time and comparison of reaction rate among NPls and nano cubes is depicted in Fig. 10C–E.

Thin film technology has advanced in the field of PV, the importance and potential of low dimension perovskites can also not be denied in this field. The significance of low dimensional perovskites for PV applications

Two dimensional perovskites 163

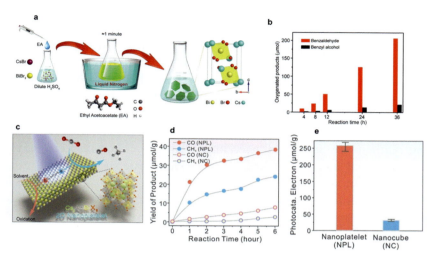

Fig. 10 (A) Schematic representation of synthesis of $Cs_3Bi_2Br_9$ NPls in presence of EA and H_2SO_4 followed by cooling treatment with liquid N_2 for 1 min. (B) Photooxidation of toluene (10 mL) with $Cs_3Bi_2Br_9$ NPls (5 mg) as catalyst under Xe arc lamp. (C) Schematic cartoon for $Cs_2AgBiBr_6$ NPls catalyzed CO_2 reduction reaction (D) the yield of photocatalytic reaction, for CO (red) and CH_4 (blue) for NPls and nano cubes. (E) Comparison of photocatalytic electrons consumed by NPls and nano cubes (time = 6 h). *(A, B) Reproduce with permission from Dai, Y., Tüysüz, H., 2021. Rapid acidic media growth of $Cs_3Bi_2Br_9$ halide perovskite platelets for photocatalytic toluene oxidation. Solar RRL 5(7), 2100265. Copyright 2021, WILEY. (C–E) Reproduce with permission from Liu, Z., Yang, H., Wang, J., Yuan, Y., Hills-Kimball, K., Cai, T., et al., 2021. Synthesis of lead-free Cs_2AgBiX_6 (X = Cl, Br, I) double perovskite nanoplatelets and their application in CO_2 photocatalytic reduction. Nano Lett. 21(4), 1620–1627. Copyright 2021, American Chemical Society.*

was highlighted by Quan et al. in 2016 (Quan et al., 2016). As the system under consideration moves from 2D to 3D the stability of the system reduces and material starts to deteriorate quickly. Recently Li et al. have achieved 18.34% power conversion efficiency (PCE) for quasi 2D RP/DJ phase heterojunction solar cells (Li et al., 2022).

5. Conclusions and perspective

The venture and research for low dimensional materials have paved a new and exciting direction in different scientific fields like optoelectronics, photocatalysis and photovoltaics. The low dimensional materials can be categorized as nanowires, nanorods, nanosheets, nanoplates and quantum dots etc. But this chapter focused specifically on non-layered or mono-layered 2D perovskites, their properties and applications.

Till now researchers have progressed a lot and developed various synthetic approaches ranging from solution based to dry approaches. Despite so many achievements and developments researchers are still struggling to get stable, single PL emissive, high yield and monodispersed nanoplatelets. Post synthesis treatments still need improvement because they can lead to calescence and incomplete removal of residual materials. These materials are getting an overwhelming amount of attention to be deployed in multiple applications. Uniform, continuous and crystalline films are pre-requisite for optoelectronic applications. It is still challenging to obtain high quality thin films.

In nutshell, the research on 2D and 2D/heterojunction materials is at an infant stage. Their astonishing unique optoelectronic properties like tunable optical properties, structural dynamics make them potential for various applications. Better control over morphology, thickness, quantum confinement, stability and homogeneity are required for further improvement in results. For heterojunction materials better interfacial engineering and understanding is required.

References

Ahmed, G.H., Yin, J., Bose, R., Sinatra, L., Alarousu, E., Yengel, E., et al., 2017. Pyridine-induced dimensionality change in hybrid perovskite nanocrystals. Chem. Mater. 29 (10), 4393–4400.

Ahmad, S., Fu, P., Yu, S., Yang, Q., Liu, X., Wang, X., et al., 2019. Dion-Jacobson phase 2D layered perovskites for solar cells with ultrahigh stability. Joule 3 (3), 794–806.

Akkerman, Q.A., Motti, S.G., Srimath Kandada, A.R., Mosconi, E., D'Innocenzo, V., Bertoni, G., et al., 2016. Solution synthesis approach to colloidal cesium lead halide perovskite nanoplatelets with monolayer-level thickness control. J. Am. Chem. Soc. 138 (3), 1010–1016.

Almeida, G., Goldoni, L., Akkerman, Q., Dang, Z., Khan, A.H., Marras, S., et al., 2018. Role of acid–base equilibria in the size, shape, and phase control of cesium lead bromide nanocrystals. ACS Nano 12 (2), 1704–1711.

Bai, W., Xuan, T., Zhao, H., Dong, H., Cheng, X., Wang, L., et al., 2023. Perovskite light-emitting diodes with an external quantum efficiency exceeding 30%. Adv. Mater. 2302283.

Berhe, T.A., Su, W.-N., Chen, C.-H., Pan, C.-J., Cheng, J.-H., Chen, H.-M., et al., 2016. Organometal halide perovskite solar cells: degradation and stability. Energy Environ. Sci. 9 (2), 323–356.

Blancon, J.-C., Tsai, H., Nie, W., Stoumpos, C.C., Pedesseau, L., Katan, C., et al., 2017. Extremely efficient internal exciton dissociation through edge states in layered 2D perovskites. Science 355 (6331), 1288–1292.

Blancon, J.-C., Stier, A.V., Tsai, H., Nie, W., Stoumpos, C.C., Traore, B., et al., 2018. Scaling law for excitons in 2D perovskite quantum wells. Nat. Commun. 9 (1), 2254.

Bohn, B.J., Tong, Y., Gramlich, M., Lai, M.L., Döblinger, M., Wang, K., et al., 2018. Boosting tunable blue luminescence of halide perovskite nanoplatelets through post-synthetic surface trap repair. Nano Lett. 18 (8), 5231–5238.

Chen, D., Chen, X., Li, J., Li, X., Zhong, J., 2018. Ultrathin $CsPbX_3$ (X = Cl, Br, I) nanoplatelets: solvothermal synthesis and optical spectroscopic properties. Dalton Trans. 47 (29), 9845–9849.

Chen, C., Zheng, S., Song, H., 2021. Photon management to reduce energy loss in perovskite solar cells. Chem. Soc. Rev. 50 (12), 7250–7329.

Dai, Y., Tüysüz, H., 2021. Rapid acidic media growth of $Cs_3Bi_2Br_9$ halide perovskite platelets for photocatalytic toluene oxidation. Sol. RRL 5 (7), 2100265.

De Graef, M., McHenry, M.E., 2012. Structure of Materials: An Introduction to Crystallography, Diffraction and Symmetry. Cambridge University Press.

Dey, A., Ye, J., De, A., Debroye, E., Ha, S.K., Bladt, E., et al., 2021. State of the art and prospects for halide perovskite nanocrystals. ACS Nano 15 (7), 10775–10981.

Dong, Q., Fang, Y., Shao, Y., Mulligan, P., Qiu, J., Cao, L., et al., 2015. Electron-hole diffusion lengths > 175 μm in solution-grown CH3NH3PbI3 single crystals. Science 347 (6225), 967–970.

Fang, H., Deng, W., Zhang, X., Xu, X., Zhang, M., Jie, J., et al., 2019. Few-layer formamidinium lead bromide nanoplatelets for ultrapure-green and high-efficiency light-emitting diodes. Nano Res. 12, 171–176.

Ferrando, A., Martinez Pastor, J.P., Suárez, I., 2018. Toward metal halide perovskite nonlinear photonics. J. Phys. Chem. Lett. 9 (18), 5612–5623.

Grancini, G., Nazeeruddin, M.K., 2019. Dimensional tailoring of hybrid perovskites for photovoltaics. Nat. Rev. Mater. 4 (1), 4–22.

Green, M.A., Ho-Baillie, A., Snaith, H.J., 2014. The emergence of perovskite solar cells. Nat. Photonics 8 (7), 506–514.

Hautzinger, M.P., Pan, D., Pigg, A.K., Fu, Y., Morrow, D.J., Leng, M., et al., 2020. Band edge tuning of two-dimensional Ruddlesden–Popper perovskites by A cation size revealed through nanoplates. ACS Energy Lett. 5 (5), 1430–1437.

Hintermayr, V.A., Polavarapu, L., Urban, A.S., Feldmann, J., 2018. Accelerated carrier relaxation through reduced coulomb screening in two-dimensional halide perovskite nanoplatelets. ACS Nano 12 (10), 10151–10158.

Hoye, R.L., Lai, M.-L., Anaya, M., Tong, Y., Gałkowski, K., Doherty, T., et al., 2019. Identifying and reducing interfacial losses to enhance color-pure electroluminescence in blue-emitting perovskite nanoplatelet light-emitting diodes. ACS Energy Lett. 4 (5), 1181–1188.

Huang, P., Kazim, S., Wang, M., Ahmad, S., 2019a. Towardphase stability: Dion–Jacobson layered perovskite for solar cells. ACS Energy Lett. 4 (12), 2960–2974.

Huang, H., Li, Y., Tong, Y., Yao, E.P., Feil, M.W., Richter, A.F., et al., 2019b. Spontaneous crystallization of perovskite nanocrystals in nonpolar organic solvents: a versatile approach for their shape-controlled synthesis. Angew. Chem. Int. Ed. 58 (46), 16558–16562.

Huo, C., Fong, C.F., Amara, M.-R., Huang, Y., Chen, B., Zhang, H., et al., 2020. Optical spectroscopy of single colloidal $CsPbBr_3$ perovskite nanoplatelets. Nano Lett. 20 (5), 3673–3680.

Ji, H., Liu, X., Li, L., Zhang, F., Qin, L., Lou, Z., et al., 2023. Two-dimensional layered Dion–Jacobson phase organic–inorganic tin iodide perovskite field-effect transistors. J. Mater. Chem. A 11 (14), 7767–7779.

Kattan, N.A., Rouf, S.A., Sfina, N., Mana Al-Anazy, M., Ullah, H., Hakamy, A., et al., 2023. Tuning of band gap by anion variation of double perovskites K_2AgInX_6 (X = Cl, Br) for solar cells and thermoelectric applications. J. Solid. State Chem. 319, 123820.

Kieslich, G., Sun, S., Cheetham, A.K., 2014. Solid-state principles applied to organic–inorganic perovskites: new tricks for an old dog. Chem. Sci. 5 (12), 4712–4715.

Kim, G., Min, H., Lee, K.S., Lee, D.Y., Yoon, S.M., Seok, S.I., 2020. Impact of strain relaxation on performance of α-formamidinium lead iodide perovskite solar cells. Science 370 (6512), 108–112.

Kostopoulou, A., Konidakis, I., Stratakis, E., 2023. Two-dimensional metal halide perovskites and their heterostructures: from synthesis to applications. Nanophotonics-Berlin 12 (9), 1643–1710.

Kumar, S., Jagielski, J., Yakunin, S., Rice, P., Chiu, Y.-C., Wang, M., et al., 2016. Efficient blue electroluminescence using quantum-confined two-dimensional perovskites. ACS Nano 10 (10), 9720–9729.

Levchuk, I., Herre, P., Brandl, M., Osvet, A., Hock, R., Peukert, W., et al., 2017a. Ligand-assisted thickness tailoring of highly luminescent colloidal $CH_3NH_3PbX_3$ (X = Br and I) perovskite nanoplatelets. Chem. Commun. 53 (1), 244–247.

Levchuk, I., Osvet, A., Tang, X., Brandl, M., Perea, J.D., Hoegl, F., et al., 2017b. Brightly luminescent and color-tunable formamidinium lead halide perovskite $FAPbX_3$ (X = Cl, Br, I) colloidal nanocrystals. Nano Lett. 17 (5), 2765–2770.

Li, S.-L., Tsukagoshi, K., Orgiu, E., Samorì, P., 2016. Charge transport and mobility engineering in two-dimensional transition metal chalcogenide semiconductors. Chem. Soc. Rev. 45 (1), 118–151.

Li, X., Yu, D., Chen, J., Wang, Y., Cao, F., Wei, Y., et al., 2017. Constructing fast carrier tracks into flexible perovskite photodetectors to greatly improve responsivity. ACS Nano 11 (2), 2015–2023.

Li, K., Yue, S., Li, X., Ahmad, N., Cheng, Q., Wang, B., et al., 2022. High efficiency perovskite solar cells employing quasi-2D Ruddlesden-Popper/Dion-Jacobson heterojunctions. Adv. Funct. Mater. 32 (21), 2200024.

Ling, T., Wang, J.J., Zhang, H., Song, S.T., Zhou, Y.Z., Zhao, J., et al., 2015. Freestanding ultrathin metallic nanosheets: materials, synthesis, and applications. Adv. Mater. 27 (36), 5396–5402.

Liu, J., Xue, Y., Wang, Z., Xu, Z.-Q., Zheng, C., Weber, B., et al., 2016. Two-dimensional $CH_3NH_3PbI_3$ perovskite: synthesis and optoelectronic application. ACS Nano 10 (3), 3536–3542.

Liu, X., Yu, D., Cao, F., Li, X., Ji, J., Chen, J., et al., 2017. Low-voltage photodetectors with high responsivity based on solution-processed micrometer-scale all-inorganic perovskite nanoplatelets. Small 13 (25), 1700364.

Liu, Z., Yang, H., Wang, J., Yuan, Y., Hills-Kimball, K., Cai, T., et al., 2021. Synthesis of lead-free Cs_2AgBiX_6 (X = Cl, Br, I) double perovskite nanoplatelets and their application in CO_2 photocatalytic reduction. Nano Lett. 21 (4), 1620–1627.

Mandal, A., Ghosh, A., Ghosh, D., Bhattacharyya, S., 2021. Interfaces, Photodetectors with high responsivity by thickness tunable mixed halide perovskite nanosheets. ACS Appl. Mater. 13 (36), 43104–43114.

Niu, L., Zeng, Q., Shi, J., Cong, C., Wu, C., Liu, F., et al., 2016. Controlled growth and reliable thickness-dependent properties of organic–inorganic perovskite platelet crystal. Adv. Funct. Mater. 26 (29), 5263–5270.

Otero-Martínez, C., García-Lojo, D., Pastoriza-Santos, I., Pérez-Juste, J., Polavarapu, L., 2021. Dimensionality control of inorganic and hybrid perovskite nanocrystals by reaction temperature: from no-confinement to 3D and 1D quantum confinement. Angew. Chem. Int. Ed. 60 (51), 26677–26684.

Otero-Martínez, C., Ye, J., Sung, J., Pastoriza-Santos, I., Pérez-Juste, J., Xia, Z., et al., 2022. Colloidal metal-halide perovskite nanoplatelets: thickness-controlled synthesis, properties, and application in light-emitting diodes. Adv. Mater. 34 (10), 2107105.

Pan, Q., Hu, H., Zou, Y., Chen, M., Wu, L., Yang, D., et al., 2017. Microwave-assisted synthesis of high-quality "all-inorganic" $CsPbX_3$ (X = Cl, Br, I) perovskite nanocrystals and their application in light emitting diodes. J. Mater. Chem. C 5 (42), 10947–10954.

Peng, S., Wen, Z., Ye, T., Xiao, X., Wang, K., Xia, J., et al., 2020. interfaces, Effective surface ligand-concentration tuning of deep-blue luminescent $FAPbBr_3$ nanoplatelets with enhanced stability and charge transport. ACS Appl. Mater. 12 (28), 31863–31874.

Qammar, M., Zia, A., Adil, O.J., 2023. Emerging trends of MXenes in supercapacitors. Handbook of Functionalized Nanostructured MXenes: Synthetic Strategies Applications from Energy to Environment Sustainability 83–95.

Qin, X., Yao, Y., Dong, H., Zhen, Y., Jiang, L., Hu, W., 2016. Perovskite photodetectors based on $CH_3NH_3PbI_3$ single crystals. Chem. Asian J. 11 (19), 2675–2679.

Quan, L.N., Yuan, M., Comin, R., Voznyy, O., Beauregard, E.M., Hoogland, S., et al., 2016. Ligand-stabilized reduced-dimensionality perovskites. J. Am. Chem. Soc. 138 (8), 2649–2655.

Roy, S., Mandal, A., Raj R, A., Bhattacharyya, S., Pal, B., 2020. Thermal nonlinear refraction in cesium lead halide perovskite nanostructure colloids. J. Phys. Chem. C 124 (28), 15558–15564.

Schmidt, L.C., Pertegás, A., González-Carrero, S., Malinkiewicz, O., Agouram, S., Minguez Espallargas, G., et al., 2014. Nontemplate synthesis of $CH_3NH_3PbBr_3$ perovskite nanoparticles. J. Am. Chem. Soc. 136 (3), 850–853.

Shamsi, J., Kubicki, D., Anaya, M., Liu, Y., Ji, K., Frohna, K., et al., 2020. Stable hexylphosphonate-capped blue-emitting quantum-confined $CsPbBr_3$ nanoplatelets. ACS Energy Lett. 5 (6), 1900–1907.

Sheng, X., Chen, G., Wang, C., Wang, W., Hui, J., Zhang, Q., et al., 2018. Polarized optoelectronics of $CsPbX_3$ (X = Cl, Br, I) perovskite nanoplates with tunable size and thickness. Adv. Funct. Mater. 28 (19), 1800283.

Sichert, J.A., Tong, Y., Mutz, N., Vollmer, M., Fischer, S., Milowska, K.Z., et al., 2015. Quantum size effect in organometal halide perovskite nanoplatelets. Nano Lett. 15 (10), 6521–6527.

Song, J., Xu, L., Li, J., Xue, J., Dong, Y., Li, X., et al., 2016. Monolayer and few-layer all-inorganic perovskites as a new family of two-dimensional semiconductors for printable optoelectronic devices. Adv. Mater. 28 (24), 4861–4869.

Stoumpos, C.C., Cao, D.H., Clark, D.J., Young, J., Rondinelli, J.M., Jang, J.I., et al., 2016. Ruddlesden–Popper hybrid lead iodide perovskite 2D homologous semiconductors. Chem. Mater. 28 (8), 2852–2867.

Tan, Z.-K., Moghaddam, R.S., Lai, M.L., Docampo, P., Higler, R., Deschler, F., et al., 2014. Bright light-emitting diodes based on organometal halide perovskite. Nat. Nanotechnol. 9 (9), 687–692.

Vybornyi, O., Yakunin, S., Kovalenko, M.V., 2016. Polar-solvent-free colloidal synthesis of highly luminescent alkylammonium lead halide perovskite nanocrystals. Nanoscale 8 (12), 6278–6283.

Wang, F., Wang, Z., Shifa, T.A., Wen, Y., Wang, F., Zhan, X., et al., 2017. Two-dimensional non-layered materials: synthesis, properties and applications. Adv. Funct. Mater. 27 (19), 1603254.

Wang, Q., Liu, X.-D., Qiu, Y.-H., Chen, K., Zhou, L., Wang, Q.-Q., 2018. Quantum confinement effect and exciton binding energy of layered perovskite nanoplatelets. AIP Adv. 8 (2).

Weidman, M.C., Seitz, M., Stranks, S.D., Tisdale, W.A., 2016. Highly tunable colloidal perovskite nanoplatelets through variable cation, metal, and halide composition. ACS Nano 10 (8), 7830–7839.

Weidman, M.C., Goodman, A.J., Tisdale, W.A., 2017. Colloidal halide perovskite nano-platelets: an exciting new class of semiconductor nanomaterials. Chem. Mater. 29 (12), 5019–5030.

Wibowo, A., Sheikh, M.A.K., Diguna, L.J., Ananda, M.B., Marsudi, M.A., Arramel, A., et al., 2023. Development and challenges in perovskite scintillators for high-resolution imaging and timing applications. Commun. Mater. 4 (1), 21.

Wu, W., Lu, H., Han, X., Wang, C., Xu, Z., Han, S.T., et al., 2023. Recent progress on wavelength-selective perovskite photodetectors for image sensing. Small Methods 7 (4), 2201499.

Wu, Y., Wang, P., Zhu, X., Zhang, Q., Wang, Z., Liu, Y., et al., 2018a. Composite of CH3NH3PbI3 with reduced graphene oxide as a highly efficient and stable visible-light photocatalyst for hydrogen evolution in aqueous HI solution. Adv. Mater. 30 (7), 1704342.

Wu, Y., Wei, C., Li, X., Li, Y., Qiu, S., Shen, W., et al., 2018b. In situ passivation of $PbBr_6^{4-}$ octahedra toward blue luminescent $CsPbBr_3$ nanoplatelets with near 100% absolute quantum yield. ACS Energy Lett. 3 (9), 2030–2037.

Xing, G., Mathews, N., Sun, S., Lim, S.S., Lam, Y.M., Grätzel, M., et al., 2013. Long-range balanced electron-and hole-transport lengths in organic-inorganic CH3NH3PbI3. Science 342 (6156), 344–347.

Xu, H., Liang, W., Zhang, Z., Cao, C., Yang, W., Zeng, H., et al., 2023. 2D perovskite Mn^{2+}-doped $Cs_2CdBr_2Cl_2$ scintillator for low-dose high-resolution X-ray imaging. Adv. Mater. 2300136.

Yang, D., Zou, Y., Li, P., Liu, Q., Wu, L., Hu, H., et al., 2018. Large-scale synthesis of ultrathin cesium lead bromide perovskite nanoplates with precisely tunable dimensions and their application in blue light-emitting diodes. Nano Energy 47, 235–242.

Yin, W., Li, M., Dong, W., Luo, Z., Li, Y., Qian, J., et al., 2021. Multidentate ligand polyethylenimine enables bright color-saturated blue light-emitting diodes based on $CsPbBr_3$ nanoplatelets. ACS Energy Lett. 6 (2), 477–484.

You, J., Meng, L., Song, T.-B., Guo, T.-F., Yang, Y., Chang, W.-H., et al., 2016. Improved air stability of perovskite solar cells via solution-processed metal oxide transport layers. Nat. Nanotechnol. 11 (1), 75–81.

Yuan, Z., Shu, Y., Xin, Y., Ma, B., 2016. Highly luminescent nanoscale quasi-2D layered lead bromide perovskites with tunable emissions. Chem. Commun. 52 (20), 3887–3890.

Zhai, W., Lin, J., Li, Q., Zheng, K., Huang, Y., Yao, Y., et al., 2018. Solvothermal synthesis of ultrathin cesium lead halide perovskite nanoplatelets with tunable lateral sizes and their reversible transformation into Cs_4PbBr_6 nanocrystals. Chem. Mater. 30 (11), 3714–3721.

Zhang, Y., Ummadisingu, A., Shivanna, R., Tjhe, D.H.L., Un, H.I., Xiao, M., et al., 2023a. Direct observation of contact reaction induced ion migration and its effect on non-ideal charge transport in lead triiodide perovskite field-effect transistors. Small 2302494.

Zhang, C., Zhang, Y., Nie, Z., Wu, C., Gao, T., Yang, N., et al., 2023b. Double per-ovskite La_2MnNiO_6 as a high-performance anode for lithium-ion batteries. Adv. Sci. 2300506.

Zhao, F., Li, J., Gao, X., Qiu, X., Lin, X., He, T., et al., 2019. Comparison studies of the linear and nonlinear optical properties of $CsPbBr_x I_{3-x}$ nanocrystals: the influence of dimensionality and composition. J. Phys. Chem. C 123 (14), 9538–9543.

Zhao, X., Liu, T., Loo, Y.L., 2022. Advancing 2D perovskites for efficient and stable solar cells: challenges and opportunities. Adv. Mater. 34 (3), 2105849.

CHAPTER SIX

CVD growth of 2D non layered materials

Shumaila Karamat* and Shabeya Kanwal

Department of Physics, COMSATS University, Islamabad, Pakistan
*Corresponding author. e-mail address: shumailakaramat@comsats.edu.pk

Contents

1. Introduction	170
2. Growth of 2D non-layered materials	170
3. Chemical vapor deposition (CVD)	171
3.1 Basic overview: How does CVD work?	171
4. Growth of phosphorene	172
5. Two-dimensional selenium nanoflakes (SeNFs)	173
6. Growth of di-indium tri-sulfide (In$_2$S$_3$)	175
7. Growth of β-Ga$_2$O$_3$	176
8. Growth of cadmium sulfide (CdS)	177
9. Growth of hematite (α-Fe$_2$O$_3$)	179
10. Growth of ε-Fe$_2$O$_3$	180
11. Growth of lead sulfide (PbS)	181
12. Growth of titanium dioxide (TiO$_2$)	183
13. Conclusion	184
References	184

Abstract

The shrinkage in the size of materials harnessed the world due to their unique and novel properties, which emerge at low dimensions. Among low geometry, 2D non-layered materials are enthusiastically pursued by the scientific community due to their novel properties. Various techniques have been developed so far to prepare 2D materials and chemical vapor deposition (CVD) is considered the well-suited growth technique for the non-layered growth of 2D materials. CVD deposits materials as a result of a reaction occurring in the gaseous phase or at the desired substrate surface. Different factors such as reactions of precursor gases occurring at the surface, diffusion of the required species, mass transfer of gaseous species, and desorption reaction highly influenced the growth of material and can be controlled by optimizing temperature, pressure, and geometrical features of the system. In this chapter, the CVD synthesis of 2D non-layered materials is presented. The synthesis routes mainly follow chemical vapor deposition or its types i.e., thermal CVD, plasma CVD, metal-oxide CVD, self-limited epitaxial growth, Vander waal epitaxy, and much more rather than using other techniques of deposition. The synthesis and characterization of monolayer materials like phosphorene, cadmium sulfide, selenium nanoflakes, titanium oxide, *Indium sulphide*, and their applications will be discussed.

1. Introduction

In 2D non-layered material atoms or molecules form a single-layer structure instead of a head-up arrangement (Pal et al., 2022), unlike 2D layered materials results in enhanced expertise in catalysis, sensing, and carrier transport. These materials contain unsaturated surface atoms with dangling bonds in between them (Zheng et al., 2020). Although 2D materials lack the layered structure of graphene, they still possess extraordinary capabilities due to their nanostructure. Chemisorption of reactants is promoted by exposed surface atoms in non-layered 2D materials. Structural engineering can be utilized to enhance the catalytic performance of these material's surfaces. The defects present on the surface of these materials are more beneficial to surface electronic structure and for charge transport mechanism. However, most conventionally used electrocatalysis which exhibits good quality are also non-layered in their crystal structure. 2D non-layered $CoSe_2$ exhibits good OER performance while MoS_2 is a layered material and it exhibits limited OER performance in alkaline conditions (Wang et al. 2020). Due to the unique atomic thickness and atomic organization, these non-layered 2D materials usually possess distinctive optical, electrical, and mechanical capabilities. Examples of non-layered 2D materials include β-Ga_2O_3, In_2S_3, CdS, phosphorene, selenium nanoflakes, titanium oxide, hematite, PbS, cobalt phosphide. Non-layered materials were previously thought to be non-exfoliated due to the lack of anisotropy in their 3D bonding network. Later, influenced by the exfoliation of layered hematite research along the non-layered orientation, the interest in exploring whether non-layered materials can also be exfoliated or not was developed. Therefore, it was investigated that non-layered materials having cleavage planes could be exfoliated along cleavage directions (Liu et al., 2019). The active 2D nanosheets were synthesized by non-layered and un-exfoliative natural materials. Recently Jiang et al. reported the synthesis of ultra-thin nickel-doped cobalt oxide single crystal through one-pot vander waal epitaxy (Jiang et al., 2023).

2. Growth of 2D non-layered materials

These materials have many potential applications, so these materials are now of great interest. The methods that are commonly used for the synthesis of 2D non-layered materials (Qin et al., 2023). The choice of technique depends on the required material properties. Several techniques such as molecular beam epitaxy (MBE), liquid phase exfoliation, hydrothermal synthesis, physical vapor deposition (PVD), CVD, and sol–gel synthesis are used to synthesize non-layered materials. CVD is considered as most appropriate technique for the synthesis of 2D non-layered materials (Fig. 1).

3. Chemical vapor deposition (CVD)

The CVD technique is used by most industries to deposit thin films or coatings of materials onto a substrate surface (Han, 2019). In CVD, a chemical reaction at or close to the substrate's surface occurs to create a solid material that condense on the surface. Precursor gases containing the individual elements of the desired materials are introduced at high-temperature furnaces/chambers during CVD (Qin et al., 2023). The general CVD principle is not enough for the synthesis of all type of materials, optimized CVD methods could be used to obtain specific 2D non-layered materials via van der Waal epitaxy growth (vWDE), space-confined CVD, self-limited epitaxial growth (Zhou et al., 2019).

3.1 Basic overview: How does CVD work?

The CVD process begins with precursor gases. The reactor chamber is filled with these gases where frequently volatile molecules decompose under high temperature and, then deposited as a thin coating on substrate.

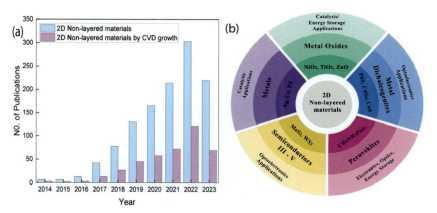

Fig. 1 (a) Comparative bar graph of 2D non-layered materials and their growth by CVD. Data were taken from Scopus by searching the word 2D non-layered materials from 2014 to 2023. (b) Applications of 2D non-layered materials.

A chemical reaction takes place when these precursor gases meet the heated substrate. The atoms or molecules of the target substance are released in this reaction by dissolving precursor molecules. The released atoms or molecules of the desired substance are now deposited on the substrate's surface. Metals, ceramics, and semiconductors can also be deposited by CVD (El Hammoumi et al., 2022) (Fig. 2).

4. Growth of phosphorene

Phosphorene is a monolayer of black phosphorous and was successfully exfoliated in 2014 (Yu et al., 2016). Synthesis of high-quality phosphorene depends on the growth parameters such as temperature, substrate orientation, cooling rate. Blue phosphorene is a monolayer allotrope of black phosphorene. The synthesis demonstration of blue phosphorene is reported due to its promising applications in optoelectronics (Tchoffo et al.) (Fig. 3).

Phosphorene possesses a puckered structure. Phosphorene can be synthesized by top-down or bottom-up approaches. Exfoliation of the parent atom of black phosphorous is done in the top-down approach. At the bottom up assembling of phosphorous's small precursor is done. To improve the surface transfer, surface modification is done with different elements. It also increases the stability when black phosphorous is exposed to oxygen. Ultrathin black phosphorous nanosheets having single atomic thickness have the greatest surface area (Chakraborty et al., 2023). The non-layered puckered structure of phosphorene is due to its sp_3 orbital hybridization (Pang et al., 2018). For monolayers, if the number of layers is reduced, the band gap tends to increase from 1.3 eV to approximately 1.5 eV. It has a characteristics of a tunable band gap, strong in-plane

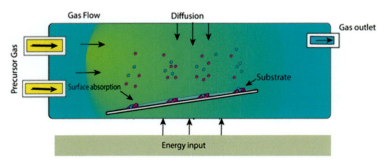

Fig. 2 Schematic of diffusion-controlled CVD setup for thin film growth.

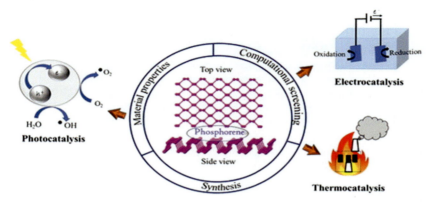

Fig. 3 Applications of phosphorene. *Reprinted with permission of Christopher Igwi Idumah with Copyrights (2022) (Idumah, 2022).*

anisotropy, and greater charge mobility (Akhtar et al., 2017). Nowadays, CVD is the most trustable technique for the synthesis of monolayer 2D material, due to precise control over the morphology, defects, etc. (Zhou et al., 2019). Smith et al. reported that, in Kimble conical bottom-glass centrifuge, substrate i.e., the red amorphous phosphorous thin film was transferred. The conical bottom glass contained mineralizing agents like purified Sn (20 mg) and SnI_4 (10 mg). In the pressure vessel reactor centrifuge tubes were placed, then the vessel was sealed, evacuated and backfilled with argon. The final adjusted pressure of the vessel was 400 psi. The vessel was heated at an external temperature of 950 °C in tube furnace. After reaching 950 °C the temperature of vessel starts to decrease at the rate of 50 °C at every 30 min. Then the temperature was kept in between 700 and 600 °C for 1 h, until it was completely off. The temperature and pressure and monitored and recorded at every cycle (Smith et al., 2016). The obtained sample was 6000 nm thick roughly, whereas thin sample were less than 3.4 nm. By using polarized Raman experiments, the visibility at 365 cm^{-1} band verifies SBP as BP (Smith et al., 2016). Nehra et al. reported systematics of phosphorene crystal structure via ab initio method DFT calculations (Nehra et al., 2023) (Fig. 4).

5. Two-dimensional selenium nanoflakes (SeNFs)

Selenium is a member of Group VI from chalcogen's family with band gap of 0.71 eV at room temperature. Selenium is also a promising

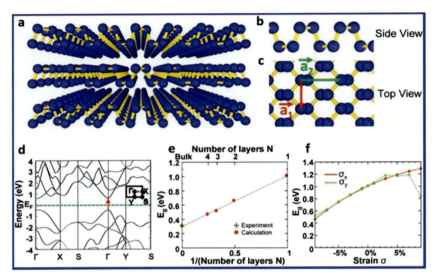

Fig. 4 Systematic of the crystal structure of phosphorene via ab initio DFT calculations (a) preceptive view, (b, c) top view, (d) band structure of monolayer phosphorene and (e, f) shows energy band gap dependence on the number of layers and the strain. Figure a–f is reprinted with copyright permission (2023) of Nehra et al.

photovoltaic material widely used in solar cells. The key ingredients affecting the performance of innovative selenium solar cells has to the shape of light absorber and crystallinity (Nielsen et al., 2023). Crystal symmetry tyrannizes the optical transition of SeNFs under illumination. Hussain et al. reported photoluminescence, room temperature functionality and phase-changing characteristics of SeNFs (Hussain et al., 2022). 2D SeNFs are good enough due to their excellent thermoelectric applications, because they possess the largest see beck coefficient (+1250 µV/K) among all the others. Selenium grows in 1D strongly due to its anisotropic atomic structure instead of 2D (Shi et al., 2022). Among all the forms of selenium square selinene, monoclinic selenium (α-Se, β-Se, γ-Se) are present, triangular selenium dominates due to its excellent electro-optic properties. Fan et al. reported that 2D SeNFs can be synthesized from bulk selenium. The SeNFs were stored away from sunlight, at −20 °C. From their results that the size of 2D SeNFs was 50–130 nm and their thickness was in the range of 5–10 nm. From the results of X-ray diffraction pattern, it is supposed that the obtained SeNFs are triangular SeNFs. For 2D Selenium the Raman shift was 144.3 cm^{-1}. The PEC method was also used to check the photo-response of them, which proves them good for self-powered photodetectors (Fan et al., 2019). Sun et al. reported that Transition metal selenides (TMSes) gain a prominent attention against noble metal based electrocatalyst. These electrocatalysts are mostly used water splitting process, the compatibility of selenides other than TMSes, for water splitting is also durable (Sun et al., 2022) (Fig. 5).

6. Growth of di-indium tri-sulfide (In$_2$S$_3$)

According to different morphological structures 2D transition metal chalcogenides can be synthesized by using different synthesizing methods. There are bulk number of unit cells in In$_2$S$_3$ which leads to many dangling bonds and results in enhanced superficial sites (Cheng et al., 2023). 2D non-layered In$_2$S$_3$ (AC-In$_2$S$_3$) contains rectification characteristics with Asymmetrical contact. In order to attain high response and low dark current, self-powdered 2D AC-In$_2$S$_3$ and traditional Si were combined to work in photo-gain mode (Lu et al., 2020). In$_2$S$_3$ exists in three crystallographic structures i.e. α-In$_2$S$_3$, β-In$_2$S$_3$ where this crystallographic structure is mostly abundant at room temperature with chemical conductivity as compared to others. β-In$_2$S$_3$ possesses intermediate band gap, due to misbalance between indium and sulfur. Kaur et al. reported that in growth of β-In$_2$S$_3$ with CVD method at optimized flow rate, in single zone furnace Ar is introduced as an inert gas. Substrate like SiO$_2$, F-mica and ZnO, TiO$_2$ layers are deposited. The temperature is maintained near to the melting point of sulfur of 150 °C. In$_2$S$_3$ layers were synthesized depending on the time at 750 °C on both SiO$_2$ and F-mica substrates. A good

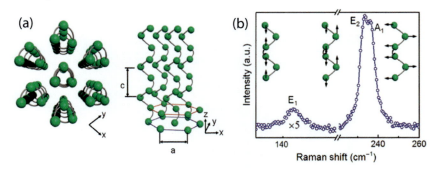

Fig. 5 (a) Atomic structure of 2D selenium and (b) Raman spectral results of 2D selenium respectively (reprinted from nano photonics CC open access journal). *Reprinted with permission, copyright 2022 by Elsevier (Sun et al., 2022).*

triangular morphology was observed in FESEM results for both substrates SiO$_2$ and F-mica on temperature 750 °C. At this temperature for SiO$_2$ the size of triangular nanoflakes was approx. 300 nm and for F-mica it was about approx. It is also observed that the increase in the deposition time or 5–15 min, the average thickness of nanoflakes also increases. Yu Zhao et al. also reported same results (Zhao, 2019). By studying Raman spectra, the characteristics peaks show phase pure tetragonal at 750 °C for both substrates. XRD patterns confirmed the triangular phase transformation at 27.52° with hkl along the direction (113) (Kaur et al., 2021) (Fig. 6).

7. Growth of β-Ga$_2$O$_3$

Gallium oxide (β-Ga$_2$O$_3$) possess a band gap of 4.9 eV. It is an appreciable material for usage in solar blind photodetector. Gallium oxide also have applications in gate dielectric, UV detectors, power electronics and much more. Suarez et al. reported the synthesis of highly pure β-Ga$_2$O$_3$ through hydrothermal method (Suárez, 2022). Ga$_2$O$_3$ shows n-type conductivity. The reported thermal conductivity of Ga$_2$O$_3$ is low in range of 10–30 W/mK. Regardless of a few studies like single crystal, thin films, and nanowire not consistently luminescence is observed in raw (powder) material. So, Chromium doped β-Ga$_2$O$_3$ has attracted most of the attention due to its enhanced thermal conductive behavior and vast applications in drug delivery due to luminescence property (Bhattacharjee and Singh, 2022). Gallium oxide (Ga$_2$O$_3$) is synthesized by direct oxidation of 2D GaSe nanosheets(Feng, 2014) (Fig. 7).

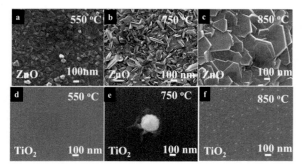

Fig. 6 (a–f) FESEM results of β-In$_2$S$_3$ on SiO$_2$/Si substrate and F-Mica substrate (Kaur et al., 2021). Figure is reprinted with permission of copyrights 2021.

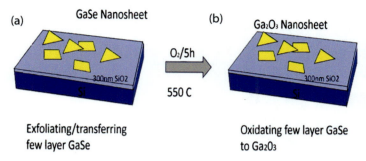

Fig. 7 Synthesis process of β-Ga$_2$O$_3$ from GaSe by direct oxidation on SiO$_2$/Si substrate.

Feng at al. reported that the diffraction patterns through XRD indexed the monoclinic phase of gallium oxide and after oxidation diffraction peaks of GaSe recedes from view. Through XRD and EDS it was ensured that the GaSe is fully transformed into β-Ga$_2$O$_3$. The responsivity, detectivity and EQE of photodetector was reported as 3.3 A/W, 4.0 × 10^{12} Jones and 1600% respectively (Fig. 8).

8. Growth of cadmium sulfide (CdS)

Cadmium sulfide (CdS) is non-layered material, where the chemical bonds are present in 3-directions. CdS is a semiconducting material, possesses a band gap of 2.42 eV (Alam et al., 2022a). Due to the existence of surface wide 2D structures and impressive anisotropy, these materials can easily be tuned. CdS possess many applications in optoelectronics, catalysis, sensing, nanoelectronics, good piezoelectric properties and energy storage (ALKATHIRI, 2022). Singh et al. reported synthesis of zinc-doped cadmium sulfide with different chemical compositions of Zn, Cd, and sulfide (Singh et al., 2022). Alam et al. reported that for preparation of CdS electrolyte, it consists of 0.24 M cadmium acetate in 75 mL of ethylene glycol and ethanol (1:2). A molybdenum substrate was polished with 600 grit carborundum paper and then rinsed with hot soap with distilled water. A yellow color film of CdS is deposited on molybdenum substrate, without any pinhole imperfection and defect. From XRD results it was observed that as the deposited film is hexagonal in structure and polycrystalline in nature. The deposition conditions can also alter the size of crystallites. According to energy dispersive spectroscopy it is observed that

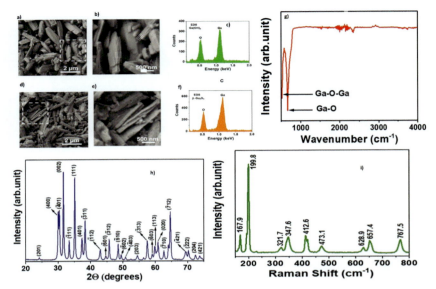

Fig. 8 (a) SEM image of $Ga(OH)_3$ powder, (b) overlapping lamellas, (c, f) shows EDS data of $Ga(OH)_3$ and $β-Ga_2O_3$ respectively, (d) SEM image after calcination at 1000 °C of $β-Ga_2O_3$, (e) magnification of selected area nanopores of image (d), (g) FTIR of $β-Ga_2O_3$, (h) XRD of $β-Ga_2O_3$ and (i) Raman spectrum of $β-Ga_2O_3$ (Suárez, 2022). For figure a–i got copyright permission 2022.

cadmium and sulfur grow towards the spectra which confirms the successful deposition of CdS on molybdenum substrate. According to the results of FTIR all peaks in the spectrum were identified (Singh et al., 2022). The peaks at 3279 cm^{-1}, 2352 cm^{-1}, 1615 cm^{-1}, 1387 cm^{-1}, 113 cm^{-1}, 739 cm^{-1} shows existence O–H stretching, C–H group, vibrational mode of O–H bending, asymmetrical stretching, Co stretching vibrations, CdS stretching respectively (Alam et al., 2022a). Jasim et al. reported the synthesis of CdS have been deposited on glass substrate from 45 mL of solution including 0.792 g of cadmium sulfate $CdSO_4$ dissolve in 25 mL of DI water acting as a source of Cd ions, and 0.511 g of thiourea $(Cs(NH_2)_2)$ in 20 mL of DI water as a source of sulfur ions. Then it was used in CVD to deposit CdS thin film (Jasim and Alfaidhi). Jassim et al. also reported the chemical vapor deposition of cadmium sulfide at low temperature from cadmium ethyl xanthate (Jassim et al., 2021). Naidoo et al. Buffer layer for heterojunction solar cell window layer were fabricated by CdS with bismuth dopes zin oxide (Naidoo et al., 2022) (Fig. 9).

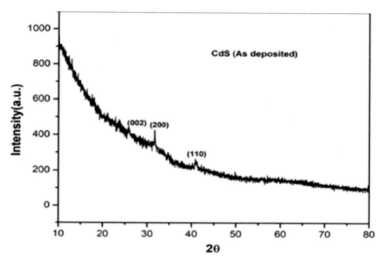

Fig. 9 XRD spectra of as deposited cadmium sulfide. Figure is reprinted with permission of Alam et al. with copy rights of Elsevier 2022 (Alam et al., 2022a).

9. Growth of hematite (α-Fe₂O₃)

Generally, there are four different phases of iron oxide (Fe$_3$O$_4$, γ, ε, α-Fe$_2$O$_3$) are reported. The structure of α-Fe$_2$O$_3$ is hexagonal where iron atoms are surrounded by six oxygen atoms with space group R-3c, lattice parameters a = 5.03 nm, c = 13.74 nm and six formula units per unit cell. The hematite shows C_{3v} symmetry. α-Fe$_2$O$_3$ is an ecofriendly friendly n-type semiconductor with energy band gap of 2.1 eV, widely used in photocatalysis, lithium-ion batteries, water treatments and gas sensor. Mishra et al. also reported the photocatalytic properties of α-Fe$_2$O$_3$ (Mishra and Chun, 2015). Recently, Chu et al. reported the synthesis of good quality α-Fe$_2$O$_3$ nanosheets on SiO$_2$/Si with Se powder by following Gib's free energy design (Chu et al., 2023). (Fig. 10).

Kozlovskly et al. recently reported the method for preservation of α-Fe$_2$O$_3$ properties by modification of these nanoparticles after electron beam irradiation (Kozlovskiy et al., 2023). Lu et al. reported recently the fabrication of α-Fe$_2$O$_3$ by high temperature thermal decomposition (Lu et al., 2023). Most of the synthesis methods for α-Fe$_2$O$_3$ were reported but Aadenan et al. reported the most influential one recently i.e., aerosol assisted chemical vapor deposition (AACVD). Fluorine doped tin-oxide substrate, 0.1 M precursor solution of FeCl$_3$.6H$_2$O was dissolved in 50 mL methanol and 50 mL

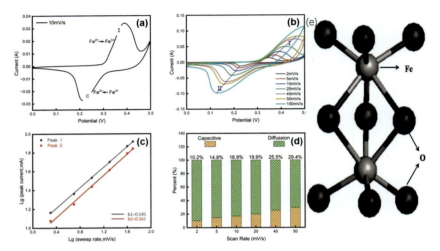

Fig. 10 (a) CV curves of α-Fe$_2$O$_3$ at 10 mV/s, (b) combination of CV curves scan results at 0–0.5 V, (c) peak current versus scan rate and (d) at different scan rate capacitance contribution ratios. These results are reported by Lu et al. (Got reprint permission through copyright access 2023). (e) Crystal structure of α-Fe$_2$O$_3$. Figure a–e is reprinted with permission copyrights 2015 (Lu et al., 2023; Mishra and Chun, 2015).

absolute ethanol separately and stirred. After that the samples was fabricated at 550 °C for 2, 5, 10 min of deposition time. The diffraction patterns obtained from both ethanol and methanol solvents with rhombohedral crystal structure of hematite, matched to JCPDS file # 00-002-0919. The strongest diffraction peak is observed at 2Θ = 35.7° originated from (110) crystal plane. According to SEM results the sample prepared by using methanol shows uneven flakes, while other shows irregular granules like structure. AFM results state that the sample produced by using methanol possess greater average surface roughness as compared to other. So, it is concluded that the choice of precursor opens different options of morphology change with variation in crystal growth (Aadenan et al., 2023).

10. Growth of ε-Fe$_2$O$_3$

ε-Fe$_2$O$_3$ is a crystalline polymorph of Fe$_2$O$_3$. Interestingly, just single phase ε-Fe$_2$O$_3$ had been present in Chinese sauce glaze porcelain (Wang, 2019). Li et al. reported the existence of highly cross-linked 3D ε-Fe$_2$O$_3$ managed by ultrathin nanosheets as high-performance anode material for LIBs (Li et al., 2022). Wang et al. reported that ε-Fe$_2$O$_3$ is a material with excellent potential because of its super-exchange intercalation between Fe^{+3} ions. But still to obtain the controlled ultra-thin $\varepsilon\text{-Fe}_2O_3$ crystal is in its infancy. The 3D layer by layer exfoliation is hindered by 2D anisotropic growth and inherent 3D chemical bonded nature. To obtain 2D $\varepsilon\text{-Fe}_2O_3$ space-confined chemical vapor deposition in a controllable manner on mica, was used. To obtain a thin rectangular $\varepsilon\text{-Fe}_2O_3$ single crystal for the very first time, carrier gases flow rate was controlled. The under-control synthesis environment reduced precursor concentrations and ultra-thin crystals were obtained. So, a space confined CVD provides the best alternative for the synthesis of ultra-thin crystals. X-ray photoelectron spectroscopy (XPS) dictates that the binding energies at 723.6 and 710.2 eV provides attributes of Fe^{+3}, while the peak of 532.4 eV was assigned to O^{-2}. XRD results confirms the formation of $\varepsilon\text{-Fe}_2O_3$. Raman intensity plotting reveals uniform color contrast. To check morphology and thickness of sample OM and AFM was done (Wang et al., 2023). Tansakenan et al. reported atomic layered deposition for $\varepsilon\text{-Fe}_2O_3$ tin film deposition (Tanskanen et al., 2017).

11. Growth of lead sulfide (PbS)

To transform lead sulfide nanoparticles into PbS nanosheets, nanoparticles are oriented in specific crystallographic directions (Zheng et al., 2020). Most of metal thiolates are allowed to serve as a precursor due to their layered structure, in synthesis of non-layered metal sulfide (Tao et al., 2019) By using 2D oriented PbS nanoparticles in presence of compounds of chlorine, ultrathin PbS nanosheets were reported (Khan et al., 2017a). PbS possess direct and narrow band gap (0.41 eV), high static dielectric constant (17.3), good carrier mobility and small effective electron mass ($<0.1m^*$) (Mamiyev and Balayeva, 2023). The PbS nanoplates were synthesized in quartz tube with diameter of 1 in. argon/hydrogen gases were used as carrier gases at atmospheric pressure. 50 mg of $PbCl_2$ was placed in center of quartz tube, 0.5 g of S was placed in ceramic boat of 4 cm length, located approximately 16 cm up streamed from the heating center. The system was ramped to 700 °C with the speed of 40 °C/min and kept for 155 min under argon/hydrogen gas at rate of 50sccm. After the growth of PbS was completed, the furnace was cooled down to room temperature. The substrate region was reported as sharp temperature gradient region where the temperature decreases 150 °C in 4 cm, that leads to different morphological behavior of PbS nanoplates (Gu et al., 2020). Gu et al. reported in results and

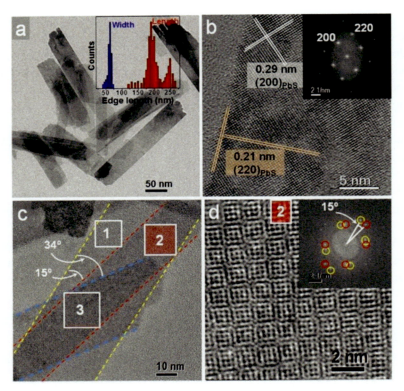

Fig. 11 (a) TEM image of PbS nanocrystals shows rectangular morphologies, the inset image red for length and blue for width distribution histograms, (b) HRTEM edge and lattice spacing with the inset FFT patterns, (c) SEM image of overlapped NCs show number of layers and their orientation and (d) HRTEM images and their corresponding FFT patterns. The number of layers and FFT spots correlates with the corresponding color codes. Figure is Reprinted with permission (copyrights 2017, ACS Publications) (Khan et al., 2017).

discussion that the crystal structure peaks were well defined. The XRD spectra exhibits the presence of cubic crystal structure in PbS nanoplates by comparing them with standard card JCPDS card of 65–9496. The lattice parameter can be calculated by peak at 30.1° was 5.94 Å (Gu et al., 2020). Khan et al. fabricated free standing single crystalline2D ultra-thin PbS of 2 nm thickness by high reaction yield at low temperature, the size of nano crystals was reported as approximately 200 nm of length and 500 nm of width, results of Raman spectroscopy show the formation of nano crystals by single seed growth mechanism. Synthesis route of these free-standing nanocrystals of PbS was reported by Khan et al. (Khan et al., 2017a) (Fig. 11).

CVD growth of 2D non layered materials   183

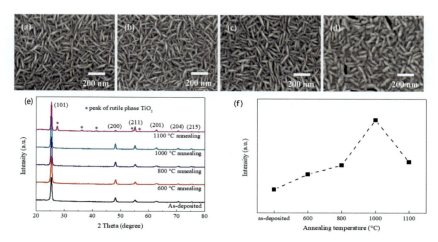

Fig. 12 (a) FESEM results of as deposited TiO$_2$, (b, c, d) TiO$_2$ at different calcination temperatures i.e., 600, 800, 1000 respectively, (e) XRD results of TiO$_2$ as deposited and with different annealing temperatures, (f) (101) peak intensities of XRD results. Figure a–e is reprinted with copyright permission 2023 (Yang et al., 2023).

12. Growth of titanium dioxide (TiO$_2$)

Titanium dioxide is always a promising nanoparticle due to its photocatalytic bacterial activity, due to its cheap cost, good chemical stability and natural abundance. Non-Layered 2D titanium dioxide has high dielectric permittivity (κ = 50–80). It is also known as titania. Titanium dioxide is a n-type semiconductor due to the availability of oxygen vacancies. These vacancies promote the charge flow and electron–hole recombination by trapping charge carriers in defects sites (Liao et al., 2020). Titanium dioxide is widely used in cosmetics as ultraviolet filter for sunscreen. Titanium dioxide is also used for incidental oral exposure in lip balms. So, it is observed that, titanium dioxide does not adversely effects on human health (Dréno et al., 2019). Shi et al. reported the synthesis of titanium nanorods by pulse chemical vapor deposition (Shi and Wang, 2011). Zhang et al. reported the synthesis of titanium dioxide through mist CVD. The precursor solution is prepared by dissolving titanium tetraisopropoxide (TTIP) in ethanol. The concentration of 0.10 mol/L. by using transducers of 2.4 MHz, the solution was atomized ultrasonically. The mist droplets of precursor were transferred to reaction chamber, using dilution gas (4.5 L/min) and compressed air (2.5 L/min). Then in a pre-heated chamber the substrate was kept at 400 °C

(Zhang and Li, 2020). Zhang et al. annealed the Titanium dioxide films to check thermal stability at temperature range 600–1100 °C for 1 h. XRD results exhibits the synthesis of pure anatase phase through mist chemical vapor deposition. The peaks of XRD at 1100 °C shows the incomplete phase transformation of anatase to rutile phase transformation. According to Raman results all phases are pure anatase and no anatase to rutile phase transformation is noticed during annealing at temperature 600, 800, and 1000 °C (Zhang and Li, 2020) (Fig. 12).

13. Conclusion

2D non-layered materials have demonstrated outstanding chemical and physical properties, including strong electrical conductivity, mechanical toughness, and optical qualities. 2D materials perform better than conventional materials when used in electronic and optoelectronic devices like transistors and photodetectors. They have been used into nano-composites to enhance the material's qualities, like how TMDCs or graphene have been incorporated into polymers to boost their heat conductivity and mechanical strength. Energy storage (such as supercapacitors) and conversion (such as fuel cells and solar cells) applications due to their high surface area and conductivity. In future there is a need to modify the properties of colloidal quantum dots (CQDs) during solid films. Now, introduction of different metal atoms during colloidal synthesis and post preparation is a main concern. In the present era, it is essential to synthesis more 2D materials with unique structure through feasible synthesis methods. In future it is required to build up interest in new emerging fields manipulate magnetism of 2D metal oxides by substitutional doping.

References

Aadenan, A., et al., 2023. Influence of solvent on accelerating deposition time of α-Fe$_2$O$_3$ photoanode via AACVD. IOP Conf. Ser. Mater. Sci. Eng. 1278 (1), 012006. https://doi.org/10.1088/1757-899x/1278/1/012006.

Akhtar, M., et al., 2017. Recent advances in synthesis, properties, and applications of phosphorene. npj 2D Mater. Appl. 1 (1), 5. https://doi.org/10.1038/s41699-017-0007-5.

Alam, A., Kumar, S., Singh, D.K., 2022a. Cadmium sulphide thin films deposition and characterization for device applications. Mater. Today Proc. 62 (P10), 6102–6106. https://doi.org/10.1016/j.matpr.2022.04.1018.

Alkathiri, T.A., 2022. Ultrathin two-dimensional oxide and oxysulphide nanomaterials and their potential applications. Doctoral dissertation, RMIT University.

Bhattacharjee, J., Singh, S.D., 2022. Temperature dependence of red luminescence in pure β-Ga$_2$O$_3$: An estimation of electron-phonon interaction. Solid State Commun 352. https://doi.org/10.1016/j.ssc.2022.114831.

Chakraborty, G., Padmashree, R., Prasad, A., 2023. Recent advancement of surface modification techniques of 2-D nanomaterials. Mater. Sci. Eng.: B 297, 116817. https://doi.org/10.1016/j.mseb.2023.116817.

Cheng, Y., et al., 2023. rGO spatially confined growth of ultrathin In$_2$S$_3$ nanosheets for construction of efficient quasi-one-dimensional Sb2Se3-based heterojunction photocathodes. Sci. China Mater. 66 (4), 1460–1470. https://doi.org/10.1007/s40843-022-2267-7.

Chu, W., et al., 2023. Synthesis of nonlayered 2D α-Fe$_2$O$_3$ nanosheets by ultralow concentration precursor with Se catalysts design. Phys. Status Solidi - Rapid Res. Lett. https://doi.org/10.1002/pssr.202300102.

Dréno, B., Alexis, A., Chuberre, B., Marinovich, M., 2019. Safety of titanium dioxide nanoparticles in cosmetics. J. Eur. Acad. Dermatol. Venereol. 33 (S7), 34–46. https://doi.org/10.1111/jdv.15943.

Fan, T., Xie, Z., Huang, W., Li, Z., Zhang, H., 2019. Two-dimensional non-layered selenium nanoflakes: facile fabrications and applications for self-powered photo-detector. Nanotechnology 30 (11), 114002. https://doi.org/10.1088/1361-6528/aafc0f.

Feng, W., et al., 2014. Synthesis of two-dimensional β-Ga$_2$O$_3$ nanosheets for high-performance solar blind photodetectors. J. Mater. Chem. C Mater. 2 (17), 3254–3259. https://doi.org/10.1039/c3tc31899k.

Gu, Y.Y., Wang, Y.F., Xia, J., Meng, X.M., 2020. Chemical vapor deposition of two-dimensional PbS nanoplates for photodetection. Chin. Phys. Lett. 37 (4), 048101. https://doi.org/10.1088/0256-307X/37/4/048101.

El Hammoumi, M., Chaudhary, V., Neugebauer, P., El Fatimy, A., 2022. Chemical vapor deposition: a potential tool for wafer scale growth of two-dimensional layered materials. J. Phys. D: Appl. Phys. 55 (47). https://doi.org/10.1088/1361-6463/ac928d.

Han, W., et al., 2019. Salt-assisted chemical vapor deposition of two-dimensional materials. Sci. China Chem. 62 (10), 1300–1311. https://doi.org/10.1007/s11426-019-9525-y.

Hussain, N., et al., 2022. Ultra-narrow linewidth photo-emitters in polymorphic selenium nanoflakes. Small 18 (52), e2204302. https://doi.org/10.1002/smll.202204302.

Idumah, C.I., 2022. Phosphorene polymeric nanocomposites for biomedical applications: a review. Int. J. Polym. Mater. Polym. Biomater. https://doi.org/10.1080/00914037.2022.2158333.

Jasim, S.A., Alfaidhi, A. Effect of Annealing Period and Temperature on the Optical and Structural Properties of CdS Thin Films.

Jassim, S., Abbas, A.M., Al-Shakban, M., Ahmed, L.M., 2021. Chemical vapour deposition of cds thin films at low temperatures from cadmium ethyl xanthate. Egypt. J. Chem. 64 (5), 2533–2538. https://doi.org/10.21608/EJCHEM.2021.60695.3451.

Jiang, J., et al., 2023. Tuning 2D magnetism in cobalt monoxide nanosheets via in situ nickel-doping. Adv. Mater. 35 (22), e2301668. https://doi.org/10.1002/adma.202301668.

Kaur, N., Sharma, D., Mehta, B.R., 2021. Growth of In$_2$S$_3$ nanolayers on F-Mica, SiO$_2$, ZnO, and TiO$_2$ substrates using chemical vapor deposition. Mater. Sci. Eng. B Solid. State Mater Adv. Technol. 264, 114889. https://doi.org/10.1016/j.mseb.2020.114889.

Khan, A.H., Pal, S., Dalui, A., Pradhan, J., Sarma, D.D., Acharya, S., 2017a. Solution-processed free-standing ultrathin two-dimensional PbS nanocrystals with efficient and highly stable dielectric properties. Chem. Mater. 29 (3), 1175–1182. https://doi.org/10.1021/acs.chemmater.6b04508.

Kozlovskiy, A.L., Rusakov, V.S., Fadeev, M.S., 2023. The influence of electron irradiation on the stability of α-Fe$_2$O$_3$ nanoparticles to natural aging processes. Crystallogr. Rep. 68 (3), 487–494. https://doi.org/10.1134/S1063774523700207.

Li, D., Liang, J., Song, S., Li, L., 2022. Highly cross-linked 3D ϵ-Fe_2O_3 networks organized by ultrathin nanosheets as high-performance anode materials for lithium-ion storage. ACS Appl Nano Mater 6, 2356–2365. https://doi.org/10.1021/acsanm.2c04359.

Liao, C., Li, Y., Tjong, S.C., 2020. Visible-light active titanium dioxide nanomaterials with bactericidal properties. Nanomaterials 10 (1). https://doi.org/10.3390/nano10010124.

Liu, S., Xie, L., Qian, H., Liu, G., Zhong, H., Zeng, H., 2019. Facile preparation of novel and active 2D nanosheets from non-layered and traditionally non-exfoliable earth-abundant materials. J. Mater. Chem. A Mater. 7 (25), 15411–15419. https://doi.org/10.1039/c9ta04442f.

Lu, J., et al., 2020. An asymmetric contact-induced self-powered 2D In_2S_3 photodetector towards high-sensitivity and fast-response. Nanoscale 12 (13), 7196–7205. https://doi.org/10.1039/d0nr00517g.

Lu, P.A., et al., 2023. Synthesis, analysis and characterization of alpha-Fe_2O_3 nanoparticles and their applications in supercapacitors. J. Mater. Sci.: Mater. Electron. 34 (9). https://doi.org/10.1007/s10854-023-10246-8.

Mamiyev, Z., Balayeva, N.O., 2023. PbS nanostructures: a review of recent advances. Mater. Today Sustain. 21. https://doi.org/10.1016/j.mtsust.2022.100305.

Mishra, M., Chun, D.M., 2015. α-Fe_2O_3 as a photocatalytic material: a review. Appl. Catal. A: General 498, 126–141. https://doi.org/10.1016/j.apcata.2015.03.023.

Naidoo, J., Fangsuwannarak, T., Laohawiroj, S., Rattanawichai, P., Limsiri, W., Phatthanakun, R., 2022. Cadmium sulphide thin film with ZnO:Bi buffer layer for heterojunction solar cell window layer applications. Integr. Ferroelectr. 225 (1), 124–138. https://doi.org/10.1080/10584587.2022.2054062.

Nehra, M., et al., 2023. Catalytic applications of phosphorene: computational design and experimental performance assessment. Crit. Rev. Environ. Sci. Technol. 1–25. https://doi.org/10.1080/10643389.2023.2224614.

Nielsen, R., Hemmingsen, T.H., Bonczyk, T.G., Hansen, O., Chorkendorff, I., Vesborg, P.C.K., 2023. Laser-annealing and solid-phase epitaxy of selenium thin-film solar cells. ACS Appl. Energy Mater. 6 (17), 8849–8856. https://doi.org/10.1021/acsaem.3c01464.

Pal, A., et al., 2022. Quantum-engineered devices based on 2D materials for next-generation information processing and storage. Adv. Mater. https://doi.org/10.1002/adma.202109894.

Pang, J., et al., 2018. Applications of phosphorene and black phosphorus in energy conversion and storage devices. Adv. Energy Mater. 8 (8). https://doi.org/10.1002/aenm.201702093.

Qin, B., et al., 2023. General low-temperature growth of two-dimensional nanosheets from layered and nonlayered materials. Nat. Commun. 14 (1). https://doi.org/10.1038/s41467-023-35983-6.

Shi, J., Wang, X., 2011. Growth of rutile titanium dioxide nanowires by pulsed chemical vapor deposition. Cryst. Growth Des. 11 (4), 949–954. https://doi.org/10.1021/cg200140k.

Shi, Z., Zhang, H., Khan, K., Cao, R., Xu, K., Zhang, H., 2022. Two-dimensional selenium and its composites for device applications. Nano Res. 15 (1), 104–122. https://doi.org/10.1007/s12274-021-3493-x.

Singh, A.K., Vijayashri, K.M., Singh, S.P., Mishra, M.K., 2022. Synthesis and characterization of zinc doped cadmium sulphide nanoparticles. Mater. Today Proc. 66, 2017–2027. https://doi.org/10.1016/j.matpr.2022.05.484.

Smith, J.B., Hagaman, D., Ji, H.F., 2016. Growth of 2D black phosphorus film from chemical vapor deposition. Nanotechnology 27 (21). https://doi.org/10.1088/0957-4484/27/21/215602.

Suárez, M., et al., 2022. Ultrahigh purity beta gallium oxide microstructures. Ceram. Int. 48 (17), 25322–25325. https://doi.org/10.1016/j.ceramint.2022.05.205.

Sun, J., Zhao, Z., Li, J., Li, Z., Meng, X., 2022. Recent advances in transition metal selenides-based electrocatalysts: rational design and applications in water splitting. J. Alloy. Compd. 918. https://doi.org/10.1016/j.jallcom.2022.165719.

Tanskanen, A., Mustonen, O., Karppinen, M., 2017. Simple ALD process for ϵ-Fe_2O_3 thin films. APL Mater. 5 (5). https://doi.org/10.1063/1.4983038.

Tao, P., Yao, S., Liu, F., Wang, B., Huang, F., Wang, M., 2019. Recent advances in exfoliation techniques of layered and non-layered materials for energy conversion and storage. J. Mater. Chem. A 7 (41), 23512–23536. https://doi.org/10.1039/c9ta06461c.

Tchoffo, B.D., Benabdallah, I., Aberda, A., Neugebauer, P., Belhboub, A., El Fatimy, A. Towards Large-area and Defects-free Growth of Phosphorene on Nickel.

Wang, Y., Zhang, Z., Mao, Y., Wang, X., 2020. Two-dimensional nonlayered materials for electrocatalysis. Energy Environ. Sci. 13 (11), 3993–4016. https://doi.org/10.1039/d0ee01714k.

Wang, L., et al., 2019. Three-dimensional microstructure of ϵ-Fe_2O_3 crystals in ancient Chinese sauce glaze porcelain revealed by focused ion beam scanning electron microscopy. Anal. Chem. 91 (20), 13054–13061. https://doi.org/10.1021/acs.analchem.9b03244.

Wang, Y., et al., 2023. Room-temperature magnetoelectric coupling in atomically thin ϵ-Fe_2O_3. Adv. Mater. 35 (7). https://doi.org/10.1002/adma.202209465.

Yang, Y., Jia, L., Wang, D., Zhou, J., 2023. Advanced strategies in synthesis of two-dimensional materials with different compositions and phases. Small Methods 7 (4). https://doi.org/10.1002/smtd.202201585.

Yu, X., Zhang, S., Zeng, H., Wang, Q.J., 2016. Lateral black phosphorene P-N junctions formed via chemical doping for high performance near-infrared photodetector. Nano Energy 25, 34–41. https://doi.org/10.1016/j.nanoen.2016.04.030.

Zhang, Q., Li, C., 2020. High temperature stable anatase phase titanium dioxide films synthesized by mist chemical vapor deposition. Nanomaterials 10 (5). https://doi.org/10.3390/nano10050911.

Zheng, Z., Yao, J., Li, J., Yang, G., 2020. Non-layered 2D materials toward advanced photoelectric devices: progress and prospects. Mater. Horiz. 7 (9), 2185–2207. https://doi.org/10.1039/d0mh00599a.

Zhou, N., Yang, R., Zhai, T., 2019. Two-dimensional non-layered materials. Mater. Today Nano 8. https://doi.org/10.1016/j.mtnano.2019.100051.

Zhou, X., et al., 2018. 2D layered material-based van der Waals heterostructures for optoelectronics. Adv. Funct. Mater. 28 (14). https://doi.org/10.1002/adfm.201706587.

Zhao, Y., et al., 2019. Thickness-dependent optical properties and in-plane anisotropic Raman response of the 2D β-In2S3. Adv. Opt. Mater. 7 (22). https://doi.org/10.1002/adom.201901085.

CHAPTER SEVEN

2D non-layered materials for energy applications

Harish Somala, Muzammil Mushtaq, and Uma Sathyakam Piratla[*]

School of Electrical Engineering, Vellore Institute of Technology, Vellore, Tamil Nadu, India
[*]Corresponding author. e-mail address: umasathyakam.p@vit.ac.in

Contents

1. Introduction — 189
2. 2D materials — 190
   2.1 2D layered materials — 191
   2.2 2D non-layered materials — 191
3. 2D non layered materials for energy application — 193
4. Supercapacitors — 193
   4.1 Electric double-layer capacitor — 193
   4.2 Pseudocapacitor — 195
   4.3 Hybrid capacitor — 196
5. Lithium-ion batteries — 200
6. Sodium-ion batteries — 205
7. Fuel cells — 209
8. Other energy storage — 213
9. Conclusions — 213
References — 213

Abstract

Two-dimensional (2D) materials with meritorious characteristics like unique layered structure, good surface reactions, flexibility and high electrochemical reactivity makes them trending materials for energy applications. Among them, energy storage devices like batteries and supercapacitors, and fuel cells are explored extensively due to their high performance dependance on the porosity, and active area. In this chapter, different types of 2D non layered materials and their applications in energy storage devices and fuel cells with their working mechanisms of charge storage and reaction kinetics, along with illustration of recently published works are outlined.

1. Introduction

The emergence of nanomaterials has caused a significant paradigm shift in various science and technology domains. Their nanoscale dimensions have resulted in unique properties that have garnered widespread attention. 2D nanomaterials have been particularly noteworthy due to their exceptional characteristics and immense application potential. Unlike three-dimensional bulk materials, 2D nanomaterials are only a few atomic layers thick, with the other two dimensions that can be extended indefinitely. This confinement of material in a single layer leads to extraordinary properties that differ from their bulk counterparts, opening exciting possibilities. 2D nanomaterials are a fascinating class of materials that have captured the attention of materials scientists and engineers worldwide. These materials consist of a single layer or a few layers of atoms, forming a two-dimensional structure often referred to as a nanosheet or nanoflake. Graphene (a single layer of carbon atoms arranged in a honeycomb lattice) is the most well-known and widely studied 2D nanomaterial. However, researchers have discovered and explored numerous other 2D materials with diverse elemental compositions and structures. One of the most intriguing aspects of 2D nanomaterials is their unique properties, which arise from their reduced dimensionality and high surface-to-volume ratio. These materials exhibit exceptional mechanical, electrical, thermal, optical, and chemical properties that differ from their bulk counterparts. For example, 2D materials often possess extraordinary mechanical strength and flexibility, high electrical conductivity, and excellent thermal conductivity.

Additionally, their large surface area enables enhanced interactions with their surrounding environment, making them suitable for sensing, catalysis, and energy storage applications. The mechanical properties of 2D nanomaterials are particularly noteworthy. Due to their atomically thin nature, these materials can bend and flex to an extraordinary degree without breaking, which makes them ideal for applications where flexibility and durability are essential, such as in flexible electronic devices. In addition, the high surface area of 2D nanomaterials makes them highly reactive, which makes them useful for catalytic applications such as fuel cells. Overall, 2D nanomaterials are a fascinating and promising research area with great potential for various electronics, energy, and medical applications. As researchers continue to explore these materials and uncover new properties and applications, we will likely see even more exciting developments soon.

2. 2D materials

2D materials have one dimension in the range of nanoscale (1–100 nm), and the remaining two dimensions may not be restricted to the nanoscale. The most common 2D materials reported these days are graphene and MXene with plate and sheet-like nanostructures. 2D materials are classified broadly into two types, as mentioned in Fig. 1.

2.1 2D layered materials

These materials consist of stacked atomic layers held together by weak Van der Waals forces. These materials possess atomic thickness (<5.0 nm) and a large specific surface area. The most well-known example of a layered 2D material is graphene, a single layer of carbon atoms arranged in a honeycomb lattice. Other layered materials include transition metals. Layered 2D materials possess several distinct characteristics:

> Anisotropic properties: These materials show different properties between the planes of layers compared to perpendicular to the layers.
> High mechanical strength: Graphene, for instance, is exceptionally strong, with a tensile strength of around 130 GPa.
> High electrical conductivity: Graphene is an excellent conductor of electricity due to its unique band structure.
> Tunable band gap: TMDCs and other layered materials can exhibit tunable band gaps, making them suitable for various electronic applications.

2.2 2D non-layered materials

2D non-layered materials are a class of 2D materials that are not composed of stacked layers. Instead, they are formed by chemical bonding in three dimensions (3D), resulting in unsaturated dangling bonds on the surface and inducing a

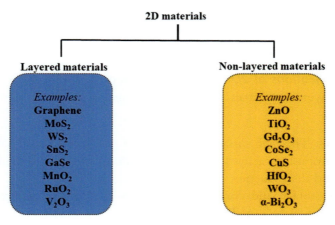

Fig. 1 Types of 2D materials.

high–activity and high-energy surface. These materials, also known as single-layer materials, comprise a single layer of atoms without a layered structure. One prominent example is black phosphorous (phosphorene), which consists of a single layer of phosphorous atoms. Other non-layered materials include silicone and germanene. These materials possess a unique set of properties:

Layer-dependent properties: These materials exhibit layer-dependent properties due to their specific atomic arrangement.
Layer-edge effects: These materials often exhibit distinct properties at their edges, leading to potential edge-related applications.
Variable band structures: The band structures of non-layered materials can be modified by strain or applied electric fields, allowing tuning properties.

Some of the most common 2D non-layered materials include:

1. Metal chalcogenides: These materials comprise a metal atom bonded to one or more chalcogen atoms (such as sulphur, selenium, or tellurium). Examples of metal chalcogenides include molybdenum disulphide (MoS_2), tungsten disulphide (WS_2), and tungsten diselenide (WSe_2).
2. Metal oxides: These materials comprise a metal atom bonded to one or more oxygen atoms. Examples of metal oxides include indium oxide (In_2O_3), tin oxide (SnO_2), and zinc oxide (ZnO).
3. III-V semiconductors: These materials comprise a group III element (such as aluminium, gallium, or indium) bonded to a group V element (such as phosphorus, arsenic, or antimony). Examples of III-V semi-conductors include gallium arsenide (GaAs) and indium phosphide (InP).
4. Organic-inorganic perovskites: These materials comprise an organic molecule bonded to an inorganic oxide. Examples of organic-inorganic perovskites include methyl ammonium lead iodide ($MAPbI_3$) and for-mamidinium lead iodide ($FAPbI_3$).

2.2.1 Applications of 2D non layered materials

2D non-layered materials have a wide range of potential applications, including:

1. Catalysis: The high surface activity of 2D non-layered materials makes them promising catalysts for various reactions, such as the oxidation of CO_2 and the reduction of H_2O.
2. Energy storage and conversion: 2D non-layered materials can make batteries, fuel cells, and solar cells.

3. Optoelectronics: 2D non-layered materials can make light-emitting diodes, lasers, and sensors.
4. Topological insulators: 2D non-layered materials can be used to make topological insulators, which conduct electricity on their surface but insulate in bulk.

3. 2D non layered materials for energy application

Energy plays a significant role in improving the development and quality of human life as we are greatly dependent on electrical and electronic gadgets and machinery, which is possible by consuming much energy. To make this happen, we need a proper energy source which is continuous and more reliable. This opens a great interest and opportunities in the energy applications like energy storage. Supplying an uninterrupted power supply is a big issue, which is possible using energy storage technologies. This chapter concentrates on energy storage applications like supercapacitors, rechargeable batteries, and fuel cells.

4. Supercapacitors

Supercapacitors (SCs) are electrochemical energy storage devices with greater power and lower energy densities than rechargeable batteries; they take significantly less time to charge, and have good cycle life, which makes them an attractive option for applications that need high power, less time to charge, and are portable.

Firstly, SCs covered a significant space in the Ragone plot, as shown in Fig. 2, that fills the huge technological gap between capacitors and batteries. This shows that SCs have hybrid characteristics of both capacitors and batteries. SCs are classified based on the type of energy storage mechanism, as depicted in the block diagram in Fig. 3 (Harish and Sathyakam, 2022).

4.1 Electric double-layer capacitor

The charge storage mechanism in electric double-layer capacitors (EDLC) is electrostatic in nature. The charge and discharge process happens in the electric double layer formed on a porous electrode under an applied electric field; Fig. 4 illustrates an EDLC capacitor while charging and discharging.

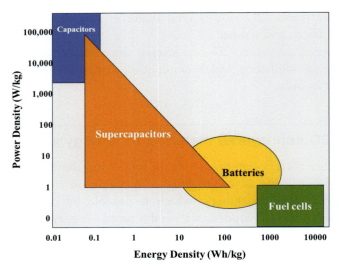

Fig. 2 Ragone plot.

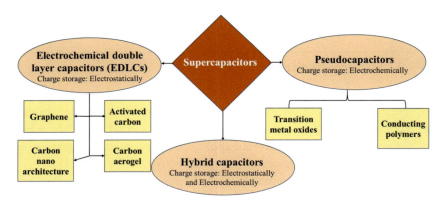

Fig. 3 Classification of supercapacitors.

As an electric field is applied across the electrodes, as shown in Fig. 4B, the solvated anions and cations are attracted to their respective opposite polarity electrodes and form a double layer called as Helmholtz layer. While discharging, the solvated ions are attracted towards the separator, as seen in Fig. 4C.

EDLC is reversible in operation and has higher cyclability and faster operation than other energy storage devices, due to the charge-discharge on the electrode surface by physical ion transfer alone. The most commonly used 2D non layered electrode materials for EDLCs are metal carbon-based materials like graphene, CNTs, aero-gels etc.

2D non-layered materials for energy applications

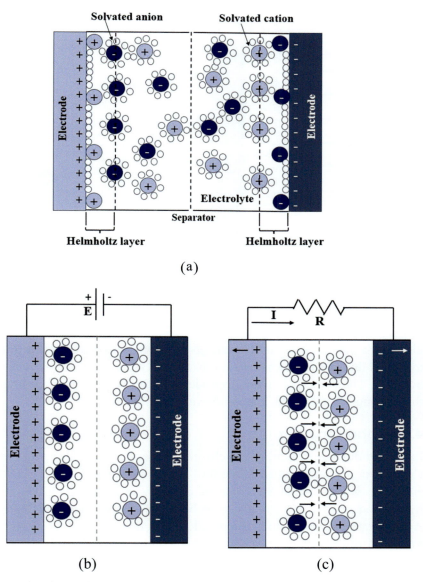

Fig. 4 (A) Electric double layer capacitor, (B) During charging, (C) During discharging.

4.2 Pseudocapacitor

The charge storage mechanism in pseudocapacitor is electrochemical in nature. The charge and discharge process happens electrochemically. The oxidation and reduction is over the surface of the electrode. Fig. 5

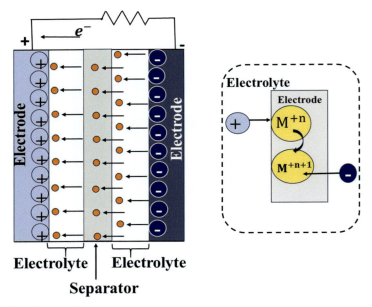

Fig. 5 Pseudocapacitor.

illustrates a pseudocapacitor. The charge-discharge process is slow compared to EDLC due to a chemical reaction that occurs at the electrode-electrolyte interface. The commonly used electrode materials for pseudocapacitors are the transition metal oxides and conducting polymers.

4.3 Hybrid capacitor

In the hybrid-type capacitors, the charge storage mechanism happens both electrostatically and electrochemically, where it consists of one EDLC electrode and the other electrode of pseudo nature, which makes it a hybrid and shows the capacitance of both types as illustrated in Fig. 6. A chemical reaction occurs at the pseudo electrode, and the double layer is formed at the EDLC electrode by applying an electric field.

Researchers are exploring 2D nanomaterials to address the issue of substantially enhancing the double-layer capacitance and pseudocapacitance. By optimising the structure of these materials through techniques like pore creation and heterostructure introduction, it may be possible to improve their capacitive charge storage even further.

Scientists have been exploring the potential of ultrathin transition metal oxides and chalcogenides as materials for supercapacitors. These materials offer a high theoretical capacitance, are environmentally friendly, and are

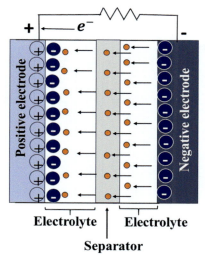

Fig. 6 Hybrid capacitor.

widely available. One good example is the ultrathin mesoporous GO-TiO$_2$ composite, which has been shown to produce high-performance supercapacitors (Abdallah et al., 2023). Its capacitance reaches 617 F/g at a specific density of 1 cm^3/g, and remains stable after 1000 continuous CV cycles indicating excellent cycling stability.

This impressive performance is attributed to the in-situ growth strategy for enhanced electrical conductivity, high specific surface area of TiO$_2$ nanosheets for many active sites, and the mesoporous structure for efficient electrolyte diffusion and charge transfer. A study was conducted to improve the pseudocapacitance of nanocomposites (SnO$_2$ quantum dots and Au nanoparticles (SQD-Au)) (Babu et al., 2022). The CV curves of an asymmetric supercapacitor, which used the functionalised QD's nanosheet cathode and 2D anode, maintained similar shapes across a wide range of scan rates. This indicated excellent pseudocapacitive behaviour with fast and reversible charge storage capability.

Moreover, the supercapacitor showed excellent cycling stability with a 98% capacitance retention after 5000 galvanostatic charge/discharge cycles. The surface chemical reactivity sparked by phosphate ion functionalisation is critical to the supercapacitor's fast and efficient faradaic pseudocapacitive reactions. Various 2D non layered nanomaterials and composites used as electrode materials in supercapacitors are tabulated in Table 1.

Table 1 2D non layered materials in supercapacitors.

Material	Method	IPCE	Types of capacitance	Specific capacitance/ rate	Cycles	References
$Ni_{1.9}Mo_{0.08}CuO_{0.20}$	Hydrothermal	89.1%	Hybrid	1136.02/2	2000	Mala et al. (2023)
$GO-TiO_2$	Tour's method	80.12%	EDLC	617/1	1000	Abdallah et al. (2023)
$Zn/Co-S@CeO_2/NF$	Solvothermal	91.1%	Pseudo	198.6/1	8000	Xu et al. (2023)
Eu_2O_3/MnO_3	Template method	91.8%	EDLC	186.25/1	7700	Ding et al., 2022
SnO_2 QD's–Au	Soft chemical method	~98%	Pseudo	87/1	5000	Babu et al. (2022)
Gd_2O_3/NiS_2 nanospheres	Solution mixing technique	82%	Hybrid	354/0.5	5000	Dhanalakshmi et al. (2020)
$ZnCo_2O_4/ZnO$ heterostructure	Hydrothermal	91.04%	Hybrid	538/1	10,000	Sivakumar et al. (2023)
$CdS@Bi_2Se_3$	SILAR	80%	Pseudo	116/1	2000	Mendhe et al. (2023)

MXene/CuS heterostructure	Hydrothermal	93.5%	Pseudo	2569.3/1	10,000	Chen et al. (2023a)
CdO/CuSe nanocomposite	Hydrothermal	94.8%	Pseudo	187/1	9000	Khan et al. (2023)
ZnSe/CdO	Hydrothermal	92%	Pseudo	1006.8/1	5000	Althubiti et al. (2023)
$CoSe_2$/NC nanocomposite	Co-precipitation	91.53%	Pseudo	554.4/2	6000	Wang et al. (2023)

Hydrothermal was adopted to prepare Mo–Cu dual-doped NiO nanorods. A single phase, with the face-centred cubic geometry, was confirmed by an X-ray diffraction study in all the nanostructures. The electrochemical analysis revealed the pseudocapacitive nature based on the faradaic redox mechanism. A maximum specific capacitance of 1136 F/g was obtained at a 2 mV/s scan rate for $Ni_{1.9}Mo_{0.08}Cu_{0.02}O$ nanorods. Electrochemical Impedance Spectroscopy (EIS) established the perfect capacitive nature in the low-frequency region, with a non-uniform distribution of ions. In conclusion, the high specific capacitance (502 F/g) of $Ni_{0.9}Mo_{0.02}Cu_{0.08}O$ electrode at 1 A/g, with energy density (21.10 Wh/kg) and power density (3.44 kW/kg) make it a fascinating candidate for the energy storage devices (Mala et al., 2023).

Bimetallic sulphides with superior electrochemical activity are extensively explored as electrode materials for electrochemical energy-related applications. Asymmetric supercapacitor (ASC) based on $Zn/Co-S@CeO_2/NF$ and activated carbon (AC) exhibits high energy storage capability (42.4 Wh/kg) and outstanding operating durability (91.1% after 8000 cycles). These results demonstrate that the as-prepared electrode is a superior candidate for electrochemical devices (Xu et al., 2023).

EV-HNSs show high areal capacity (186.25 mF/cm^2 at 1 mA/cm^2), excellent cycling stability, and good rate capability under the voltage from -0.9 to -0.1 V (vs. Ag/AgCl), superior to their bulk counterparts. By choosing the electrodeposited MnO_2 as a positive electrode, the EV-HNSs//MnO_2 ASCs are further fabricated, which manifest satisfactory operating voltage (1.55 V) and high areal energy density (9.4 $\mu Wh/cm^2$ at 775 $\mu W/cm^2$), outperforming most of reported ASCs devices. This work may shed light on developing hierarchical MOFs-nanosheets-based multifunctional negative electrodes and promote their applications in smart energy storage devices or other clean energy options (Ding et al., 2022).

5. Lithium-ion batteries

A Lithium-ion battery, commonly referred to as a LIB, is a rechargeable battery that has become increasingly popular in recent years. This battery type comprises several key components, including a positive electrode, a negative electrode, an electrolyte, a membrane that allows only lithium ions to pass through, and a battery shell. The negative electrode of a LIB is typically made of carbon material, while the positive electrode consists of lithium-containing compounds such as $LiFePO_4$, $LiMn_2O_4$, and $Li(NiCoMn)O_2$. The primary function of a LIB is to move lithium ions between the positive and negative poles. During discharging, the lithium ions detach from the negative electrode and pass through the electrolyte to reach the positive electrode. This movement of ions produces energy and heat, which can be harnessed to power various applications.

The charging process for a LIB is the opposite of the discharging process. In this case, lithium ions are moved from the positive electrode back to the negative electrode through the electrolyte. This electrochemical reaction is highly complex and takes time to complete. Fig. 7 illustrate the electrode reaction in the charge and discharge processes, we can use the example of a lithium iron phosphate battery.

During the discharging process, the lithium ions move from the negative electrode, composed of carbon, to the positive electrode, made up of lithium iron phosphate. This movement of ions releases energy and heat, which can be used to power various electronic devices. When the battery is recharged, the opposite reaction occurs, with the lithium ions moving back to the negative electrode, ready for the next cycle of use. Overall, Lithium-ion batteries are a vital component of modern technology, powering everything from smartphones to electric vehicles. Understanding their composition and function is essential for anyone interested in developing or using these powerful energy sources.

The positive electrode reaction equation for the discharging of LIB is:

$$LiFeO_4 \rightarrow Li_{1-x}FePO_4 + xLi + xe \qquad (1)$$

The negative reaction equation is:

$$6C + xLi^+ + xe^- \rightarrow Li_xC_6 \qquad (2)$$

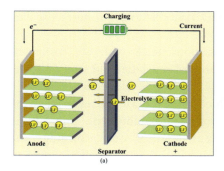 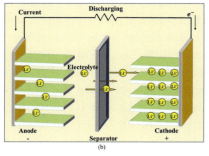

Fig. 7 Lithium-Ion battery.

The overall reaction equation is:

$$Li_x C_6 + Li_{1-x}FePO_4 \rightarrow 6C + LiFeO_4 \tag{3}$$

There are four primary sources of heat inside a battery:

1. The reaction heat from a reversible reaction
2. The by-reaction heat generated by electrolyte decomposition during overcharging or over-discharge
3. Joule heat due to the internal resistance of the cell
4. The heat of polarisation generated by the polarisation reaction

A battery with a higher discharge current will have a lower discharge capacity and experience a faster voltage drop. Similarly, if the charging current increases, the charging speed will also increase, leading to more heat generation by the battery. Generally, the standard operating temperature for LIBs is between 15 °C and 40 °C.

Using 2D non layered nanomaterials can significantly improve the energy density of supercapacitors and the power density and cycling stability of rechargeable batteries. These nanomaterials have been shown to have many advantages in improving the performance of high-performance LIBs. Firstly, their atomic thickness allows for a higher fraction of surface atoms and greater surface area. This means the electrode is in complete contact with the electrolyte, resulting in shorter paths for Li+ diffusion in and out of the crystals. Secondly, their high elasticity and flexibility prevent volume expansion and structural degradation during the charge/discharge process. Thirdly, the increased DOS near the Fermi level improves electrical conductivity and carrier mobility, a common challenge for non-carbonaceous electrodes. Fourthly, ultrathin 2D nanosheets can easily be used to construct nanocomposites, greatly enhancing the synergistic coupling effect due to vast interfacial areas. Overall, the use of 2D nanomaterials promises excellent lithium storage performance.

Nanosheets made of ultrathin transition metal oxide are commonly used in LIBs because of their high theoretical capacities and safety features. For instance, Fernando et al (Wang et al., 2023). developed a composite of graphitic nanostructure and NiO, which achieved a high reversible capacity of 752 mAh/g at 1.2 C and excellent rate capability with high capacity retention of 88.7% at significant current densities, outperforming other options (Fernando et al., 2023). Another team of researchers created a TiO_2/CNT nanostructure as a cathode material for LIBs, which delivered high specific capacities of 275.14 at 0.10 C and showed long-term cycling stability with a capacity retention of 72% after 150 cycles at 2.25 C, surpassing bare TiO_2 nanosheets and any previously reported TiO_2 composites (Liu et al., 2023). In another study conducted by Sun et al. Al_2O_3 nanosheets, after being coated with a CuO-ZnO, exhibit excellent lithium storage performance, including long cycle life (capacity retention of 86.9% at 0.02 C after 100 cycles) (Sun et al., 2023). The remarkable lithium storage performance is mainly due to nanosheets' atomic thickness and mesoporosity, as well as graphene's electrical conductivity and flexibility, which enhance charge transportation and structural stability. Additionally, nanomaterials and graphene's highly exposed surface atoms and defects provide extra lithium storage and the large interfacial areas between nanomaterials and composites. Various other nanostructures and their performance are tabulated in Table 2.

Self-supportive integrated thick electrodes with dual hierarchical carbon skeleton of CMFs and CNTs and TiO_2 are synthesised via vacuum filtration-casting technique. By tuning the ratio of both CMF and CNT, the optimised electrode achieves a relatively high reversible areal capacity of 5.32 mAh/cm^2 @ 0.0125 A/g, attractive rate performance of 0.91 mAh/cm^2 @ 0.50 A/g and impressive cyclic stability areal capacity of 2.75 mAh/cm^2 @ 0.10 A/g up to 150 cycles as anode material for LIBs (Fernando et al., 2023).

Transition metal oxides are said to have the ability to enhance the gravimetric capacity of lithium-ion batteries by two or three times that of current state-of-the-art graphite anodes. novel CuO-ZnO@Al_2O_3 submicroflakes were prepared by magnetron sputtering deposition of Cu-Zn alloy films on a removable sacrificial substrate, followed by a facile thermal oxidation treatment and Al_2O_3 coating procedure by particle atomic layer deposition. The CuO-ZnO@Al_2O_3 submicroflakes electrodes deliver a high reversible capacity of 814 mAh/g at a current density of 50 mA/g, and still maintain a moderate capacity of 447 mAh/g at a high current density of 1000 mA/g. Pairing with a commercial $LiFePO_4$ cathode, the full cell shows high capacity retention and stable cycling performance, suggesting the feasibility of the CuO-ZnO@Al_2O_3 submicroflakes in practical energy storage applications (Liu et al., 2023).

$LiMn_2O_4$ (LMO) electrode materials were prepared via sol–gel method with island-like CeO_2 coating (CeLMO) for lithium electrochemical adsorption from simulated brine. CeO_2 helps inhibit the agglomeration of active metal oxides and promotes the contact and electron transfer between metal oxides. Based on X-ray diffraction, X-ray photoelectron, and Raman spectroscopy results, the fate of lithium intercalation and deintercalation in CeLMO

Table 2 2D nanostructure performance in Li-ion storage batteries.

Material	Method	ICE	Capacity/rate	Cycles	References
NiO/Graphite	Hydrothermal	88.7%	752/1.2	370	Fernando et al. (2023)
TiO_2/CNT	Vacuum filtration testing technique	72%	275.14/0.10	150	Liu et al. (2023)
CeO_2/LMO	Sol-gel method	60%	50/0.0016	30	Luo et al. (2023)
SnO_2/NC	Facile template method	76.68%	869.8/1.01	60	Xin et al. (2023)
CuO-ZnO@Al_2O_3	Magnetron sputtering method	86.9%	447/0.02	100	Sun et al. (2023)
PbS	Electrochemical synthesis	92.4%	239.75/1.20	13,000	Zhao et al. (2023a)
PbSe@C	Solvothermal	70.47%	600.1/1	100	Lu et al. (2023a)
CdS	Oil bath method	76.1	854/0.50	300	Qing et al. (2019)
CuS	Hydrothermal	60%	528/3.0	10	Han et al. (2011)
CuSe	Hydrothermal	55%	264.9/1	500	Xiao et al. (2020)
ZnSe	Hydrothermal	71%	433/3	50	Fu et al. (2015)
$CoSe_2$	Hydrothermal	51.4%	638.3/0.5	100	Yu et al. (2019)

is illuminated. CeLMO-2//Ag electrochemical system can produce 96% pure Li^+ from Li_2CO_3 precipitated mother liquor, and the adsorption capacity is 36.52 mg/g. The capacity retention after 30 cycles was 60% at 50 mA/g, and the Mn dissolution loss rate was 0.016%. Island-like oxide coating improves electrochemical lithium adsorption efficiency and is a promising technology for producing battery-grade lithium products (Luo et al., 2023).

6. Sodium-ion batteries

The sodium–ion battery, or NIB or SIB, is a rechargeable battery that uses sodium ions (Na+) as its charge carriers. Similar to lithium–ion batteries (LIB), the working principle and cell construction of SIBs can be relatively similar. However, what sets SIBs apart is the use of sodium in place of lithium as the cathode material. Sodium is in the same group as lithium on the periodic table and shares similar chemical properties. SIBs have garnered significant attention from academics and commercial industries in recent years. This is due to the high cost, environmental impact, and uneven geographic distribution of materials required for lithium–ion batteries. Commonly used materials in LIBs, such as lithium, cobalt, copper, and nickel, are only sometimes necessary for many SIBs.

The most significant advantage of SIBs is the abundance of sodium, which can be easily obtained from saltwater. Despite their many benefits, SIBs do face particular challenges. One such challenge is their lower energy density compared to LIBs. Additionally, SIBs have limited charge-discharge cycles, which can limit their overall lifespan. However, researchers remain optimistic about the potential of sodium-ion batteries and are working towards developing and improving their performance. The Group I alkali elements found on the periodic table also referred to as Group IA, are known for their highly reactive nature. These elements tend to lose their valence electrons, transforming into cations with a charge, allowing them to form ionic bonds with other elements. The alkali elements in this group share similarities, and the mechanism of rocking chair SIBs (to and fro movement of charge carriers between anode and cathode) is comparable to LIBs. To develop new electrode materials that possess specific structures, it is crucial to have a clear understanding of the sodium transport properties (Fig. 8).

Initially, there was a belief that the larger radius of Na+ ions (97 p.m.) would result in slow transport. Still, later discoveries revealed that the

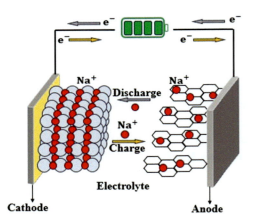

Fig. 8 Sodium-Ion battery.

crystal morphology (A-O coordination, octahedral, or prismatic) plays a significant role in the transport properties. Studies have shown that in some cases, the Na+ ion migration barrier can be lower than that of Li+ ions in layered structures. Sodium-based layered electrode materials can be categorised into O3-type or P2-type, where the sodium ion occupies octahedral or prismatic sites, respectively. O3-type $NaFeO_2$ is highly electrochemically active, and its reaction mechanism with O3-type $NaMnO_2$ and graphitic carbon electrode materials helps comprehend the general working mechanism as discussed below:

During the charging process:

$$\text{Cathode:} -NaMnO_n \rightarrow Na_{1-x}MnO_2 + xN^a + xe^- \quad (4)$$

$$\text{Anode:} -C + xN^a + xe^- \rightarrow Na_x C \quad (5)$$

$$\text{Overall:} -C + NaMnO_2 \rightarrow Na_{1-x}MnO_2 + Na_x C \quad (6)$$

During discharging process:

$$\text{Cathode:} Na_{1-x}MnO_2 + xNa + xe^- \rightarrow NaMnO_2 \quad (7)$$

$$\text{Anode:} Na_x C \rightarrow C + xNa + xe^- \quad (8)$$

$$\text{Overall:} Na_{1-x}MnO_2 + Na_x C \rightarrow C + NaMnO_2 \quad (9)$$

Sodium-ion batteries (SIBs) often face challenges as electrode materials due to the larger radius of Na^+ compared to Li^+. This can result in low specific capacity, poor rate capability, and short cycle life. To address this, 2D non layered nanomaterials with confined thicknesses and large specific surface areas are crucial for adequate Na^+ storage. These materials' atomic thickness helps shorten the Na^+ diffusion length and improve Na^+ insertion and desertion kinetics. Meanwhile, the large surface areas allow for efficient capacitive storage, leading to reversible capacity, excellent rate capability, and long-term cycling stability.

In one experiment, TiO_2/CFC (CFC is a type of flexible carbon cloth) nanocomposites were created and tested as anodes for SIBs by Liang's team (Liu et al., 2017). These composites showed high rate capability with 148.7 mAh/g capacity retention at 73.8 mA/g and a cycle life of up to 2000 cycles with minimal capacity fading. By analysing CV curves and separating the capacitive storage from the total charge storage, they discovered that the capacitive contribution increased as the scan rate increased and reached a maximum of 90% at 1 mV/s. First principles DFT calculations showed that the improved capacitive storage was mainly due to intercalation pseudocapacitive behaviour at the interface of partly bonded carbon-TiO_2, which created feasible channels for Na^+ insertion/extraction with low energy barriers. In another study, researchers grew mesoporous $CoSe_2$ as an anode material for SIBs (Yang et al., 2019). This anode showed an average capacity of 363 mA/hg for the first 240 cycles at 5 mA/g, excellent rate capability with an average capacity of 83 mAh/g at 0.2 mA/g, and good cycling stability. The kinetic analysis revealed an improved capacitive contribution with an increasing scan rate. The excellent electrochemical performance was attributed to the atomic thickness of mesoporous $CoSe_2$ and the direct growth method for electrode processing, which led to remarkably enhanced surface redox pseudocapacitance and interfacial double-layer capacitance. Other nanomaterials used for Na-ion storage are tabulated in Table 3.

A very simple co-precipitation approach is adopted to prepare ZrO_2-coated NiO on MWCNTs nanocomposites with NiO nanoparticles within 10–15 nm size. The XPS studies confirm the presence of Zr, Ni, C, and O elements in the sample, while the BET and BJH analyses reveal a typical surface area of 204.44 m^2/g with pores between 10 and 15 nm. The electrochemical performance studies of the ZrO_2-coated nanocomposite electrode show a higher charge/discharge capacity of 688.3/688.7 mAh/g after 200 cycles with excellent retention capacity (96%) and cycling stability. A minor capacity fading has been observed in the rate performance of the electrodes at currents ranging from 100 to 5000 mA/g. It also reveals that coin cells tend to maintain their maximum Columbic efficiency of 99.9% at a low current density, revealing a notable reversible capacity (Subhan et al., 2023).

Table 3 2D nanostructure performance in Na-ion storage batteries.

Material	Method	ICE	Capacity/rate	Cycles	References
ZrO_2-coated NiO	Co-precipitation	96%	688.7/1.5	200	Subhan et al. (2023)
TiO_2/CFC	Hydrothermal	90%	148.7/1	2000	Liu et al. (2017)
ZnSe/CeO_2	Hydrothermal	69.8%	113,2/2	2000	Dong et al. (2021)
SnO_2 @MWCNT	Solvothermal	72%	839/0.1	50	Wang et al. (2013)
CuO-on Cu foil	Sol-Gel	51.8%	580/0.2	200	Chen et al. (2016)
PbSe@CNT	Ball milling	88%	458.9/0.75	200	Zhao et al. (2023b)
CdS	Oil bath method	98.2%	236/0.5	600	Ding et al. (2023)
CuS-NTAs@GDY	Template synthesis	78.7%	165.2/2	1000	Zhai et al. (2022)
MOF-derivedCu_2Se@C	Hydrothermal	76.5%	276/0.5	500	Kotova et al. (2019)
ZnSe/MWNT	Hydrothermal	88%	382/0.5	180	Tang et al. (2018)
Nano-Te@C	Hydrothermal	90%	410/0.01	1000	Zhang et al. (2015)
Mesoporous $CoSe_2$	Selenization	80%	363/5	240	Yang et al. (2019)

ZnSe/CeO$_2$/RGO displays excellent sodium and lithium-storage performances. It reveals a sodium storage capacity of 113.2 mAh/g after 2000 cycles at 2.0 A/g. Meanwhile, ZnSe/CeO$_2$/RGO as an anode for LIBs delivers a reversible capacity of 675 mAh/g at 0.1 A/g after 200 cycles. The electrochemical kinetic analysis reveals that the redox reactions are dominated by pseudocapacitive behaviour. The excellent performance reveals a bright prospect of the bimetallic organic frameworks derived inorganic material for energy storage (Dong et al., 2021).

SnO$_2$ @MWCNT nanocomposite was synthesised by the Solvothermal method. SEM and TEM results showed the uniform distribution of SnO$_2$ nanoparticles on CNTs. SnO$_2$ @MWCNT nanocomposite exhibited a high sodium storage capacity of 839 mAh/g in the first cycle when used as anode materials in Na-ion batteries. SnO$_2$ @MWCNT nanocomposite also demonstrated much better cycling performance than bare SnO$_2$ nanoparticles and bare MWCNTs (Wang et al., 2013).

7. Fuel cells

Fuel cells are highly efficient devices that generate electricity by converting chemical energy into electrical energy without combustion. Unlike traditional power sources, fuel cells produce minimal waste heat and exhaust gases, making them an environmentally friendly alternative. Additionally, co-generation systems can generate additional electricity from the waste heat produced by the fuel cells.

One of the ways to generate electricity is by utilising the reaction between hydrogen and oxygen through a fuel cell. This process involves transferring the hydrogen and oxygen through carbon electrodes into a concentrated sodium hydroxide solution as illustrated in Fig. 9.

The cell's response can be expressed through the following equation:

Reaction at cathode: $O_2 + 2H_2O + 4e^- \rightarrow 4OH^-$
Reaction at anode: $2H_2 + 4OH^- \rightarrow 4H_2O + 4e^-$
The net reaction of cell: $2H_2 + O_2 \rightarrow 2H_2O$.

The electrochemical response observed in this scenario is characterised by a relatively low reaction rate, which can present a significant challenge. However, it is possible to overcome this obstacle using a catalyst, such as platinum or palladium. Incorporating the catalyst into the electrodes in a finely divided form increases the effective surface area, resulting in a more efficient electrochemical

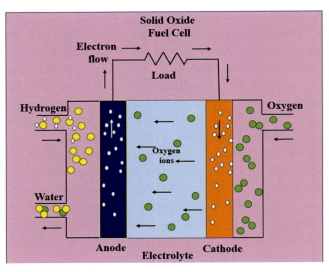

Fig. 9 Schematic illustration of SOFC.

response. This approach can significantly enhance the performance of electrochemical processes that low reaction rates would otherwise limit.

There are six different types of fuel cells; each categorised based on the type of electrolyte and fuel used (Fig. 10).

A recent study by scientists from the University of Manchester and Rice University yielded promising results in fuel cell technology. In a groundbreaking experiment, researchers incorporated 2D materials - specifically, graphene and hexagonal boron nitride - into the membrane of fuel cells for the first time. The study revealed a significant reduction in crossover without any adverse effects on proton conductivity. The outcome was a marked performance boost of up to 50%. These findings could have far-reaching implications for developing miniaturised fuel cells with increased efficiency. The studies carried by various researchers and their corresponding findings is tabulated in Table 4.

$La_{0.6}Sr_{0.4}Fe_{0.9}Sc_{0.1}O_{3-\delta}$ (LSFSc)-YSZ/YSZ/CuNi–CeO$_2$-YSZ was fabricated by tape casting, co-sintering and impregnation technologies. The single cell was evaluated at fuel cell (FC) and electrolysis cell (EC) modes. Significant maximum power density of 436.0 and 377 mW/cm^2 was obtained at 750 °C in H$_2$ and CH$_4$ fuel atmospheres, respectively. At electrolysis voltage of 1.3 V and 50% steam content, a current density of −0.718, −0.397, −0.198 and −0.081 A/cm^2 is obtained at 750, 700, 650 and 600 °C respectively. Much higher electrolysis performance than FC mode is

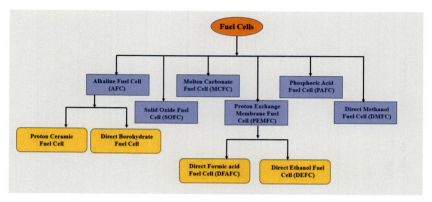

Fig. 10 Types of fuel cells.

exhibited, probably due to the optimised electrodes with increased triple phase boundary (TPB) area and faster gas diffusion (oxygen and steam) and electrochemical reactions for water splitting (Chen et al., 2023b).

Free-standing Pd/SnO$_2$/CP electrodes are prepared by a simple hydrothermal process combined with electrodeposition. Pd nanoparticles are modified on the top of SnO$_2$ nanorods, showing excellent mass normalised activity (8885 A g$_{Pd}^{-1}$) owing to the corresponding low Pd loadings. The Mg-H$_2$O$_2$ semi-fuel cell with Pd/SnO$_2$/CP as cathode displays a high peak power density (192 mW/cm^2 at 60 °C) and excellent stability during the discharge process of 50 h. The microscopic 3D structure constructed by SnO$_2$ interlayer growing on carbon paper accelerates the electrolyte delivery. The synergistic effect between SnO$_2$ and Pd ensures the efficient utilisation of Pd and provides more Pd active sites for enhancing the electrode performance further (Wen et al., 2023).

Fe-doped Gd$_2$O$_3$ with varying compositions was successfully synthesised to function as electrolytes for SOFCs. The as-prepared material, 10% Fe-doped Gd$_2$O$_3$ (Fe$_{0.1}$Gd$_{1.9}$O$_3$), exhibited an excellent peak power density of 1352 mW/cm^2 at 550 °C, while the ionic conductivity reached 0.25 S/cm. Various spectroscopic measurements, such as X-ray photoelectron spectroscopy, ultraviolet-visible (UV–vis) spectroscopy, and density functional theory calculations, were employed to understand the enhanced ion transportation mechanism and the improved performance of Fe-doped Gd$_2$O$_3$. The results showed that the Fe-doped Gd$_2$O$_3$ and energy bandgap tuning of electrolytes can significantly improve fuel cell performance at low temperatures, which is of great significance for the future development of low-temperature ceramic fuel cells (Lu et al., 2023b).

Table 4 2D materials performance for different fuel cells.

Material	Method	Current density	Peak voltage	Max. power density	Type of fuel cell	References
Ni/NiO/MWCNT	Combustion	15.94 mA/cm^2	0.43 mV	41.8 mW/cm^2	DMFC	Askari et al. (2019)
FTO/TiO$_2$ seed layer/ TiO$_2$ nanorods	Hydrothermal method	30 mA/cm^2	41 mV	0.15 μW/cm^2	PMFC	Li et al. (2023)
CuNi-CeO$_2$	Type casting/ co-sintering	0.718 A/cm^2	1.3 V	41.8 mW/cm^2	SOFC	Chen et al. (2023b)
Pd/SnO$_2$/CP	Hydrothermal	50 mA/cm^2	2.1 V	192 mW/cm^2	Semi-fuel cell	Wen et al. (2023)
Fe-doped Gd$_2$O$_3$	Sol-gel	100 mA/cm^2	0.9 V	1352 mW/cm^2	SOFC	Lu et al. (2023b)
Pt@Al-Ti doped ZnO	Solvothermal	83.4 mA/cm^2	0.2 V	15.1 mW/cm^2	DMFC	Amirinejad and Parsa (2023)
CuS/NiFe-LDH/NF	Hydrothermal	10 mA/cm^2	1.57 V	—	SOFC	Sarfraz et al. (2023)
CdS	Solvothermal	655.2 μA/cm^2	79.7 mV	144.7 μW/cm^2	SOFC	Qing et al. (2022)
Metal chalcogenides	Hydrothermal/ solvothermal	47.5 mA/cm^2	0.2 V	800 mW/cm^2	PEMFC	Eisa et al. (2023)

8. Other energy storage

The energy field can significantly benefit from 2D nanomaterials, particularly in solar cells, formic acid oxidation, oxygen reduction catalysis, and photodetectors. A notable example is the work of Zheng and colleagues, who developed freestanding hexagonal palladium nanosheets with less than ten atomic layers thick (Huang et al., 2011). These nanosheets displayed impressive electrocatalytic activity towards formic acid oxidation, surpassing commercial palladium black catalysts by 2.5 times. Furthermore, Pd@Pt monolayer nanosheets with core-shell structures were also created and evaluated for ORR catalysis (Wang et al., 2015). They exhibited around a seven-fold increase in mass activity and much better durability than commercial Pt/C catalysts. Another research group designed ultraviolet photodetectors that are transparent and flexible by depositing atomically thin transition metal oxides (TiO_2, ZnO, Co_3O_4, WO_3) on a single-layer graphene-backed electrode (Sun et al., 2014) . These photodetectors showed excellent stability and quick response, and the I-V characteristics demonstrated ideal linear responses and good ohmic behaviours (except for WO_3). These developments indicate significant promise for photoelectric or photochemical devices.

9. Conclusions

Having novel physical and electrochemical properties of 2D materials makes it promising in energy applications like supercapacitors, Fuel cells, lithium-Ion and sodium-Ion batteries. In the illustration and the working mechanism of these energy applications, it is observed that the 2D non layered materials are efficient, cost-effective and feasible electrode materials. Various 2D non layered materials and their composites are tabulated based on their application with synthesis methods, ICE, capacity, and cyclabilityin supercapacitors and rechargable batteries along with methods, current density, peak voltage, and maximun power density in fuel cells.

References

Abdallah, F., et al., 2023. Electrochemical performance of corncob-derived activated carbon-graphene oxide and TiO_2 ternary composite electrode for supercapacitor applications. J. Energy Storage, vol. 68. https://doi.org/10.1016/j.est.2023.107776.

Althubiti, N.A., et al., 2023. Fabrication of novel zinc selenide/cadmium oxide nanohybrid electrode via hydrothermal route for energy storage application. J. Energy Storage, vol. 70, 108154. https://doi.org/10.1016/j.est.2023.108154.

Amirinejad, S., Parsa, J.B., 2023. Promotion of catalytic activity of Pt@Al-Ti doped ZnO nanostructured anodes for direct methanol fuel cells. J. Solid State Electrochem. https://doi.org/10.1007/s10008-023-05491-0.

Askari, M.B., Salarizadeh, P., Seifi, M., Rozati, S.M., 2019. Ni/NiO coated on multi-walled carbon nanotubes as a promising electrode for methanol electro-oxidation reaction in direct methanol fuel cell. Solid State Sci. 97. https://doi.org/10.1016/j.solidstatesciences.2019.106012.

Babu, B., Kim, J., Yoo, K., 2022. Nanocomposite of SnO_2 quantum dots and Au nanoparticles as a battery-like supercapacitor electrode material. Mater Lett. 309. https://doi.org/10.1016/j.matlet.2021.131339.

Chen, X., Ge, H., Yang, W., Liu, J., Yang, P., 2023a. Construction of high-performance solid-state asymmetric supercapacitor based on Ti_3C_2Tx MXene/CuS positive electrode and Fe_2O_3@rGO negative electrode. J. Energy Storage, vol. 68. https://doi.org/10.1016/j.est.2023.107700.

Chen, C., et al., 2016. Rapid synthesis of three-dimensional network structure CuO as binder-free anode for high-rate sodium ion battery. J. Power Sources, vol. 320, 20–27. https://doi.org/10.1016/j.jpowsour.2016.04.063.

Chen, T., et al., 2023b. Application of CuNi–CeO_2 fuel electrode in oxygen electrode supported reversible solid oxide cell. Int. J. Hydrogen Energy 48 (26), 9565–9573. https://doi.org/10.1016/j.ijhydene.2022.11.236.

Dhanalakshmi, S., Mathi Vathani, A., Muthuraj, V., Prithivikumaran, N., Karuthapandian, S., 2020. Mesoporous Gd_2O_3/NiS_2 microspheres: a novel electrode for energy storage applications. J. Mater. Sci.: Mater. Electron. 31 (4), 3119–3129. https://doi.org/10.1007/s10854-020-02858-1.

Ding, Y., et al., 2023. Carbon quantum dots modified small molecular quinone salt as cathode materials for sodium-ion batteries. J. Electroanal. Chem. 928. https://doi.org/10.1016/j.jelechem.2022.117054.

Ding, Z., Cheng, Z., Shi, N., Guo, Z., Ren, Y., Han, M., ... Huang, W., 2022. Dual-electroactive metal–organic framework nanosheets as negative electrode materials for supercapacitors. Chemical Engineering Journal 450, 137193.

Dong, C., et al., 2021. Zn-Ce based bimetallic organic frameworks derived ZnSe/CeO_2 nanoparticles encapsulated by reduced graphene oxide for enhanced sodium-ion and lithium-ion storage. J. Alloys Compd. 875. https://doi.org/10.1016/j.jallcom.2021.159903.

Eisa, T., et al., 2023. Critical review on the synthesis, characterization, and application of highly efficient metal chalcogenide catalysts for fuel cells. Progress in Energy and Combustion Science, vol. 94 Elsevier Ltd, https://doi.org/10.1016/j.pecs.2022.101044.

Fernando, N., Kannan, H., Robles Hernandez, F.C., Ajayan, P.M., Meiyazhagan, A., Abdelkader, A.M., 2023. Graphitic nanostructure integrated NiO composites for high-performance lithium-ion batteries. J. Energy Storage 71, 108015. https://doi.org/10.1016/j.est.2023.108015.

Fu, Y., Zhang, Z., Du, K., Qu, Y., Li, Q., Yang, X., 2015. Spherical-like ZnSe with facile synthesis as a potential electrode material for lithium ion batteries. Mater Lett. 146, 96–98. https://doi.org/10.1016/j.matlet.2015.02.019.

Han, Y., et al., 2011. Synthesis of novel CuS with hierarchical structures and its application in lithium-ion batteries. Powder Technol. 212 (1), 64–68. https://doi.org/10.1016/j.powtec.2011.04.028.

Harish, S., Sathyakam, P.U., 2022. A review of tin selenide-based electrodes for rechargeable batteries and supercapacitors. J. Energy Storage 52 (PB), 104966. https://doi.org/10.1016/j.est.2022.104966.

Huang, X., et al., 2011. Freestanding palladium nanosheets with plasmonic and catalytic properties. Nat. Nanotechnol. 6 (1), 28–32. https://doi.org/10.1038/nnano.2010.235.

Khan, A.U., et al., 2023. A new cadmium oxide (CdO) and copper selenide (CuSe) nanocomposite: an energy-efficient electrode for wide-voltage hybrid supercapacitors. Colloids Surf. A Physicochem. Eng. Asp. 656. https://doi.org/10.1016/j.colsurfa.2022. 130327.

Kotova, S.P., Mayorova, A.M., Prokopova, D.V., Samagin, S.A., 2019. Tunable liquid crystal astigmatic plate. Journal of Physics: Conference SeriesInstitute of Physics Publishing. ⟨10.1088/1742-6596/1368/2/022018⟩.

Liu, L., Ouyang, T., Jiang, Z., Ogundare, S.A., Balogun, M.-S., 2023. Vacuum filtration-casted hierarchical carbon skeleton/TiO_2 thick electrodes for lithium-ion batteries with high areal capacity. J. Energy Storage 70, 108026. https://doi.org/10.1016/j.est.2023. 108026.

Liu, S., et al., 2017. TiO_2 nanorods grown on carbon fiber cloth as binder-free electrode for sodium-ion batteries and flexible sodium-ion capacitors. J. Power Sources 363, 284–290. https://doi.org/10.1016/j.jpowsour.2017.07.098.

Li, X., et al., 2023. A portable photocatalytic fuel cell based on TiO_2 nanorod photoanode for wastewater treatment and efficient electricity generation. Ceram Int. 49 (16), 26665–26674. https://doi.org/10.1016/j.ceramint.2023.05.201.

Luo, G., et al., 2023. Island-like CeO_2 decorated $LiMn_2O_4$: surface modification enhancing electrochemical lithium extraction and cycle performance. Chem. Eng. J. 455. https:// doi.org/10.1016/j.cej.2022.140928.

Lu, T., Zhao, J., Yuan, J., Xu, J., Jin, J., 2023a. Facile preparation of PbSe@C nanoflowers as anode materials for Li-ion batteries. Chem. Eng. Sci. 265. https://doi.org/10.1016/j. ces.2022.118220.

Lu, Y., Liu, Y., Yousaf, M., Shah, M.A.K.Y., Yan, S., Lu, C., 2023b. Efficient ion conductivity enhancement mechanism induced by metal ion diffusion of SOFCs based on Fe-doped Gd_2O_3 electrolyte. Electrochim. Acta 458. https://doi.org/10.1016/j. electacta.2023.142481.

Mala, N.A., et al., 2023. Unravelling the structure and electrochemical performance of Mo–Cu dual-doped NiO nanorod shaped electrodes for supercapacitor application. Int. J. Hydrogen Energy. https://doi.org/10.1016/j.ijhydene.2023.05.068.

Mendhe, A.C., Deshmukh, T.B., Soni, V., Sankapal, B.R., Jang, S.H., 2023. Facile three-step strategy to design $CdS@Bi_2Se_3$ core-shell nanostructure: an efficient electrode for supercapacitor application. Ceram. Int. 49 (13), 21978–21987. https://doi.org/10. 1016/j.ceramint.2023.04.022.

Qing, M., et al., 2019. Building nanoparticle-stacking MoO_2-CDs via in-situ carbon dots reduction as high-performance anode material for lithium ion and sodium ion batteries. Electrochim. Acta 319, 740–752. https://doi.org/10.1016/j.electacta.2019.07.017.

Qing, S., Wang, L., Jiang, L., Wu, X., Zhu, J., 2022. Live microalgal cells modified by L-cys/ Au@carbon dots/bilirubin oxidase layers for enhanced oxygen reduction in a membrane-less biofuel cell. SmartMat 3 (2), 298–310. https://doi.org/10.1002/smm2.1100.

Sarfraz, B., Bashir, I., Rauf, A., 2023. CuS/NiFe-LDH/NF as a bifunctional electrocatalyst for hydrogen evolution (HER) and oxygen evolution reactions (OER). Fuel 337. https://doi.org/10.1016/j.fuel.2022.127253.

Sivakumar, P., Kulandaivel, L., Park, J.W., Raj, C.J., Manikandan, R., Jung, H., 2023. MOF-derived flower-like $ZnCo_2O_4$/ZnO nanoarchitecture as a high-performance battery-type redox-active electrode material for hybrid supercapacitor applications. J. Alloys Compd. 952. https://doi.org/10.1016/j.jallcom.2023.170042.

Subhan, F., et al., 2023. Nanometer-thin ZrO_2 coating for NiO on MWCNTs as anode for improved performance of sodium-ion batteries. ACS Appl. Nano Mater. 6 (4), 2507–2516. https://doi.org/10.1021/acsanm.2c04860.

Sun, Z., et al., 2014. Generalized self-assembly of scalable two-dimensional transition metal oxide nanosheets. Nat. Commun. 5. https://doi.org/10.1038/ncomms4813.

Sun, X., Jing, M., Dong, H., Xie, W., Luo, F., 2023. CuO-ZnO submicroflakes with nanolayered Al_2O_3 coatings as high performance anode materials in lithium-ion batteries. J. Alloys Compd. 953. https://doi.org/10.1016/j.jallcom.2023.170137.

Tang, C., et al., 2018. ZnSe microsphere/multiwalled carbon nanotube composites as high-rate and long-life anodes for sodium-ion batteries. ACS Appl. Mater. Interfaces 10 (23), 19626–19632. https://doi.org/10.1021/acsami.8b02819.

Wang, X., et al., 2023. Dodecahedral NC-doped $CoSe_2$ nanoparticles with excellent stability for high-performance flexible solid-state supercapacitors. J. Electroanal. Chem. 943. https://doi.org/10.1016/j.jelechem.2023.117612.

Wang, Y., Su, D., Wang, C., Wang, G., 2013. SnO_2@MWCNT nanocomposite as a high capacity anode material for sodium-ion batteries. Electrochem. Commun. 29, 8–11. https://doi.org/10.1016/j.elecom.2013.01.001.

Wang, W., Zhao, Y., Ding, Y., 2015. 2D ultrathin core–shell Pd@Pt $_{monolayer}$ nanosheets: defect-mediated thin film growth and enhanced oxygen reduction performance. Nanoscale 7 (28), 11934–11939. https://doi.org/10.1039/C5NR02748A.

Wen, F., et al., 2023. Free-standing $Pd/SnO_2/CP$ cathode for high-efficiency magnesium-hydrogen peroxide semi-fuel cell. Mater. Sci. Eng. B Solid State Mater. Adv. Technol. 287. https://doi.org/10.1016/j.mseb.2022.116105.

Xiao, J., Liu, H., Huang, J., Lu, Y., Zhang, L., 2020. Decahedron Cu1.8Se/C nanocomposites derived from metal–organic framework Cu–BTC as anode materials for high performance lithium-ion batteries. Appl. Surf. Sci. 526. https://doi.org/10.1016/j.apsusc.2020.146746.

Xin, Y., et al., 2023. Engineering amorphous SnO_2 nanoparticles integrated into porous N-doped carbon matrix as high-performance anode for lithium-ion batteries. J. Colloid Interface Sci. 639, 133–144. https://doi.org/10.1016/j.jcis.2023.02.065.

Xu, D., Xue, Z., Han, L., Tao, K., 2023. Interface engineered $Zn/Co-S@CeO_2$ heterostructured nanosheet arrays as efficient electrodes for supercapacitors. J. Alloys Compd. 946. https://doi.org/10.1016/j.jallcom.2023.169399.

Yang, S.H., Park, S.K., Kang, Y.C., 2019. Mesoporous $CoSe_2$ nanoclusters threaded with nitrogen-doped carbon nanotubes for high-performance sodium-ion battery anodes. Chem. Eng. J. 370, 1008–1018. https://doi.org/10.1016/j.cej.2019.03.263.

Yu, N., Zou, L., Li, C., Guo, K., 2019. In-situ growth of binder-free hierarchical carbon coated $CoSe_2$ as a high performance lithium ion battery anode. Appl. Surf. Sci. 483, 85–90. https://doi.org/10.1016/j.apsusc.2019.03.258.

Zhai, X., Zuo, Z., Xiong, Z., Pan, H., Gao, X., Li, Y., 2022. Large-scale CuS nanotube arrays@graphdiyne for high-performance sodium ion battery. 2D Mater. 9 (2). https://doi.org/10.1088/2053-1583/ac5d84.

Zhang, J., Yin, Y.X., Guo, Y.G., 2015. High-capacity Te anode confined in microporous carbon for long-life Na-Ion batteries. ACS Appl. Mater. Interfaces 7 (50), 27838–27844. https://doi.org/10.1021/acsami.5b09181.

Zhao, J., Lu, H., Peng, J., Li, X., Zhang, J., Xu, B., 2023a. Establishing aqueous zinc-ion batteries for sustainable energy storage. Energy Storage Mater. 60. https://doi.org/10.1016/j.ensm.2023.102846.

Zhao, D., et al., 2023b. A novel PbSe@CNTs anode material based on dual conversion-alloying mechanism for sodium-ion batteries. Sci. China Mater. 66 (1), 61–68. https://doi.org/10.1007/s40843-022-2129-1.

CHAPTER EIGHT

Sensing applications of non-layered 2D materials

Tuan Sang Tran[a,b] and Dzung Viet Dao[c,d,*]

[a]School of Engineering, RMIT University, Melbourne, Victoria, Australia
[b]School of Chemical Engineering, The University of New South Wales, Sydney, NSW, Australia
[c]School of Engineering and Built Environment, Griffith University, Gold Coast, QLD, Australia
[d]Queensland Micro, and Nanotechnology Centre, Griffith University, Brisbane, QLD, Australia
*Corresponding author. e-mail address: d.dao@griffith.edu.au

Contents

1. Introduction	218
2. Structure and synthesis of non-layered 2D materials	219
2.1 Structure	219
2.2 Synthesis	220
2.3 Wet chemical synthesis	221
2.4 Optimised chemical vapour deposition	223
2.5 Liquid metals synthesis	225
3. Sensing mechanisms of non-layered 2D materials	227
3.1 Piezoresistive mechanism	227
3.2 Piezoelectric mechanism	229
3.3 Electrochemical mechanism	229
3.4 Optoelectronic mechanism	230
3.5 Thermoresistive mechanism	231
3.6 Fluorescence resonance mechanism	231
4. Sensing applications of non-layered 2D materials	232
4.1 Nanoarchitecture designs for sensing applications	232
4.2 Non-layered 2D materials for solute sensing	235
4.3 Non-layered 2D materials for gas sensing	238
4.4 Non-layered 2D materials for mechanical sensing	242
4.5 Photodetectors and radiation sensing	243
5. Conclusions	245
References	246

Abstract

The growing demand for real-time monitoring and ultrasensitive sensors has driven the development of innovative smart materials and nanostructures for sensing applications. Among various cross-functional platforms, non-layered 2D (2-dimenstional) materials have emerged as an outstanding class of sensing materials due to its atomically thin nanostructures and tuneable functionalities. In this chapter, we explore the technical principles that underpin the synthesis and properties of non-layered 2D materials, while emphasising the design of components and sensing mechanisms governing various sensor technologies. The component design and sensing mechanism governing various sensor technologies are highlighted. An in-depth examination of the fundamental engineering strategies for non-layered 2D materials across various sensing applications is explicated, elucidating the structure-interaction-sensitivity correlations between the sensing materials and the analytes. Finally, we discuss the challenges associated with non-layered 2D materials for sensing applications and possible solutions.

1. Introduction

Monitoring hazardous chemicals, gases, microbes, radiation, and other physical quantities has been a major challenge for the scientific community to remedy the environment and improve human health. The ability of a sensor to accurately resolve and process a signal is heavily dependent on the manufacturing processes and the properties of its constituent materials (Wilson, 2004). Advanced nanofabrication techniques for traditional semiconductor-based sensors have been the subject of intensive study for the past several decades in an effort to improve their sensing performance (Mittal et al., 2021). However, when it comes to design complexity, selectivity, and sensitivity, conventional silicon-based sensing technology has hit its physical limits. Therefore, the development of new sensing materials with excellent properties rather than relying on the progress of nanofabrication techniques is expected to be the most effective way for the advancement of next-generation sensing technologies.

Two-dimensional (2D) nanosheets, an emerging class of ultrathin nanomaterials, whose remarkable properties have opened exciting new opportunities in the field of chemistry and condensed matter physics. 2D materials are defined as crystalline materials composed of a single or a few layers of atoms, in which the in-plane interatomic interactions are significantly stronger than those along the stacking direction (Tran et al., 2018). Owning to their nanometre-scale thickness, novel physical properties, high flexibility, and strong integration capability with Si-based electronics, 2D materials have attracted a great deal of interest for advanced optoelectronics and sensing applications.

The history of 2D materials is dated back to 2004 when the first monolayer graphene was isolated from graphite by A. Geim and K. Novoselov (Novoselov et al., 2004), which motivated scientists to devise new ways for synthesising hundreds of others, each with a wide range of useful and interesting properties.

Sensing applications of non-layered 2D materials   219

The configuration of 2D nanomaterials can be fine-tuned to alter their bandgaps and conductivities (Tran et al., 2023a). Compared to its bulk counterparts, 2D materials have a greater surface-to-volume ratio. Such a high ratio generates a substantial amount of surface area for interactions between the 2D materials and the analyte, making 2D materials particularly well-suited for the development of high-performance and ultrasensitive sensors.

1wOn the basis of their crystal structures, 2D materials can generally be divided into two categories: layered and non-layered 2D materials. Graphene, black phosphorus, and transition metal dichalcogenides (TMDCs) are typical examples of layered 2D materials, in which the in-plane atoms are linked by a strong chemical bond, and the layers stacked on top of one another are tied together by the weak Van der Waals interactions. In contrast, noble metals, metal oxides, and metal chalcogenides are non-layered 2D materials whose crystal structures are formed by chemical bonding in three-dimensional (3D) directions (Zhou et al., 2019). Due to their obvious structural distortion and abundant unsaturated surface dangling bonds, the discovery of non-layered nanomaterials has not only greatly enriched the 2D material family, but also provided new means for the fabrication of devices with new physical effects and novel capabilities. While the properties and applications of layered 2D materials have been studied extensively for many years, the development of novel devices based on non-layered 2D materials has only recently emerged (Zhou et al., 2019, Tran et al., 2023b). In this chapter, we discuss major innovations in the synthesis and properties of non-layered 2D materials, with a special focus placed on the design and fabrication of their nanoarchitectures for sensing applications. Current problems of non-layered 2D materials-based sensors, and the strategies for overcoming them are also discussed.

2. Structure and synthesis of non-layered 2D materials

2.1 Structure

Materials with layered structures make up only a small portion of all materials, while the overwhelming majority are non-layered materials. To date, a variety of non-layered materials have been successfully synthesised into 2D crystals (Zhou et al., 2019; Dou et al., 2017). Atomically thin metal materials (e.g., Au, Ag, Co, Pt and Rh) are usually synthesised on graphene template, which fit hexagonal close-packed *(hcp)* crystal structure along $[110]_h$ direction (Huang et al., 2011). Metal oxides (e.g., In_2O_3, CeO_2,

Co_3O_4, and Fe_2O_3) and metal chalcogenides (M_aX_b, where M = In, Ni, Co, Zn, Cd, Pd, Cu, etc., and X = S, Te, Se) also form non-layered 2D structures, where metal layer and oxygen (or chalcogenide) layer are alternately arranged in a cubic crystal phase (Kumbhakar et al., 2021). Transition metal dichalcogenides (TMDCs, such as $CoSe_2$, FeS_2, and CoS_2) form thermodynamically stable 2D phases in either octahedral (1T) or trigonal prismatic (2H) coordination of metal atoms, which are different stacking orders of the atomic planes (Manzeli et al., 2017). Perovskites (e.g., $CsPdBr_3$ and $CH_3NH_3PdI_3$) can also be crystallised into non-layered 2D crystal structures, where an equatorial halide atom is shared by two neighbour octahedrons within the same layer (octahedral sheets) and sandwiched between two layers of organic spacer cations (Chen et al., 2018). Owning to their unique atomically thin geometry, non-layered 2D nanomaterials offer unique and exotic structural and electronic properties with enhanced carrier mobility, adjustable band gaps, numerous surface-active sites, and short ion-diffusion paths, which have greatly promoted their use in sensing applications. A brief summary of the classifications and chemical structures of non-layered 2D materials is provided in Fig. 1.

2.2 Synthesis

The increasing popularity of non-layered 2D nanomaterials also stimulates the rapid development of new synthetic techniques. Due to the inherent isotropic chemical bonding in three-dimensional directions, non-layered crystals typically manifest cubic crystal structure to achieve the lowest system energy and maintain a thermodynamic equilibrium state (Dou et al., 2017). When a bulk crystal becomes atomically thin, its surface is exposed to unsaturated suspended bonds, which must be compensated to satisfy the surface electrostatics via surface reconstruction, electron redistribution, or adsorption of charged species (Zhou et al., 2019). In contrast to 2D layered materials whose crystal structure matches the substrate, non-layered 2D materials typically exhibit a distinct crystal orientation compared to bulk materials (Gupta et al., 2015). Hence, the synthesis of atomically thin non-layered 2D materials is fraught with immense difficulties and obstacles. Numerous techniques have been devised thus far for synthesising non-layered 2D nanomaterials. This section focuses on the most recent developments in the synthesis strategies, such as wet chemical synthesis, optimised chemical vapour deposition (CVD), and liquid metals printing. The corresponding mechanism, advantages, and limitations are also thoroughly discussed.

Sensing applications of non-layered 2D materials   221

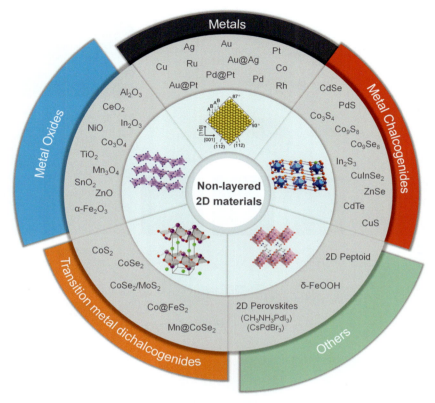

Fig. 1 Classification and chemical structures of non-layered 2D materials.

2.3 Wet chemical synthesis

Wet chemical synthesis is a process by which substances are produced in a liquid medium with the aid of surfactants or polymers through a chemical reaction. Wet chemistry has become a popular way to synthesise various types of non-layered 2D materials due to its high yield, low cost, versatility, and controllability.

Surface-energy-controlled synthesis (SECS) can be used to create 2D metals like Ag, Pd, Rh, and Ru. Surfactants such as poly (vinylpyrrolidone) (PVP) and cetyltrimethylammonium bromide (CTAB) are commonly used to limit the growth directions and favour the formation of 2D crystals (Fig. 2A) (Zhang et al., 2010). Similarly, confined space synthesis (CSS) creates a 2D space to constrain the reaction to reduce their surface energy, where the reactive precursor ions arrange themselves in a crystallographic face and form the desired 2D nanosheets (Wang et al., 2016a).

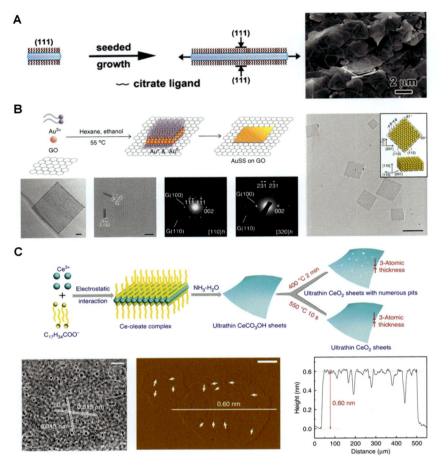

Fig. 2 Wet chemical synthesis of non-layered 2D materials. (A) Surface-energy-controlled synthesis of 2D Ag nanoplates based on selective ligand adhesion. (B) Template-directed synthesis of 2D close-packed hexagonal Au nanostructures on GO sheets. (C) Solvothermal synthesis of ultrathin CeO_2 nanosheets. *(A) Adopted with permission from Zhang et al. (2010). (B) Adopted with permission from Huang et al. (2011). (C) Adopted with permission from Sun et al. (2013).*

Unlike the confined growth method, template-directed synthesis (TDS) provides 2D templates to aid in the growth of 2D structures. Various non-layered 2D crystals have been recently synthesised using this method. Graphene oxide (GO) has long been utilised as a template to drive the formation of non-layered 2D crystals such as metals, metal oxides, and metal chalcogenides. Fig. 2B illustrates the synthesis of gold (Au) nanosheets employing GO as templates (Huang et al., 2011). Another template-directed method is colloidal templated synthesis (CTS), which uses acid, octylamine, and oleylamine surfactants (instead of existing nanosheets) as colloidal templates to direct anisotropic crystal growth by suppressing vertical growth and selectively assembling the precursor along one direction (Du et al., 2012).

Solvothermal synthesis is also a popular methodology to prepare ultrathin non-layered 2D nanosheets, where solvents are sealed and heated above their boiling point to drive the crystallisation reaction under high pressure (Fig. 2C) (Tran et al., 2017; Sun et al., 2013). When water is utilised as a solvent, the solvothermal process is also known as the hydrothermal process. This technique is commonly used to prepare non-layered 2D metal oxides (e.g., CeO_2, In_2O_3, and SnO_2) and metal chalcogenides (e.g., ZnS, ZnSe, and CdS) (Sun et al., 2013).

In general, wet chemical synthesis provides an efficient route for synthesising non-layered 2D crystals, but it has certain drawbacks. Since polymers and organic solvents are used throughout the process, the resulting 2D nanosheets usually contain residual surfactants and impurities, which negatively affect the properties of the 2D crystals. Moreover, they can be easily folded or crumpled during the transfer process, creating challenges for characterisation and device fabrication.

2.4 Optimised chemical vapour deposition

CVD is the most popular technique for producing 2D materials with controlled morphology, tuneable composition, and high crystal quality. However, due to intrinsic isotropic chemical bonding, conventional CVD methods cannot be used to synthesise non-layered 2D crystals (Zhou et al., 2019). This obstacle can be eliminated by incorporating epitaxial substrates with planar and chemically inert surfaces into the growth process. Thus, optimised CVD methods, such as self-limited epitaxial growth, space-confined CVD, and van der Waals epitaxy (vdWE), were developed for growing non-layered 2D crystals.

The vdWE growth method has enormous potential for producing non-layered 2D materials by introducing suitable epitaxial substrates and precisely adjusting growth parameters. Unlike the traditional epitaxial growth, lattice matching between 2D materials and substrates was not rigorously required, and the relatively weak interaction between them facilitated reactive molecule migration and anisotropic growth of high-crystallinity and large-size 2D crystals (Utama et al., 2013). Usually, chemically inert and atomically smooth surfaces, such as graphene, h-BN, mica, and MoS_2, can be used as the vdWE substrates (Jiang et al., 2023). The growth process includes two primary steps, where atoms first assemble into nanoislands, then they consolidate and expand into a nanoplate. The synergy of the vdWE substrates and the highly active lateral planes suppresses the growth in the vertical direction, and highly anisotropic non-layered 2D crystals can be successfully growth (Fig. 3A) (Zhu et al., 2016). However, as the growth in both the vertical and lateral directions occur simultaneously, it is very challenging to obtain ultrathin non-layered 2D crystals with large dimensions.

Space-confined CVD is another promising technique for obtaining non-layered 2D nanosheets. In principle, the bulk form of non-layered materials is thermodynamically stable. Therefore, introducing kinetic growth while inhibiting thermodynamic growth can be an effective method for the controlled synthesis of ultrathin non-layered 2D crystals (Dou et al., 2017). The space-confined CVD approach creates a confined reaction region by stacking two mica substrates, which reduces the diffusion speed and concentration of the precursors, thereby limiting the reaction rate and the vertical growth of

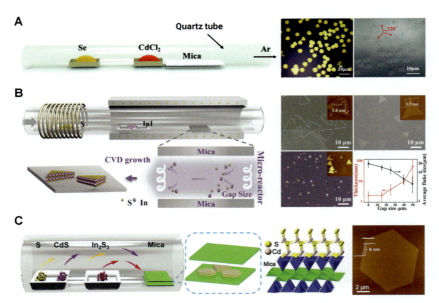

Fig. 3 Optimised chemical vapour deposition of non-layered 2D materials. (A) vdWE growth of 2D ultrathin CdSe nanoplates. (B) Space-confined CVD synthesis of 2D In$_2$S$_3$ nanoflakes within two stacked mica substrates. (C) Self-limited epitaxial growth of ultrathin CdS flakes using In$_2$S$_3$ as the passivation agent. *(A) Adopted with permission from Zhu et al. (2016). (B) Adopted with permission from Huang et al. (2017). (C) Adopted with permission from Jin et al. (2018).*

adsorbed species. When the diffusion of the precursors and reaction rates achieve equilibrium at a sufficient mica spacing, ultrathin 2D crystals are produced, as illustrated in Fig. 3B (Huang et al., 2017). In general, atomically smooth mica substrates can enhance horizontal development while suppressing vertical growth, allowing for the transition from thermodynamically to kinetically controlled growth, which aids the formation of ultrathin non-layered 2D nanosheets. In comparison to the vdWE method, the space-confined CVD is relatively simple to control the morphology and crystallinity of the 2D nanosheets.

In addition to the space-confined method, foreign elements like indium [In] or chloride [Cl] can also be introduced into the reaction system to limit the vertical growth of the crystals, allowing for the synthesis of non-layered 2D nanosheets (Jin et al., 2018; Dou et al., 2017). This self-limited epitaxial growth strategy was utilised to synthesise non-layered 2D CdS and Ge flakes by introducing [In] and [Cl] elements, respectively (Fig. 3C) (Jin et al., 2018). In summary, whereas each CVD-based technique has a unique mechanism, they all employ the same strategy: using external intervention to suppress the growth in vertical directions and kinetically promote 2D anisotropic growth.

2.5 Liquid metals synthesis

Liquid metals are metals and metal alloys exist in liquid form at temperatures below $\sim$350 °C (Daeneke et al., 2018). Under ambient conditions, most liquid metals react spontaneously with oxygen to form an ultrathin "oxide skin" at the surface via the self-limiting Cabrera-Mott reaction (Fig. 4A), and the weak forces between the solid surface and the bulk liquid of liquid metals allow for these oxide skins to be easily removed (Aukarasereenont et al., 2022). By leveraging the Cabrera-Mott reaction at the interface, an ultrathin 2D metal oxide layer can be synthesised and isolated. In recent years, several synthesis approaches based on this phenomenon have been developed to use liquid metals as a platform for producing non-layered 2D materials, including direct printing of 2D metal oxides and post-growth processing toward 2D metal chalcogenides (selenides, sulphides, and tellurides) nanosheets.

The oxide skins of liquid metals can be easily removed by various printing techniques (Fig. 4B). This is accomplished by gently touching the surface of a liquid metal droplet with a suitable substrate or squeezing the droplet between two substrates, allowing the oxide skins to be transferred. Pre-heating the substrate is essential because liquid metals may solidify on a

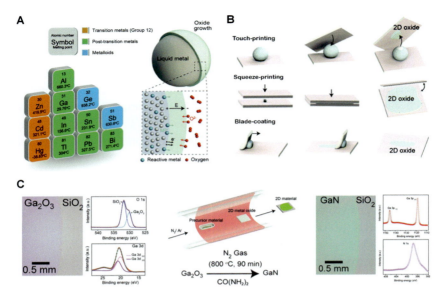

Fig. 4 Liquid metals synthesis of non-layered 2D materials. (A) Selected liquid metal base elements and illustration of the Cabrera–Mott process. (B) Various techniques for printing the 2D oxide skins out of liquid metals. (C) Vapour phase anion exchange process for converting 2D Ga_2O_3 to GaN nanosheets. *(A and B) Adopted with permission from Aukarasereenont et al. (2022). (C) Adopted with permission from Syed et al. (2019).*

cold substrate, impounding the transfer process. Other printing techniques (brushing, blade coating, and roll transfer) share a similar concept, where a liquid metal droplet is transported across the substrate, leaving behind an oxide skin that is only weakly linked to the liquid core. Various non-layered 2D metal oxide nanosheets, such as Ga_2O_3, In_2O_3, Bi_2O_3, TeO_2, and SnO, have been produced by direct printing the oxide skins out of liquid metals.

Since liquid metals can be used as a platform to produce non-layered 2D metal oxides, they also serve as an intermediate route for obtaining other non-layered 2D compounds with many exotic properties (e.g., piezoelectricity, high carrier mobilities, and tuneable bandgaps) via post-growth treatments (Aukarasereenont et al., 2022). By vapour phase anion exchange in a high-temperature furnace tube (Fig. 4C), the 2D metal oxides nanosheets derived from liquid metals can be transformed into metal sulphides (e.g., Ga_2O_3 to Ga_2S_3, In_2O_3 to In_2S_3 nanosheets) and metal nitride (e.g., Ga_2O_3 to GaN and GaP nanosheets) (Syed et al., 2019).

Wet chemical approaches have also been investigated to convert In_2O_3 to $In_2O_{3-x}S_x$ nanosheets, where the highly reactive trisulphur radical anion was used to enable sulphur insertion into the 2D metal oxides by mixing in dimethylformamide (DMF) solutions (Nguyen et al., 2021). Other reactive species, such as reactive pnictogen and chalcogen species, can be used to substantially expand the library of non-layered 2D materials derived from liquid metal synthesis (Aukarasereenont et al., 2022).

3. Sensing mechanisms of non-layered 2D materials

The remarkable properties of non-layered 2D materials are inextricably linked to their geometric dimensions. Owing to its ultrathin planar structure and quantum confinement effect, non-layered 2D materials possess compelling physical and electronic properties for sensing applications. Besides the nature of the sensing materials, the sensitivity of sensors is also heavily dependent on the transduction mechanism, which responds to various inputs and converts them into readable signals (Dral and Ten Elshof, 2018). The sensing capability of non-layered 2D nanomaterials comes from different origins. At the atomic level, any stimuli applied to the 2D crystals would alter its atomic lattice and electronic structure, generating signals in response to external inputs (Wang et al., 2019). In this section, basic principles of various transduction mechanisms, including piezoresistive, piezoelectric, electrochemical, themoresistive, optical, and fluorescence resonance, are discussed in correlation with the atomic structure and electronic properties of non-layered 2D materials employed for sensing (Fig. 5).

3.1 Piezoresistive mechanism

When non-layered materials are reduced to atomical thickness, the strong quantum confinement effect and the unsaturated dangling bonds on the surface induce interesting electrical properties and unique electronic structures that differ from their bulk counterparts, which is beneficial to sensing applications (Zhou et al., 2019). At the most basic level, the piezoresistive mechanism relies on the change in electrical resistivity of the sensing materials upon the application of mechanical stress (Koo et al., 2019). The resistance of a material under stress or strain varies based on its geometric factors and resistivity (Huo et al., 2022). For mechanical sensors, the gauge factor (GF), which is the ratio of relative change in electrical

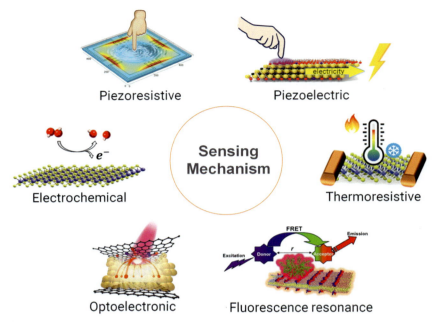

Fig. 5 Sensing mechanisms of non-layered 2D materials.

resistance to the applied mechanical strain, is used to determine the sensitivity of the sensing materials (Tong et al., 2022). Additionally, the sensitivity can be expressed as the ratio of the change in normalised resistance to external stimuli.

2D materials such as graphene have been extensively utilised in piezoresistive sensors due to its large piezoresistive properties, such as the linear change in resistance as a function of applied strain (Hu et al., 2022a). Particularly, the piezoresistivity of 2D materials is derived from three main mechanisms: (a) the deformation of crystal lattice structure, (b) the tunnelling effect between adjacent sheets, and (c) over-connection of the 2D nanosheets (Çakiroğlu et al., 2023). Similarly, the piezoresistive effect has been found on non-layered 2D materials. However, the difficulty of obtaining ultrathin non-layered 2D crystals with large lateral sizes (on the scale of centimetres) limits their widespread practical applications. There are also numerous demonstrations of piezoresistive sensors where 2D materials are incorporated in three-dimensional structures, such as 3D scaffolds, polymer microstructures, and nano/microfibers to fully exploit the piezoresistive mechanism (Cataldi et al., 2022).

3.2 Piezoelectric mechanism

In contrast to the piezoresistive effect, which is characterised by variations in the resistance of materials caused by stress or strain, the piezoelectric effect generates electricity when a force is applied (and vice versa) (Jiang et al., 2020). Therefore, the piezoelectric materials are central to the performance of piezoelectric sensors. The piezoelectricity of inorganic piezoelectric crystals is caused by the arrangement of ions in the noninversely symmetric structure of the dielectric material. Internal polarisation of the material varies linearly with applied stress, resulting in the formation of an electrical field across the material boundary (Fu et al., 2017). In organic piezoelectric polymers, the piezoelectric effect is caused by the orientation and molecular structure of the polymer. The molecular structure of these polymers should contain molecular dipoles that can be reoriented inside the bulk material and maintained in their preferred orientation state. The polarisation of these dipoles varies with the applied stress on the polymer structure and generates electricity (i.e. has a piezoelectric response) (Ramadan et al., 2014).

In the family of 2D materials, transition metal chalcogenides (TMCs) are among the most thoroughly studied 2D piezoelectric materials. Owning to its buckled and asymmetric structure, the piezoelectric effects in 2D materials are derived from the intrinsic in-plane and out-of-plane piezoelectricity (Sherrell et al., 2022). In addition, the majority of 2D crystals can assemble with other similar or dissimilar 2D crystals to form a plethora of complex 2D heterostructures whose properties are quite distinct from those of the constituents (Pham et al., 2022). The unprecedented versatility in tailoring the properties of 2D crystals, combined with their minimal thickness, makes them highly desirable for advanced piezoelectric-based sensing applications.

3.3 Electrochemical mechanism

Electrochemical sensors work by reacting with the analytes and producing electrical signals. The electrochemical mechanism is popular among various analytical techniques due to its inherent precision, specificity, sensitivity, rapid response, and user-friendliness (Varsha and Nageswaran, 2020). The electrochemical reactions that occur between the electrode surface and the analytes are detected by electrochemical sensors, where chemical concentrations are often converted to quantifiable electrical signals. In general, electrochemical sensors are usually relied on conductometry, potentiometry, and amperometry (Wang, 1996). As a result, optimal design of the working electrode modified with non-layered 2D nanosheets can improve quantitative detection of the signals (i.e., oxidation, reduction, or electron transfer of a specific analyte) in electrochemical sensors. In many cases, the inability of some analytes to trigger redox reactions requires the use of additional mediators to produce an electrochemical signal proportionate to the analyte concentration.

Two of the most common types of electrochemical sensors are voltammetric and amperometric sensors (Tajik et al., 2021). Voltammetric sensors detect redox reactions of an electroactive analyte or mediator by measuring the current as a function of the change in potential. Voltammetric sensors find widespread use in the detection of small biorelevant chemical compounds (including uric acid, ascorbic acid, and dopamine) that are essential for monitoring the health of the human body (Zheng et al., 2021). Amperometric measurements, on the other hand, are conducted at a constant potential. The analyte concentration is calculated from the constantly recorded Faradaic current, which is produced by the redox reactions of the electroactive molecules (Baracu and Gugoasa, 2021). Non-layered 2D nanomaterials were employed to construct amperometric sensors used in health, food, agriculture, and environmental monitoring (Meng et al., 2019).

3.4 Optoelectronic mechanism

Optoelectronic mechanism relies on the interaction between photons and the electronic structure of semiconductors. When photons with a suitable wavelength shine on a semiconductor, it will generate free electrons in the semiconductor. These free electrons will alter the electrical conductivity of the material, thereby enabling the measurement of the incident light (Hu et al., 2022b). Many photodetectors utilise the photovoltaic effect, which converts photon energy into electrical energy or photovoltage. To quantify various properties of light, including wavelength, frequency, intensity, scattering, and polarisation, modern photodetectors usually rely on the photoconductive effect of the sensing materials, such as a light-dependent resistor whose resistance can be controlled by the irradiated light. Non-layered 2D materials, due to their large surface area, high electron mobility, and unique optical properties, are promising for the development of highly sensitive optical sensors and photodetectors (Tan et al., 2020). Furthermore, dopants or metallic nanostructures may aid in the enhancement of sensor performance. In recent years, the research on optoelectronic properties of 2D materials is expanding rapidly in numerous crucial disciplines, including environmental monitoring, biomedicine, and food analysis (Gan et al., 2022; Lei and Guo, 2022).

3.5 Thermoresistive mechanism

Temperature sensors are useful in various applications (e.g., monitoring physiological changes of the human body, thermal changes of artificial systems, and the environment). Thermistors and Resistance Temperature Detectors (RTDs) are two common types of temperature sensors for measuring and monitoring the temperature in various applications. Both sensors rely on the principle that the electrical resistance of the sensing materials changes with temperature (Dinh et al., 2017). RTDs usually use a pure metal or metal alloy as the sensing element. As the temperature changes, the electrical resistance of the metal element also changes in a predictable and repeatable manner. This change in resistance is then measured and converted into a temperature reading (Blasdel et al., 2014). Thermistors are temperature sensors that use semiconductor materials as the sensing element. The temperature-dependent resistance of the thermistor is used to calculate the temperature based on the change in resistance (Dinh et al., 2018). The main difference between thermistors and RTDs is the temperature range. A thermistor is better for lower temperatures, whereas RTDs are more suitable for higher temperatures (Kuzubasoglu and Bahadir, 2020). The mechanical stiffness of conventional bulk materials, however, limits their applications in flexible temperature sensors. In contrast to the bulk materials, non-layered 2D materials are atomically thin and possess excellent physiochemical properties, making them ideal for fabricating flexible temperature sensors. In addition, 2D materials can be assembled into various forms (e.g., fibres, films, and composites), enabling the integration of 2D material-based thermal sensors into other wearable devices and overcoming the limitations of traditional temperature sensors.

3.6 Fluorescence resonance mechanism

The basic principle of fluorescence sensing is the absorption and emission of light produced by molecular interaction between the targets and the sensing materials. Fluorescence sensors offer simple and cost-effective designs with increased sensitivity and portability for sensing applications. Due to their photoluminescent properties, non-layered 2D materials can be utilised as a quencher or fluorophore in fluorescence-based chemical sensors (Liu et al., 2023). This principle allows for non-layered 2D nanosheets to function under fluorescence resonance energy transfer (FRET) mechanism (Hiremath et al., 2022).

FRET is a non-radiative energy transfer process that utilises a light-absorbing acceptor and a fluorescent donor. This process is predicated on the absorption of light at a lower wavelength by the fluorophore, followed by its emission at a longer wavelength. FRET occurs when the absorption and emission spectra of the donor and acceptor overlap by less than 10 nanometres (Liu et al., 2023). In general, 2D materials exhibit an excellent quenching capability. Non-layered 2D nanosheets have been used as acceptors in fluorescent sensors in a variety of applications, such as biochemical analysis, single molecule spectroscopy, cellular imaging, and environmental monitoring (Hiremath et al., 2022; Algar and Krause, 2022).

4. Sensing applications of non-layered 2D materials
4.1 Nanoarchitecture designs for sensing applications

As a class of 2D materials, non-layered 2D materials have infinite planar dimensions and molecular thickness, giving them a high proportion of surface atoms and large surface area (Zhou et al., 2019). The lattice structural deformation of 2D nanosheets, which is generally accompanied by bond rotation, contraction, elongation, and even bond breakage, leads to significant changes in electronic states compared to their bulk counterparts (Fig. 6A) (Sun et al., 2012). The variation in electronic structures (e.g., the enhanced density of states near the Fermi level) causes predictable changes in their carrier mobility, electrical conductivity, and energy band structure, which can lead to unusual features such as superconductivity, half-metallicity, topological insulator, and ferromagnetism.

The atomic structure of non-layered 2D nanomaterials establishes a direct relationship between the electronic structure and the macroscopic physiochemical properties. As a result, modulating the atomic structures of non-layered 2D materials is critical for achieving increased characteristics and unique capabilities (Dou et al., 2017). To date, various ways for manipulating the atomic structures of non-layered 2D materials have been devised, including thickness control, pit/pore generation, vacancy engineering, elemental doping, and 2D heterostructure assembly.

(a) Thickness control: Ultrathin 2D nanosheets with thickness reduced down to the atomic scale inevitably experience surface lattice distortion to preserve structural stability, resulting in differences in electronic states compared to their bulk counterparts(Knøsgaard and Thygesen, 2022). Therefore, adjusting the thickness of the nanosheets is the most reasonable

Sensing applications of non-layered 2D materials

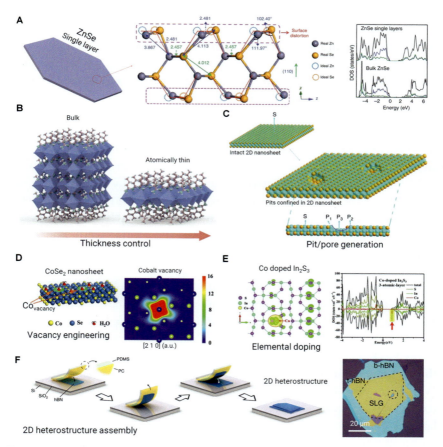

Fig. 6 Nanoarchitecture designs of non-layered 2D materials for sensing applications. (A) Structural model of a non-layered ZnSe nanosheet viewed along the (110) plane and the calculated DOS of ZnSe single layers and bulk ZnSe. (B) Thickness control. (C) Pit/pore generation. (D) Vacancy engineering. (E) Elemental doping. (F) 2D heterostructure assembly. *(A) Adopted with permission from Sun et al. (2012). (B) Adopted with permission from Leng et al. (2020). (C) Adopted with permission from Sun et al. (2013). (D) Adopted with permission from Liu et al. (2014). (E) Adopted with permission from Lei et al. (2015). (F) Adopted with permission from Purdie et al. (2018).*

option for modulating their electrical structure (Fig. 6B). According to DFT calculations, when nanosheet thickness decreases, the density of states (DOS) rises dramatically and the charge density at the conduction band edge becomes more dispersed (Gao et al., 2016). Due to the increased electrical conductivity and rapid carrier transport along the 2D nanosheets, these modifications are advantageous for enhancing sensitivity.

(b) Pit/pore generation: It is well established that because exterior atoms have a lower coordination number than inside ones, they are more disordered. On the surface of the nanosheets, one can make pits or pores that further reduce the coordination number (Fig. 6C). As a result, the electronic states of the atoms close to the pits/pores change noticeably. As an example, a mechanism called ultrafast open space transformation was used to change $CeCO_3OH$ precursor nanosheets into CeO_2 nanosheets with an extremely thin structure. On the surface of the nanosheets, a large number of pits with an occupancy percentage of 20% developed simultaneously (Sun et al., 2013). The ultrathin CeO_2 sheet pits are surrounded by a clear lattice distortion. According to the calculated DOS, the resulting pits produce defect states close to the Fermi level, which clearly increases the hole carrier concentration when compared to bulk CeO_2 and pristine CeO_2 nanosheets (Sun et al., 2013, 2014). These non-layered 2D nanosheets with surface pits have distinct surface electronic states and perform better in a variety of applications (Liu et al., 2016a; Sun et al., 2014).

(c) Vacancy engineering: Incorporating vacancies, which are considered to be more effective than pits and pores due to their high concentration and atomic scale, is another technique to modulate the electronic states of non-layered 2D materials (Liu et al., 2016b). The presence of vacancies endows the non-layered nanosheets with increased DOS at valence band maximum (VBM) (Lei et al., 2014). Ultrathin $CoSe_2$ nanosheets containing abundance of Co vacancies were created by using an intermediate-assisted lamellar exfoliation technique followed by ultrasonication, which caused a disintegration of Co atoms (Fig. 6D). Significant lattice structure distortion and exotic electrical characteristics have been achieved due to the presence of Co vacancies on the 2D nanosheets, which leads to low coordination numbers and a notably decreased Co-Se distance (Liu et al., 2014).

(d) Elemental doping: Another promising method for controlling the electronic structure of non-layered 2D nanosheets is the incorporation of heteroatoms with similar atomic configurations into its lattice structure (Lei et al., 2015). When compared to undoped nanosheets and their bulk counterpart, the DOS at the conduction band minimum is dramatically increased due to the micromechanical disturbance generated by the elemental doping, which redistributes the electron density and distorts the atomic arrangement (Fig. 6E).

(e) 2D heterostructure assembly: A heterostructure is created by establishing intimate electrical contact at the interface between two dissimilar semiconducting materials, whose physical interface is referred to as a

heterojunction (Pham et al., 2022). Due to variations in the density of states, the work function, the Fermi levels across the interface, and the positions of the conduction and valence bands, 2D heterostructures made from ultrathin non-layered 2D nanosheets (by chemical bonds or van der Waals interactions) usually display enhanced performance (Wang et al., 2013; Joseph et al., 2022). The most widely used materials for coupling with non-layered 2D nanomaterials are graphene and its derivatives, which enhance the inherent properties and electrical conductivity of the host materials (Wang et al., 2013). Combining various non-layered 2D materials into atomically thin layered heterostructures also yields a range of possible applications, especially new advancements in optical-based sensing (Fig. 6F).

4.2 Non-layered 2D materials for solute sensing

For solute sensing, non-layered 2D materials can be immobilised on a support or used in suspension. Owning to their nanometre-scale thickness and a substantial amount of surface area for interactions with the analytes, non-layered 2D crystals are especially suitable for use as active sensing elements. Adsorption-based solute sensing typically entails only turn-on sensing (i.e., the target molecules are permanently bound to the sensing platform) (Dral and ten Elshof, 2018). If the 2D crystals are immobilised on a support, desorption of the target molecules may be possible for regeneration and reuse of the sensors.

Fluorescent, plasmonic, and chemiresistive signals can be used to detect adsorption events (Fig. 7A) (Wang et al., 2020). In fluorescent sensors, light-absorbing 2D crystals typically suppress the fluorescence of adsorbed signalling species. Upon interaction with the target molecule, the signalling species desorbs and escapes the quenching effect of the 2D nanosheets, thereby emitting fluorescence (Dral and ten Elshof, 2018). Plasmonic sensors take advantage of the surface plasmons in 2D crystals with charged species, which alters the surface electron density and plasmonic absorption intensity (Yuan et al., 2014a). Plasmonic sensing benefits from 2D sensing elements due to the potential for ionic intercalation, which permits extremely high dopant concentrations (ultradoping) (Alsaif et al., 2014). Ultradoping can increase the free carrier concentration in semiconductors to such a degree that plasmonic resonances migrate towards the near-infrared and visible light regions, which is useful for various practical applications since these wavelengths are often employed in optical systems. Fluorescent sensors usually use light-absorbing materials like MoO_3 and

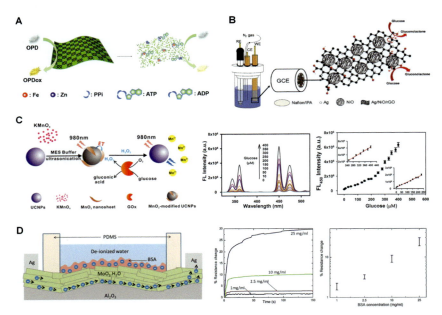

Fig. 7 Non-layered 2D materials for solute sensing. (A) Schematic illustration for adsorption-based colorimetric sensing. (B) Illustration of an electrochemical measurement systems for Glucose sensing. (C) Non-layered MnO$_2$ nanosheets for sensitive turn-on fluorescence detection of H$_2$O$_2$ and glucose. (D) Field effect biosensing based on non-layered 2D α-MoO$_3$ nanosheets. *(A) Adopted with permission from Wang et al. (2020). (B) Adopted with permission from Ngo et al. (2017). (C) Adopted with permission from Yuan et al. (2015). (D) Adopted with permission from Balendhran et al. (2013).*

MnO$_2$ to extinguish the fluorescence of signalling species. Due to their significant d-d transitions, the majority of 2D transition metal oxides absorb a significant amount of light (Dral and ten Elshof, 2018). Compared to 2D metal sulphides and selenides, non-layered 2D metal oxides are advantageous for fluorescent sensing because oxygen atoms produce stronger electrostatic interactions with signalling species, resulting in closer proximity and more effective quenching (Ren et al., 2020). In addition to their light absorption, an advantage of atomically thin MnO$_2$ nanosheets is their potent oxidation of organic compounds, which can be used in conjunction with its fluorescent property.

As an alternative to adsorption-based sensing, solutes may also be detected via electrochemical reactions (Fig. 7B). As a result of the oxidation or reduction of the target species at an electrode covered with 2D nanosheets, an electric current or potential is generated as response signals (Ngo et al., 2017). Electrochemical sensors require oxidative or reducing non-layered 2D crystals towards the target species. There are numerous non-layered 2D materials with high isoelectric points, which facilitates the immobilisation and subsequent charge transfer of a wide range of biomolecules, making them suitable for electrochemical sensors (Zhou et al., 2019). Ultrathin NiO nanosheets are usually employed for electrochemical glucose sensors due to its low cost, high electrocatalytic activity, biocompatibility, and good electron transfer capabilities. Other electrochemically active materials, such as CuO and Co_3O_4, are also widely used for solute sensing (Mahmoudian et al., 2016).

The majority of glucose sensors rely on non-enzymatic electrochemical detection using atomically thin NiO, CuO, and Co_3O_4 nanoflakes (Dral and ten Elshof, 2018). With flower-like structures of ultrathin NiO nanoflakes on a carbon core, Cui et al. reported the highest sensitivity in a glucose sensor, exceeding $30\,\text{mA mM}^{-1}\,\text{cm}^{-2}$. The glucose signal was not substantially affected by the addition of NaCl, ascorbic acid, urea, l-lysine, and l-leucine. The greatest linear response range was reported by Ngo et al. (2017) using NiO flowers on a rGO film decorated with Ag particles. In the linear response region between 50 M and 7.5 mM, the sensitivity was $1.9\,\text{mA mM}^{-1}\,\text{cm}^{-2}$, and $116\,\text{mA mM}^{-1}\,\text{cm}^{-2}$ between 10 and 25 mM. The presence of dopamine, lactose, sucrose, fructose, ascorbic acid, or uric acid did not result in any significant interference. Solute interference can be reduced not only by fine-tuning the sensor design, but also by fine-tuning the operating conditions (e.g., pH, temperature). Yuan et al. (2015) reported a fluorescent enzymatic glucose sensor based on doped sodium yttrium fluoride ($NaYF_4$) upconversion nanoparticles. In this strategy, MnO_2 nanosheets on $NaYF_4$ nanoparticles serve as a quencher. By addition of H_2O_2, which degraded the MnO_2 nanoflakes restored the fluorescence, allows for the detection of glucose (Fig. 7C).

With fluorescent sensors, DNA- or peptide-based molecular recognition of different biomolecules in suspension has been achieved. He et al. (2014) reported the use of MnO_2 nanosheets assembled into flower structures and decorated with dye-labelled ssDNA signalling species. A linear relationship between fluorescence and target molecule concentration was observed over a concentration range of 0–5 nM with a detection limit of 300 pM. To detect ochratoxin A and cathepsin D, Yuan et al. (2014b) employed single-layer MnO_2 nanoflakes coated with dye-labelled ssDNA and peptides. The reaction to ochratoxin A was linear in the $0.02–2.0\,\text{ng mL}^{-1}$ range and 1000 times more sensitive to aflatoxin B1 and B2. A linear response for cathepsin D detection was observed at values ranging from 1 to $100 \, ng \, mL^{-1}$. Yang et al. (2015) detected microRNA inside living cells using single-layer MnO_2 nanoflakes coated with dye-labelled DNA. Non-layered MnO_2 nanoflakes not only contributed to fluorescence signalling with a linear response between 0 and 100 pM and a detection limit of 1 pM, but they also aided cell internalisation and preserved the DNA label from enzymatic disintegration.

Other types of solute sensors have also been developed using non-layered 2D materials. For example, glutathione sensors are typically composed of fluorescent probes and MnO_2 nanoflakes. The non-layered MnO_2 nanoflakes function as both fluorescence quenchers and oxidising agents, where the redox reactions with glutathione deplete the nanoflakes and restore the fluorescence of the probes (Wang et al., 2015). This sensing mechanism only produces turn-on signals, but it has a high selectivity for glutathione over other macromolecules and electrolytes. By employing individually suspended non-layered MoO_{3-x} nanoflakes (4–6 nm thickness), bovine serum albumin (BSA) detection using plasmonic sensing was reported by Alsaif et al. (2016). The response factor was observed to improve with decreasing lateral flake size and thickness, as well as with decreasing free electron concentration. Balendhran et al. (2013) developed a FET-based BSA sensor using drop-cast films of non-layered 2D MoO_3 coated with an atomically thin layer of Au (1–1.5 nm), as shown in Fig. 7D. The Au coating was applied to improve BSA adsorption and facilitate charge transfer between the film and the electrodes. For concentrations of 1 and $25 \, mg \, mL^{-1}$, resistance changes of roughly 1% and 30% were obtained, respectively.

4.3 Non-layered 2D materials for gas sensing

The majority of 2D material-based gas sensors are semiconductor gas sensors, which rely on adsorption of gas molecules on the sensing elements to produce electrical signals (Tripathi et al., 2016). Thus, charge carriers in atomically thin non-layered 2D materials play a vital role in the sensing performance. The density of charge carriers can be altered in contact with the target gas under appropriate conditions, and the concentration of the target gas can be measured based on the change in the amount of charge carriers. The doping state of 2D materials can be modulated by reaction with target gases, causing change in conductivity, which is the essential principle of 2D semiconductor gas sensors (Yang et al., 2017). In general, the number of charge carriers is governed by the electron affinity of gas molecules. Oxidising gas molecules such as NO and NO_2 prefer to absorb electrons from 2D materials, whereas reducing gas molecules such as H_2, NH_3, and the majority of volatile organic compounds (VOCs) prefer to donate electrons to 2D materials (Yang et al., 2017; Dral and ten Elshof, 2018). The sensor sensitivity is dependent on the atomic thickness, charge carrier density, band bending, and relative permittivity of the 2D semiconductors.

With modest variations in chemical composition, stoichiometry, and crystal quality, electronic processes vary considerably. It is widely assumed that oxygen vacancies promote the production of oxygen species on the crystal surface and/or provide active locations for gas adsorption (Du et al., 2012). Metal cation defects are essential, particularly in p-type semi-conductors, yet research into their impact is scarce. The sensitivity of 2D NiO sensors to NO_2 has been observed to increase with increasing nickel vacancy concentrations (Zhang et al., 2016). Although the majority of non-layered 2D materials only contain a single metal species, multi-component structures can be obtained through internal doping, external decoration, or composite formation. Secondary components can enhance sensor performance by altering the concentration of charge carriers (electronic sensitisation), forming p-n junctions, catalysing chemical interactions, or increasing the concentration of adsorbed gas via the spillover effect (chemical sensitisation) (Luo et al., 2017).

Among the various varieties of gas sensors based on non-layered 2D materials, those responses to ethanol, acetone, formaldehyde, and NO_2 have been the most widely reported. Due to the inherent nonselective nature of the sensing mechanism, numerous methods to boost the sensor response to a particular gas will have qualitatively equivalent effects on other oxidising or reducing target gases. As a non-layered 2D sensing material, In_2O_3 has demonstrated the greatest sensitivity for Nitric Oxide (NO_x, including NO and NO_2) detection. Through a facile two-step synthetic method, Wang et al. (2017) developed ultrathin In_2O_3 nanosheets with uniform mesopores, which exhibit an ultrahigh response of 213 for the detection of 10 ppm NO_x at relatively low temperature (120 °C) (Fig. 8A). Hu et al. (2017) reported ultrathin In_2O_3 architectures adorned with PdO particles for gas sensors, which yielded responses as high as 4080 for 50 ppm NO_2 at the optimal operating temperature of 110 °C. The response to NO_2 was roughly 2500–4000 times greater than that to CH_4, H_2, CO, and ethanol, where gas concentrations as low as 500 ppb were detectable. This outstanding performance can be attributed to the catalytic effect of Pd, which promotes the dissociation of both ambient O_2 and targeted NO_2 on the sensor surface. This decreases the number of electrons

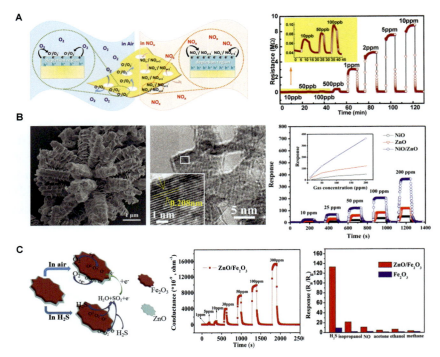

Fig. 8 Non-layered 2D materials for gas sensing. (A) Ultrathin In$_2$O$_3$ nanosheets with uniform mesopores for highly sensitive NO$_x$ detection. (B) Heterostructure of porous NiO nanosheets assembled on flower-like ZnO nanorods for high-performance gas sensor. (C) Ultrathin Fe$_2$O$_3$/ZnO heterostructure for H$_2$S gas sensor. *(A) Adopted with permission from Wang et al. (2017). (B) Adopted with permission from Lu et al. (2017). (C) Adopted with permission from Fan et al. (2017).*

in In$_2$O$_3$ and increases its electrical resistance. The undoped In$_2$O$_3$ sensor achieved a maximal NO$_2$ response of 1310 for 50 ppm, which is still superior in comparison to the other 2D materials. Xu et al. (2012) reported pure In$_2$O$_3$ nanosheets with porous petals that had a response of 1210 at room temperature for 1 ppm NO$_2$ with 30% relative humidity and could detect concentrations as low as 50 ppb. Wang et al. reported NO$_2$ sensors based on ultrathin WO$_3$ flower architectures with responses of 150–250 for 0.8 ppm NO$_2$ at 90–120 °C and the ability to detect concentrations as low as 2–40 ppb. It is unclear why non-layered In$_2$O$_3$ nanosheets outperform other 2D materials for NO$_2$ detection. Hu et al. (2022a) cite the high electrical conductivity of In$_2$O$_3$ as the reason for its widespread use in gas sensors. Nonetheless, the concentration and mobility of charge carriers in non-layered 2D semiconductors are highly dependent on crystal defects and doping levels, implying that the intrinsic conductance of the 2D crystals is of marginal significance. Alternately, the surface of In_2O_3 may be conducive to interaction with NO_2. The specific response of In_2O_3 extends to other oxidising gases, such as O_3, for which it is known that surface hydroxyl groups facilitate O_3 dissociation.

For acetone and ethanol detection, various materials and sensor designs have been reported, in which non-layered ZnO and NiO nanosheets yielded the most effective results as sensing materials. Xie et al. (2017) reported non-layered ZnO nanosheets assembled into flower architectures, capable of detecting concentrations as low as 1 ppm with a response of 362–100 ppm acetone at 300 °C. The response to acetone was approximately 12 times greater than the response to ethanol. Wang et al. (2016b) described a composite of ultrathin CoO, SnO, and SnO_2 nanosheets in a flower-like architecture. The p-n junctions between CoO (p-type) and SnO_2 (n-type) result in the depletion of charge carriers and an increase in resistance in the sensing elements. CoO was also believed to catalyse the dissociation of ethanol molecules on the surface of the sensors. Lu et al. (2017) used 2D NiO nanoflakes to embellish 1D ZnO nanorod-assembled flowers. The local p-n junctions between p-type NiO and n-type ZnO augment electron depletion and electrical resistivity in ZnO. At 240 °C, a response of 205 for 100 ppm acetone was attained, and concentrations as low as 10 ppm were detected (Fig. 8B).

Other types of gas sensors have also been developed using non-layered 2D materials. The highest responses for formaldehyde detection have been observed using non-layered ZnO and In_2O_3 nanomaterials. With a floral architecture of ZnO nanosheets, Cao et al. (2017) achieved a response of 35 for 100 ppm formaldehyde at 260 °C. The response to formaldehyde was 2.3 times that of ethanol, 4 times that of acetone, and more than ten times that of NH_3, toluene, and benzene. Guo (2016) reported ZnO flowers with Fe doping, delivering a response of 33 at 300 °C for a formaldehyde level of only 10 ppm. When Zn^{2+} is replaced with Fe^{3+} in the 2D ZnO lattice, additional electrons are released in the conduction band. This enhances the reduction of O_2 at the sensor surface, increasing the number of locations where formaldehyde can be oxidised. Additionally, Fe doping is thought to improve gas sensing by raising the oxygen vacancy concentration. Fan et al. (2017) designed an ultrathin ZnO/Fe_2O_3 heterostructure for gas sensors towards H_2S (Fig. 8C), which could realise a response of 133.1–100 ppm H_2S gas at temperatures ranging from 150 °C to 350 °C.

In terms of the relationship between the thickness of 2D nanoflakes and the sensor response, thinner nanosheets produce more surface area and, as a result, more active sensing sites per unit volume of non-layered 2D materials. However, in sensor design, the amount of non-layered 2D materials utilised is not always critical. The active surface area of a sensor can then be modified further by altering the size and design of the substrate, as well as the packing density of the 2D nanosheets. Because sensor performance reports often do not take into account the amount of sensing materials utilised in a sensor, it is important to exercise caution when evaluating correlations between response and flake size or specific surface areas. Aside from the material loadings, other fabrication and operation factors might have a major impact on the sensing performance. The electrical contact is determined by the bonding between the sensing material and the device, therefore growing 2D-based designs directly onto the sensing device could be beneficial. However, morphological evolution during post-processing can be deleterious. When it comes to operating settings, sensor responses and selectivity vary depending on factors such as gas flow rate, environmental humidity, and operation temperature.

4.4 Non-layered 2D materials for mechanical sensing

Mechanical quantities, such as stress and strain, can alter the atomic structure of non-layered 2D materials, producing signals for sensing applications (Zhou et al., 2019). Numerous studies on the mechanoelectrical properties of 2D materials have been conducted to investigate the strain-induced modification of their bandgap and electronic structure (Yang et al., 2021). Strain-induced bandgap engineering is crucial and has been a research hotspot for controlling the bandgap of atomically thin 2D semiconductors. In addition to the bandgap structure, piezoelectricity is a crucial strain-induced electrical property for non-layered 2D materials, particularly those with an asymmetrical monolayer structure. The piezoelectric effect (i.e., mechanical-to-electrical) of non-layered 2D materials is highly beneficial for developing flexible and highly sensitive mechanical sensors.

One of the most studied 2D piezoelectric materials is transition metal chalcogenides (TMCs), as their buckled and asymmetric structure induces piezoelectric properties (Sherrell et al., 2022). Recently discovered monolayer group-IV monochalcogenides (e.g., SnSe, SnS, GeSe, GeS) have been found to exhibit large theoretical d_{11} (in-plane) piezoelectric coefficients (between 75 pm V^{-1} to 250 pm V^{-1}) (Fei et al., 2015). The emergence of liquid metal synthesis techniques has also stimulated an upsurge in the production and customisation of non-layered 2D metal oxide nanocrystals. Khan et al. (2020) demonstrated a high piezoelectric response for liquid metal derived non-layered SnS nanosheets with a calculated d_{11} coefficient of $-144\,\mathrm{pm\,V^{-1}}$ and remarkable performance (output voltage of $150\,\mathrm{mV}$ at just 0.7% strain). While lead zirconate titanate (PZT) is the most commonly studied bulk piezoelectric material, little has been uncovered regarding their non-layered 2D material counterparts. Ghasemian et al. (2020) reported piezoelectric properties of 2D lead oxide (PbO) derived from liquid metals synthesis. This atomically thin PbO crystals exhibited a d_{31} (in-plane to out-of-plane) and a d_{33} (out-of-plane) piezoelectric coefficients of $18.9\,\mathrm{pm\,V^{-1}}$ and $30.41\,\mathrm{pm\,V^{-1}}$, respectively, which are among the highest values reported for 2D crystals. By measuring the generated voltage of the embedded PbO monolayers under a controlled alternating force, the piezoelectric property of the ultra-thin PbO sheets was experimentally investigated (Fig. 9A). Despite the fact that it remains substantially lower than that of PZT, this research demonstrates the ability to produce environmentally-friendly piezoelectric 2D crystals that are chemically similar to PZT.

As an emerging class of materials, 2D perovskites have been a research hotspot for optoelectronic applications. Recent research has found that strain over 2D perovskites can introduce octahedral distortion of the crystal structure and influence their carrier transport, which was utilised to develop the first 2D perovskites wearable strain sensor (Fig. 9B) (Xia et al., 2020). Stacking dissimilar 2D crystals into 2D heterostructures could further unlock unique piezoelectric outputs, enabling the transition of 2D piezoelectric devices from the lab to practical applications.

4.5 Photodetectors and radiation sensing

The fundamental principle of photodetectors is the conversion of radiation to electrical signals utilising optoelectronic mechanism. The electronic structure of non-layered 2D nanosheets is inextricably linked to their thickness, which in turn influences their optical absorption and optoelectronic properties. Hence, non-layered 2D materials have garnered great attention and attained exceptional performance in the field of photodetection and radiation sensing.

The first non-layered 2D material-based photodetector was fabricated by Weller's group in 2010 using ultrathin PbS nanosheets (Schliehe et al., 2010), which exhibits exceptional photoresponse capabilities with a responsivity of $0.472\,\mathrm{A\,W^{-1}}$ under 532 nm laser irradiation. Since then, there have been many reports of non-layered 2D nanoflakes for photodetectors and radiation sensing.

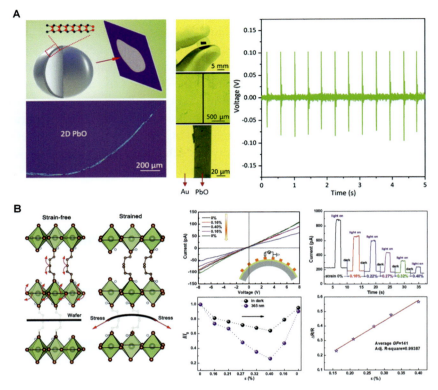

Fig. 9 Non-layered 2D materials for mechanical sensing. (A) Liquid metal-derived ultrathin PbO nanocrystals with piezoelectric properties. (B) 2D perovskites as sensitive strain sensors. *(A) Adopted with permission from Ghasemian et al. (2020). (B) Adopted with permission from Xia et al. (2020).*

In the family of non-layered 2D materials, atomically thin metal oxides, including MgO, ZnO, SnO, MoO_3, and In_2O_3, have shown tremendous potential in photodetection. Zheng et al. (2018) developed a vacuum ultraviolet (VUV) photodetector based on non-layered 2D MgO nanocrystals, which can identify extremely weak VUV signals (0.85 pW) with an external quantum efficiency of 1539% (Fig. 10A). This research lays the way for the future development of next-generation vacuum ultraviolet photodetectors. Alenezi et al. (2014) developed UV detectors with non-layered 2D ZnO nanodisks of 10–100 nm thickness using a photoelectric sensing mechanism. UV irradiation causes electron-hole pairs to form in the 2D crystal lattice. Adsorbed oxygen ions neutralise the holes as they travel to the surface, while electrons boost conductivity in the 2D

Sensing applications of non-layered 2D materials   245

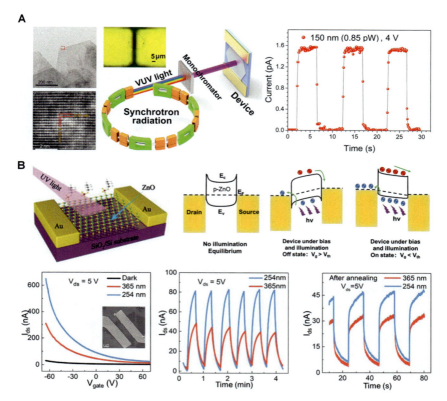

Fig. 10 Non-layered 2D materials for photodetectors and radiation sensing. (A) Vacuum ultraviolet photodetection using non-layered 2D MgO nanocrystals. (B) Atomically thin ZnO nanosheets for visible-blind ultraviolet photodetection. *(A) Adopted with permission from Zheng et al. (2018). (B) Adopted with permission from Yu et al. (2020).*

nanocrystals. The fabricated photodetectors exhibit a record external quantum efficiency value of 1.2×10^4 and photoresponsivity of $3300 \, \text{A W}^{-1}$. Similarly, Yu et al. (2020) recently reported visible-blind ultraviolet photodetectors based on ZnO nanosheets with ultrahigh performance (Fig. 10B). The highest responsivity reached 2.0×10^4 A W^{-1} and the detectivity was as high as 6.83×10^{14} Jones.

5. Conclusions

Non-layered 2D materials offer new opportunities in the whole plethora of sensing applications, including electrochemical, bio, optoelectronic, mechanical and thermal sensing. Alongside the numerous extraordinary physical and chemical properties exhibited by non-layered 2D materials, their high surface area and unique electronic structures constitute significant advantages for sensing applications. Several factors, like the layer thickness, the composition, and the quality of the nanosheets, also have an important influence on their sensing behaviours. Engineering the electronic structure of non-layered 2D nanosheets in correlation with the sensing mechanisms hold the key to enhance the response and performance of the sensing platforms. One of the greatest challenges to be tackled in the near future is the controlled production of non-layered 2D materials with desired characteristics at scale. Sensing applications of non-layered 2D materials is still in its infancy. As research continues to unfold, the potential of non-layered 2D materials in revolutionising sensing technologies appears promising and opens up exciting avenues for future advancements in this field. As the long-term goal is to commercialise sensing devices incorporating non-layered 2D materials, achieving a robust and reliable sensor design with low production cost and high scalability are the most important industrial requirements for market entry.

References

Alenezi, M.R., Alshammari, A.S., Alzanki, T.H., Jarowski, P., Henley, S.J., Silva, S.R.P., 2014. ZnO nanodisk based UV detectors with printed electrodes. Langmuir 30, 3913–3921.

Algar, W.R., Krause, K.D., 2022. Developing FRET networks for sensing. Annu. Rev. Anal. Chem. 15, 17–36.

Alsaif, M.M.Y.A., Field, M.R., Daeneke, T., Chrimes, A.F., Zhang, W., Carey, B.J., et al., 2016. Exfoliation solvent dependent plasmon resonances in two-dimensional substoichiometric molybdenum oxide nanoflakes. ACS Appl. Mater. Interfaces 8, 3482–3493.

Alsaif, M.M.Y.A., Latham, K., Field, M.R., Yao, D.D., Medehkar, N.V., Beane, G.A., et al., 2014. Tunable plasmon resonances in two-dimensional molybdenum oxide nanoflakes. Adv. Mater. 26, 3931–3937.

Aukarasereenont, P., Goff, A., Nguyen, C.K., Mcconville, C.F., Elbourne, A., Zavabeti, A., et al., 2022. Liquid metals: an ideal platform for the synthesis of two-dimensional materials. Chem. Soc. Rev. 51, 1253–1276.

Balendhran, S., Walia, S., Alsaif, M., Nguyen, E.P., Ou, J.Z., Zhuiykov, S., et al., 2013. Field effect biosensing platform based on 2D α-MoO$_3$. ACS Nano 7, 9753–9760.

Baracu, A.M., Gugoasa, L.A.D., 2021. Recent advances in microfabrication, design and applications of amperometric sensors and biosensors. J. Electrochem. Soc. 168, 037503.

Blasdel, N.J., Wujcik, E.K., Carletta, J.E., Lee, K.-S., Monty, C.N., 2014. Fabric nanocomposite resistance temperature detector. IEEE Sens. J. 15, 300–306.

Çakiroğlu, O., Island, J.O., Xie, Y., Frisenda, R., Castellanos-Gomez, A., 2023. An automated system for strain engineering and straintronics of 2D materials. Adv. Mater. Technol. 8, 2201091.

Cao, J., Wang, S., Zhang, H., 2017. Controllable synthesis of zinc oxide hierarchical architectures and their excellent formaldehyde gas sensing performances. Mater. Lett. 202, 44–47.

Cataldi, P., Liu, M., Bissett, M., Kinloch, I.A., 2022. A review on printing of responsive smart and 4D structures using 2D materials. Adv. Mater. Technol. 7, 2200025.

Chen, Y., Sun, Y., Peng, J., Tang, J., Zheng, K., Liang, Z., 2018. 2D Ruddlesden–Popper perovskites for optoelectronics. Adv. Mater. 30, 1703487.

Daeneke, T., Khoshmanesh, K., Mahmood, N., De Castro, I.A., Esrafilzadeh, D., Barrow, S.J., Dickey, M.D., et al., 2018. Liquid metals: fundamentals and applications in chemistry. Chem. Soc. Rev. 47, 4073–4111.

Dinh, T., Phan, H.-P., Nguyen, T.-K., Balakrishnan, V., Cheng, H.-H., Hold, L., et al., 2018. Unintentionally doped epitaxial 3C-SiC (111) nanothin film as material for highly sensitive thermal sensors at high temperatures. IEEE Electron. Device Lett. 39, 580–583.

Dinh, T., Phan, H.P., Qamar, A., Woodfield, P., Nguyen, N.T., Dao, D.V., 2017. Thermoresistive effect for advanced thermal sensors: fundamentals, design considerations, and applications. J. Microelectromech. Syst. 26, 966–986.

Dou, Y., Zhang, L., Xu, X., Sun, Z., Liao, T., Dou, S.X., 2017. Atomically thin nonlayered nanomaterials for energy storage and conversion. Chem. Soc. Rev. 46, 7338–7373.

Dral, A.P., Ten Elshof, J.E., 2018. 2D metal oxide nanoflakes for sensing applications: review and perspective. Sens. Actuators B: Chem. 272, 369–392.

Du, Y., Yin, Z., Zhu, J., Huang, X., Wu, X.-J., Zeng, Z., et al., 2012. A general method for the large-scale synthesis of uniform ultrathin metal sulphide nanocrystals. Nat. Commun. 3, 1177.

Fan, K., Guo, J., Cha, L., Chen, Q., Ma, J., 2017. Atomic layer deposition of ZnO onto Fe_2O_3 nanoplates for enhanced H_2S sensing. J. Alloy. Compd. 698, 336–340.

Fei, R., Li, W., Li, J., Yang, L., 2015. Giant piezoelectricity of monolayer group IV monochalcogenides: SnSe, SnS, GeSe, and GeS. Appl. Phys. Lett. 107.

Fu, Y.Q., Luo, J.K., Nguyen, N.T., Walton, A.J., Flewitt, A.J., Zu, X.T., et al., 2017. Advances in piezoelectric thin films for acoustic biosensors, acoustofluidics and lab-on-chip applications. Prog. Mater. Sci. 89, 31–91.

Gan, X., Englund, D., Van Thourhout, D., Zhao, J., 2022. 2D materials-enabled optical modulators: from visible to terahertz spectral range. Appl. Phys. Rev. 9.

Gao, S., Jiao, X., Sun, Z., Zhang, W., Sun, Y., Wang, C., et al., 2016. Ultrathin Co_3O_4 layers realizing optimized CO_2 electroreduction to formate. Angew. Chem. Int. Ed. 55, 698–702.

Ghasemian, M.B., Zavabeti, A., Abbasi, R., Kumar, P.V., Syed, N., Yao, Y., et al., 2020. Ultra-thin lead oxide piezoelectric layers for reduced environmental contamination using a liquid metal-based process. J. Mater. Chem. A 8, 19434–19443.

Guo, W., 2016. Design of gas sensor based on Fe-doped ZnO nanosheet-spheres for low concentration of formaldehyde detection. J. Electrochem. Soc. 163, B517.

Gupta, A., Sakthivel, T., Seal, S., 2015. Recent development in 2D materials beyond graphene. Prog. Mater. Sci. 73, 44–126.

He, D., He, X., Wang, K., Yang, X., Yang, X., Li, X., et al., 2014. Nanometer-sized manganese oxide-quenched fluorescent oligonucleotides: an effective sensing platform for probing biomolecular interactions. Chem. Commun. 50, 11049–11052.

Hiremath, S.D., Banerjee, M., Chatterjee, A., 2022. Review of 2D MnO_2 nanosheets as FRET-based nanodot fluorescence quenchers in chemosensing applications. ACS Appl. Nano Mater.

Hu, J., Liang, Y., Sun, Y., Zhao, Z., Zhang, M., Li, P., et al., 2017. Highly sensitive NO_2 detection on ppb level by devices based on Pd-loaded In_2O_3 hierarchical microstructures. Sens. Actuators B: Chem. 252, 116–126.

Hu, L., Kim, B.J., Ji, S., Hong, J., Katiyar, A.K., Ahn, J.-H., 2022a. Smart electronics based on 2D materials for wireless healthcare monitoring. Appl. Phys. Rev. 9, 041308.

Hu, X., Liu, K., Cai, Y., Zang, S.-Q., Zhai, T., 2022b. 2D oxides for electronics and optoelectronics. Small Sci. 2, 2200008.

Huang, W., Gan, L., Yang, H., Zhou, N., Wang, R., Wu, W., et al., 2017. Controlled synthesis of ultrathin 2D β-In_2S_3 with broadband photoresponse by chemical vapor deposition. Adv. Funct. Mater. 27, 1702448.

Huang, X., Li, S., Huang, Y., Wu, S., Zhou, X., Li, S., et al., 2011. Synthesis of hexagonal close-packed gold nanostructures. Nat. Commun. 2, 292.

Huo, Z., Wei, Y., Wang, Y., Wang, Z.L., Sun, Q., 2022. Integrated self-powered sensors based on 2D material devices. Adv. Funct. Mater. 32, 2206900.

Jiang, H., Zheng, L., Liu, Z., Wang, X., 2020. Two-dimensional materials: from mechanical properties to flexible mechanical sensors. InfoMat 2, 1077–1094.

Jiang, J., Cheng, R., Feng, W., Yin, L., Wen, Y., Wang, Y., et al., 2023. Van der Waals epitaxy growth of 2D single-element room-temperature ferromagnet. Adv. Mater. 2211701.

Jin, B., Huang, P., Zhang, Q., Zhou, X., Zhang, X., Li, L., et al., 2018. Self-limited epitaxial growth of ultrathin nonlayered CdS flakes for high-performance photodetectors. Adv. Funct. Mater. 28, 1800181.

Joseph, X.B., Stanley, M.M., Wang, S.-F., George, M., 2022. Growth of 2D-layered double hydroxide nanorods heterojunction with 2D tungsten carbide nanocomposite: improving the electrochemical sensing of norfloxacin. J. Ind. Eng. Chem. 110, 434–446.

Khan, H., Mahmood, N., Zavabeti, A., Elbourne, A., Rahman, M.A., Zhang, B.Y., et al., 2020. Liquid metal-based synthesis of high performance monolayer SnS piezoelectric nanogenerators. Nat. Commun. 11, 3449.

Knøsgaard, N.R., Thygesen, K.S., 2022. Representing individual electronic states for machine learning GW band structures of 2D materials. Nat. Commun. 13, 468.

Koo, W.-T., Jang, J.-S., Kim, I.-D., 2019. Metal-organic frameworks for chemiresistive sensors. Chem 5, 1938–1963.

Kumbhakar, P., Chowde Gowda, C., Mahapatra, P.L., Mukherjee, M., Malviya, K.D., Chaker, M., et al., 2021. Emerging 2D metal oxides and their applications. Mater. Today 45, 142–168.

Kuzubasoglu, B.A., Bahadir, S.K., 2020. Flexible temperature sensors: a review. Sens. Actuators A: Phys. 315, 112282.

Lei, F., Sun, Y., Liu, K., Gao, S., Liang, L., Pan, B., et al., 2014. Oxygen vacancies confined in ultrathin indium oxide porous sheets for promoted visible-light water splitting. J. Am. Chem. Soc. 136, 6826–6829.

Lei, F., Zhang, L., Sun, Y., Liang, L., Liu, K., Xu, J., et al., 2015. Atomic-layer-confined doping for atomic-level insights into visible-light water splitting. Angew. Chem. Int. Ed. 54, 9266–9270.

Lei, Z.L., Guo, B., 2022. 2D material-based optical biosensor: status and prospect. Adv. Sci. 9, 2102924.

Leng, K., Fu, W., Liu, Y., Chhowalla, M., Loh, K.P., 2020. From bulk to molecularly thin hybrid perovskites. Nat. Rev. Mater. 5, 482–500.

Liu, H., Li, C., Li, J., Cheng, Y., Zhao, J., Chen, J., et al., 2023. Plasmon-enhanced fluorescence resonance energy transfer in different nanostructures and nanomaterials. Appl. Mater. Today 30, 101731.

Liu, Y., Cheng, H., Lyu, M., Fan, S., Liu, Q., Zhang, W., et al., 2014. Low overpotential in vacancy-rich ultrathin CoSe2 nanosheets for water oxidation. J. Am. Chem. Soc. 136, 15670–15675.

Liu, Y., Liang, L., Xiao, C., Hua, X., Li, Z., Pan, B., et al., 2016a. Promoting photo-generated holes utilization in pore-rich WO_3 ultrathin nanosheets for efficient oxygen-evolving photoanode. Adv. Energy Mater. 6, 1600437.

Liu, Y., Xiao, C., Li, Z., Xie, Y., 2016b. Vacancy engineering for tuning electron and phonon structures of two-dimensional materials. Adv. Energy Mater. 6, 1600436.

Lu, Y., Ma, Y., Ma, S., Yan, S., 2017. Hierarchical heterostructure of porous NiO nanosheets on flower-like ZnO assembled by hexagonal nanorods for high-performance gas sensor. Ceram. Int. 43, 7508–7515.

Luo, Y., Zhang, C., Zheng, B., Geng, X., Debliquy, M., 2017. Hydrogen sensors based on noble metal doped metal-oxide semiconductor: a review. Int. J. Hydrog. Energy 42, 20386–20397.

Mahmoudian, M.R., Basirun, W.J., Woi, P.M., Sookhakian, M., Yousefi, R., Ghadimi, H., et al., 2016. Synthesis and characterization of Co_3O_4 ultra-nanosheets and Co_3O_4 ultra-nanosheet-$Ni(OH)_2$ as non-enzymatic electrochemical sensors for glucose detection. Mater. Sci. Eng. C 59, 500–508.

Manzeli, S., Ovchinnikov, D., Pasquier, D., Yazyev, O.V., Kis, A., 2017. 2D transition metal dichalcogenides. Nat. Rev. Mater. 2, 17033.

Meng, Z., Stolz, R.M., Mendecki, L., Mirica, K.A., 2019. Electrically-transduced chemical sensors based on two-dimensional nanomaterials. Chem. Rev. 119, 478–598.

Mittal, M., Sardar, S., Jana, A., 2021. Nanofabrication techniques for semiconductor chemical sensors. Handbook of Nanomaterials for Sensing Applications. Elsevier.

Ngo, Y.-L.T., Hoa, L.T., Chung, J.S., Hur, S.H., 2017. Multi-dimensional Ag/NiO/reduced graphene oxide nanostructures for a highly sensitive non-enzymatic glucose sensor. J. Alloy. Compd. 712, 742–751.

Nguyen, C.K., Low, M.X., Zavabeti, A., Jannat, A., Murdoch, B.J., Della Gaspera, E., et al., 2021. Ultrathin oxysulfide semiconductors from liquid metal: a wet chemical approach. J. Mater. Chem. C 9, 11815–11826.

Novoselov, K.S., Geim, A.K., Morozov, S.V., Jiang, D.-E., Zhang, Y., Dubonos, S.V., et al., 2004. Electric field effect in atomically thin carbon films. Science 306, 666–669.

Pham, P.V., Bodepudi, S.C., Shehzad, K., Liu, Y., Xu, Y., Yu, B., et al., 2022. 2D heterostructures for ubiquitous electronics and optoelectronics: principles, opportunities, and challenges. Chem. Rev. 122, 6514–6613.

Purdie, D.G., Pugno, N.M., Taniguchi, T., Watanabe, K., Ferrari, A.C., Lombardo, A., 2018. Cleaning interfaces in layered materials heterostructures. Nat. Commun. 9, 5387.

Ramadan, K.S., Sameoto, D., Evoy, S., 2014. A review of piezoelectric polymers as functional materials for electromechanical transducers. Smart Mater. Struct. 23, 033001.

Ren, B., Wang, Y., Ou, J.Z., 2020. Engineering two-dimensional metal oxides via surface functionalization for biological applications. J. Mater. Chem. B 8, 1108–1127.

Schliehe, C., Juarez, B.H., Pelletier, M., Jander, S., Greshnykh, D., Nagel, M., et al., 2010. Ultrathin PbS sheets by two-dimensional oriented attachment. Science 329, 550–553.

Sherrell, P.C., Fronzi, M., Shepelin, N.A., Corletto, A., Winkler, D.A., Ford, M., et al., 2022. A bright future for engineering piezoelectric 2D crystals. Chem. Soc. Rev. 51, 650–671.

Sun, Y., Gao, S., Lei, F., Liu, J., Liang, L., Xie, Y., 2014. Atomically-thin non-layered cobalt oxide porous sheets for highly efficient oxygen-evolving electrocatalysts. Chem. Sci. 5, 3976–3982.

Sun, Y., Liu, Q., Gao, S., Cheng, H., Lei, F., Sun, Z., et al., 2013. Pits confined in ultrathin cerium(IV) oxide for studying catalytic centers in carbon monoxide oxidation. Nat. Commun. 4, 2899.

Sun, Y., Sun, Z., Gao, S., Cheng, H., Liu, Q., Piao, J., et al., 2012. Fabrication of flexible and freestanding zinc chalcogenide single layers. Nat. Commun. 3, 1057.

Syed, N., Zavabeti, A., Messalea, K.A., Della Gaspera, E., Elbourne, A., Jannat, A., et al., 2019. Wafer-sized ultrathin gallium and indium nitride nanosheets through the ammonolysis of liquid metal derived oxides. J. Am. Chem. Soc. 141, 104–108.

Tajik, S., Dourandish, Z., Jahani, P.M., Sheikhshoaie, I., Beitollahi, H., Asl, M.S., et al., 2021. Recent developments in voltammetric and amperometric sensors for cysteine detection. RSC Adv. 11, 5411–5425.

Tan, T., Jiang, X., Wang, C., Yao, B., Zhang, H., 2020. 2D material optoelectronics for information functional device applications: status and challenges. Adv. Sci. 7, 2000058.

Tong, B., Nguyen, T.-H., Nguyen, H.-Q., Nguyen, T.-K., Nguyen, T., Dinh, T., et al., 2022. Highly sensitive and robust 3C-SiC/Si pressure sensor with stress amplification structure. Mater. Des. 224, 111297.

Tran, T.S., Balu, R., De Campo, L., Dutta, N.K., Choudhury, N.R., 2023a. Sulfonated polythiophene-interfaced graphene for water-redispersible graphene powder with high conductivity and electrocatalytic activity. Energy Adv. 2, 365–374.

Tran, T.S., Balu, R., Nguyen, C.K., Mata, J., Truong, V.K., Dutta, N.K., et al., 2023b. Graphene nanosheets stabilized by P3HT nanoparticles for printable metal-free electrocatalysts for oxygen reduction. ACS Appl. Nano Mater. 6, 908–917.

Tran, T.S., Dutta, N.K., Choudhury, N.R., 2018. Graphene inks for printed flexible electronics: graphene dispersions, ink formulations, printing techniques and applications. Adv. Colloid Interface Sci. 261, 41–61.

Tran, T.S., Tripathi, K.M., Kim, B.N., You, I.-K., Park, B.J., Han, Y.H., et al., 2017. Three-dimensionally assembled graphene/α-MnO$_2$ nanowire hybrid hydrogels for high performance supercapacitors. Mater. Res. Bull. 96, 395–404.

Tripathi, K.M., Kim, T., Losic, D., Tung, T.T., 2016. Recent advances in engineered graphene and composites for detection of volatile organic compounds (VOCs) and non-invasive diseases diagnosis. Carbon 110, 97–129.

Utama, M.I.B., Zhang, Q., Zhang, J., Yuan, Y., Belarre, F.J., Arbiol, J., et al., 2013. Recent developments and future directions in the growth of nanostructures by van der Waals epitaxy. Nanoscale 5, 3570–3588.

Varsha, M., Nageswaran, G., 2020. 2D layered metal organic framework nanosheets as an emerging platform for electrochemical sensing. J. Electrochem. Soc. 167, 136502.

Wang, F., Seo, J.-H., Luo, G., Starr, M.B., Li, Z., Geng, D., et al., 2016a. Nanometre-thick single-crystalline nanosheets grown at the water–air interface. Nat. Commun. 7, 10444.

Wang, F., Zhang, Y., Gao, Y., Luo, P., Su, J., Han, W., et al., 2019. 2D metal chalcogenides for IR photodetection. Small 15, 1901347.

Wang, J., 1996. Electrochemical transduction. Handb. Chem. Biol. Sens. 123–137.

Wang, L., Meric, I., Huang, P.Y., Gao, Q., Gao, Y., Tran, H., et al., 2013. One-dimensional electrical contact to a two-dimensional material. Science 342, 614–617.

Wang, Q., Li, X., Liu, F., Sun, Y., Wang, C., Li, X., et al., 2016b. Three-dimensional flake-flower Co/Sn oxide composite and its excellent ethanol sensing properties. Sens. Actuators B: Chem. 230, 17–24.

Wang, X., Jiang, X., Wei, H., 2020. Phosphate-responsive 2D-metal–organic-framework-nanozymes for colorimetric detection of alkaline phosphatase. J. Mater. Chem. B 8, 6905–6911.

Wang, X., Su, J., Chen, H., Li, G.-D., Shi, Z., Zou, H., et al., 2017. Ultrathin In$_2$O$_3$ nanosheets with uniform mesopores for highly sensitive nitric oxide detection. ACS Appl. Mater. Interfaces 9, 16335–16342.

Wang, Y., Jiang, K., Zhu, J., Zhang, L., Lin, H., 2015. A FRET-based carbon dot–MnO$_2$ nanosheet architecture for glutathione sensing in human whole blood samples. Chem. Commun. 51, 12748–12751.

Wilson, J.S., 2004. Sensor Technology Handbook. Elsevier.

Xia, M., Yuan, J.-H., Luo, J., Pan, W., Wu, H., Chen, Q., et al., 2020. Two-dimensional perovskites as sensitive strain sensors. J. Mater. Chem. C 8, 3814–3820.

Xie, X., Wang, X., Tian, J., Song, X., Wei, N., Cui, H., 2017. Growth of porous ZnO single crystal hierarchical architectures with ultrahigh sensing performances to ethanol and acetone gases. Ceram. Int. 43, 1121–1128.

Xu, X., Wang, D., Wang, W., Sun, P., Ma, J., Liang, X., et al., 2012. Porous hierarchical In2O3 nanostructures: hydrothermal preparation and gas sensing properties. Sens. Actuators B: Chem. 171, 1066–1072.

Yang, K., Zeng, M., Fu, X., Li, J., Ma, N., Tao, L., 2015. Establishing biodegradable single-layer MnO_2 nanosheets as a platform for live cell microRNA sensing. RSC Adv. 5, 104245–104249.

Yang, S., Chen, Y., Jiang, C., 2021. Strain engineering of two-dimensional materials: methods, properties, and applications. InfoMat 3, 397–420.

Yang, S., Jiang, C., Wei, S.-H., 2017. Gas sensing in 2D materials. Appl. Phys. Rev. 4, 021304.

Yu, H., Liao, Q., Kang, Z., Wang, Z., Liu, B., Zhang, X., et al., 2020. Atomic-thin ZnO sheet for visible-blind ultraviolet photodetection. Small 16, 2005520.

Yuan, J., Cen, Y., Kong, X.-J., Wu, S., Liu, C.-L., Yu, R.-Q., et al., 2015. MnO_2-nanosheet-modified upconversion nanosystem for sensitive turn-on fluorescence detection of H_2O_2 and glucose in blood. ACS Appl. Mater. Interfaces 7, 10548–10555.

Yuan, Y., Wu, S., Shu, F., Liu, Z., 2014a. An MnO_2 nanosheet as a label-free nanoplatform for homogeneous biosensing. Chem. Commun. 50, 1095–1097.

Yuan, Y., Wu, S., Shu, F., Liu, Z., 2014b. An MnO_2 nanosheet as a label-free nanoplatform for homogeneous biosensing. Chem. Commun. 50, 1095–1097.

Zhang, J., Zeng, D., Zhu, Q., Wu, J., Huang, Q., Xie, C., 2016. Effect of nickel vacancies on the room-temperature NO_2 sensing properties of mesoporous NiO nanosheets. J. Phys. Chem. C 120, 3936–3945.

Zhang, Q., Hu, Y., Guo, S., Goebl, J., Yin, Y., 2010. Seeded growth of uniform Ag nanoplates with high aspect ratio and widely tunable surface plasmon bands. Nano Lett. 10, 5037–5042.

Zheng, W., Lin, R., Zhu, Y., Zhang, Z., Ji, X., Huang, F., 2018. Vacuum ultraviolet photodetection in two-dimensional oxides. ACS Appl. Mater. Interfaces 10, 20696–20702.

Zheng, Y., Li, J., Zhou, B., Ian, H., Shao, H., 2021. Advanced sensitivity amplification strategies for voltammetric immunosensors of tumor marker: state of the art. Biosens. Bioelectron. 178, 113021.

Zhou, N., Yang, R., Zhai, T., 2019. Two-dimensional non-layered materials. Mater. Today Nano 8, 100051.

Zhu, D.-D., Xia, J., Wang, L., Li, X.-Z., Tian, L.-F., Meng, X.-M., 2016. van der Waals epitaxy and photoresponse of two-dimensional CdSe plates. Nanoscale 8, 11375–11379.

CHAPTER NINE

2D-non-layered materials: Advancement and application in biosensors, memristors, and energy storage

Zina Fredj and Mohamad Sawan[*]

CenBRAIN Neurotech, School of Engineering, Westlake University, Hangzhou, P.R. China
[*]Corresponding author. e-mail address: sawan@westlake.edu.cn

Contents

1. Introduction	254
2. Non-layered 2D materials based biosensing platforms	255
2.1 Electrochemical biosensors	255
2.2 Optical biosensors	262
2.3 Nano-FET based biosensor	264
3. Memristors based 2D non layered material	267
4. 2D non-layered materials in energy storage	270
5. Challenges and future perspectives	272
References	273

Abstract

Two-dimensional (2D) non-layered materials have received a considerable attention due to their atomically thin structures, highly active surfaces, and seamless integration with silicon-based electronic devices. In this chapter, we emphasize their applications in biosensing, including electrochemical, optical and nanoFET-based systems. Highlighting their impact, we showcase how 2D non-layered materials enhance the sensitivity and selectivity of electrochemical biosensors and enable label-free sensing in optical biosensors. Moreover, they empower ultra-sensitive detection in nanoFET-based biosensors. Beyond biosensing, we touch upon their potential in memristors and energy storage, driving novel computing devices and high-performance energy systems. Addressing challenges, we present strategies to overcome synthesis, functionalization, and integration hurdles, unlocking the practical use of 2D non-layered materials. Moreover, future research directions are outlined, envisioning advancements that can unlock the full potential of these materials in biosensing applications.

1. Introduction

The field of materials science and nanotechnology has been revolutionized by the discovery and exploration of two-dimensional materials. Among the intriguing avenues of research in this domain is the utilization of two-dimensional materials as a platform for biosensing applications. These materials, characterized by their unique structure and properties, hold great promise for the development of highly sensitive and efficient biosensors. Two-dimensional (2D) materials can be broadly categorized into two types based on their crystal structures: layered and non-layered materials. Layered 2D materials, such as transition metal organic frameworks (MOFs) (Chakraborty et al., 2021), and graphene oxide (GO) (Gutiérrez, 2020), exhibit strong chemical bonding within their in-plane atoms and weak van der Waals interactions between stacked layers. Alternatively, 2D nonlayered materials exhibit intriguing properties due to chemical bonding in three dimensions, resulting in unsaturated dangling bonds on their surfaces and creating highly active and energetically favorable surface characteristics (Zheng et al., 2020; Zhou et al., 2019). These materials encompass noble metals like Pd, Au, and Rh; metal oxides such as WO_3, CeO_2, In_2O_3, SnO_2, TiO_2, and Fe_2O_3; and metal chalcogenides including CdSe, CuS, CuSe, PbS, SnSe, ZnS, and ZnSe (Yang et al., 2012; Wang et al., 2020a). Traditionally, biosensing has relied on layered structures or hierarchical systems to achieve enhanced sensitivity and selectivity. However, the unique properties of two-dimensional non-layered materials have opened up new avenues for biosensing, challenging conventional approaches and driving researchers to explore innovative strategies for detection and analysis. These materials exhibit remarkable electrical, optical, and mechanical characteristics, as well as large surface-to-volume ratios and high surface reactivity. Moreover, their inherent biocompatibility makes them well-suited for interfacing with biological systems and detecting various analytes with high precision. The diverse applications of two-dimensional non-layered materials in biosensing are explored in this chapter, with a specific focus on electrochemical/optical biosensors and nano-FET based biosensors. These materials demonstrate great potential in enabling highly sensitive and selective detection of biomolecules, facilitating advancements in diagnostics, healthcare, and environmental monitoring. Furthermore, the potential of these materials in memristors and energy storage applications is discussed, highlighting their versatility beyond biosensing applications. However, it is important to note that the development of biosensing technologies using 2D non-layered materials is accompanied by certain challenges. Therefore, we will address these hurdles, including synthesis, functionalization, and integration into practical devices, and present strategies to overcome them. Lastly, we will discuss the future prospects of this rapidly evolving field, outlining the potential impact of two-dimensional non-layered materials on the future of biosensing. By examining the current state-of-the-art research, we aim to inspire researchers, engineers, and scientists to explore the vast opportunities within this emerging field and foster innovative solutions to address the pressing challenges in biosensing. The ultimate goal is to shed light on the exciting advances in the use of two-dimensional non-layered materials for biosensing applications and demonstrate their potential to revolutionize diagnostics, healthcare, and environmental monitoring.

2. Non-layered 2D materials based biosensing platforms

Harnessing the unique advantages of their 2D planar structure and quantum confinement effect, 2D materials have demonstrated exceptional physical properties, making them highly promising for diverse applications. Notably, 2D non-layered materials stand out due to their isotropic crystal structure, leading to exceptional surface activity and stronger quantum confinement effect compared to their layered counterparts. This intrinsic distinction results in a broader range of applications for these materials. In this section, we present cutting-edge advancements in biosensing applications of 2D non-layered materials, with a specific focus on electrochemical, optical, and FET techniques.

2.1 Electrochemical biosensors

Electrochemical biosensors have established as the most common and accessible biosensors on the current market. Their popularity is due to their portability, user-friendly and cost-effectiveness, which distinguish them from other types of biosensors (Fredj and Sawan, 2023; Fredj et al., 2023). These characteristics make electrochemical biosensors ideal for the development of point-of-care (POC) devices that can be used for rapid on-site testing (Fredj and Sawan, 2023). Over the past few years, there has been an increasing focus on boosting the efficiency of electrochemical biosensors through the integration of non-layered 2D materials. These novel substances present clear benefits compared to conventional layered materials, positioning them as highly favorable contenders for various biosensing applications. One key advantage of non-layered 2D materials is their ability to undergo significant structure distortion, resulting in chemically active surfaces with abundant surface dangling bonds. This distinguishing characteristic, which is not commonly observed in layered materials, contributes to exceptionally reactive surfaces and improved catalytic efficiency. Furthermore, the exposed surface atoms in non-layered materials possess low coordination numbers, thereby promoting the chemisorption of reactants and facilitating rapid interfacial charge transfer. Another key advantage lies in the ability to finely adjust the structural and electronic properties of non-layered materials through structure and surface engineering, enabling further customization of catalytic performance. Additionally, inherent lattice defects, such as vacancies, play a significant role in influencing the surface electronic structure and charge transport properties of these materials. The ultrathin 2D structure of these materials also presents an ideal platform for studying catalytic mechanisms at the atomic level and modeling electronic-state modulation, thus facilitating the establishment of dependable structure-property relationships. Overall, these unique characteristics make non-layered 2D materials well-suited for understanding catalytic processes and optimizing electrochemical biosensor performance (Wang et al., 2020a). Zinc oxide (ZnO) is a widely used 2D nanolayered metal oxide known for its efficiency in fabricating electrochemical sensors and biosensors to detect various analytes (Li et al., 2015a). As a metal-oxide semiconductor, ZnO serves as an excellent intracellular sensor or transducer platform. Its versatility lies in its ability to be tailored to match the size of the target biological species, allowing for precise detection. Additional key benefit is its exceptional electrical signal generation capabilities. By modifying its dimensions, ZnO can be transformed into an ideal transducer for converting biological signals into electrical signals. This flexibility in size customization allows for enhanced sensitivity and specificity in biosensing applications. In addition to its excellent transducer properties, ZnO is an attractive material for biosensor design due to its favorable characteristics including non-toxicity, cost-effectiveness, and ease of synthesis (Shetti et al., 2019). At the nanoscale, surface properties are critical, and ZnO's diverse morphologies provide a controllable platform for customizing its properties for specific sensing applications, allowing precise manipulation of ZnO's interactions with antibodies and enzymes, thus facilitating their immobilization on the surface (Krishna et al., 2023).

For this propose, Gerbreders et al. (Gerbreders et al., 2019). developed a DNA electrochemical sensor based on ZnO nanostructures modified silicon substrates. The developed biosensor demonstrated effective discrimination between complementary, non–complementary and partially complementary *T. britovi* DNA sequences. Also, it enabled accurate identification of PCR products derived from various Trichinella species. Furthermore, ZnO provides significant benefits for enzyme immobilization stability due to electrostatic interactions between ZnO and the enzyme's acidic functional groups. This interaction is a result of the disparity in isoelectric points, where ZnO possesses a high isoelectric point of 9.5. In a study conducted by Jindal and coworkers (Jindal et al., 2017), they proposed a novel enzymatic approach for real-time monitoring of uric acid. To create the bioelectrode, they carefully controlled the thickness of ZnO thin films by adjusting the number of applied laser pulses to around 90 nm. Subsequently, they employed a deposition process to place CuO onto the ZnO thin film's surface, utilizing a shadow mask with closely spaced and evenly distributed pores measuring 600 μm in diameter. The deposition resulted in nano–thin micro–clusters of CuO with a thickness of 10 nm, forming a CuO/ZnO arrayed heterostructure (Fig. 1a). The response of the biosensor was evaluated using the cyclic voltammetry (CV) technique, as illustrated in Fig. 1b. The results demonstrated a linear sensing response over a wide range of uric acid concentrations, spanning from 0.05 mM to 1.0 mM. Recently, Urbiola's group (Chavez–Urbiola et al., 2022) carried out the synthesis of thin films comprising ZnO coated with Au on a flexible substrate. These films were specifically engineered to immobilize the glucose oxidase (GOx) enzyme, in conjunction with ferrocene methanol (Fc) and multiwalled carbon nanotubes, to enable glucose detection (Fig. 1c). Following the establishment of the catalytic effect between the enzyme and the ZnO thin layer, experiments were conducted to measure the amperometric response at various glucose concentrations (Fig. 1d). The presence of the ZnO layer resulted in heightened sensitivity, with a measurement of 30.00 $\mu A/mM/cm^2$, along with significantly low limits of detection and quantification, recorded at 0.25 mM and 0.83 mM, respectively. In a subsequent study by Li et al. (2019), a novel sensing approach was developed to detect T4 Polynucleotide Kinase (T4 PNK). This innovative method involved a cosensitization strategy using CdSe as a 2D nonlayered nanostructure with an exceptionally narrow bandgap. To create an efficient sensitizer, CdSe was combined with a ternary $TiO_2/g-C_3N4$, ensuring a well-matched band-edge level. To evaluate the

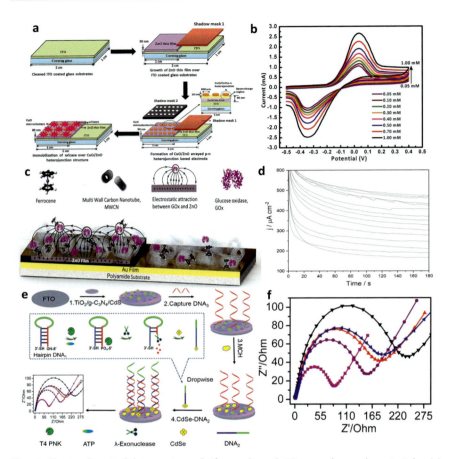

Fig. 1 Electrochemical biosensing platforms based 2D non layered materials: (a) fabrication process of the heterojunction structure based uricase/CuO/ZnO, (b) CV curves depicting the biosensor's response to a range of uric acid concentrations (0.05–1.0 mM) under a scan rate of 100 mV/s (Jindal et al., 2017), (c) Bioelectrode configuration incorporating ZnO and Au thin films, MWCNTs, and FcMeOH for the immobilization of glucose oxidase, (d) Amperometric response of the Au/ZnO/GOx-Fc/MWCNT structure at various glucose concentrations (Chavez-Urbiola et al., 2022), (e) Illustration of the fabrication process for the sensing platform tailored for AA detection, (f) Nyquist plots showcasing the impedance measurements obtained during various modification steps (Li et al., 2019).

T4 PNK activity, the phosphorylation process was employed, catalyzing a 5′-hydroxyl group terminal hairpin DNA1 to produce a phosphate group (Fig. 1e). Subsequently, electrochemical impedance spectroscopy (EIS) was used to examine the interfacial properties of the fabricated electrode as presented in Fig. 1f. The results demonstrated that the proposed biosensing platform exhibited exceptional analytical performance. Notably, it featured a broad linear range from 0.0001 to 0.02 U/mL and an impressively low detection limit of 6.9×10^{-5} U/mL.

Furthermore, Lincy et al. (2023) demonstrated that 2D non-layered metal oxide Fe_2O_3 holds promise for enhancing the bioavailability and effectiveness of antibiotics. Whereas, it exhibits excellent stability, tunable properties such as size, shape, and porosity, making it suitable for biosensor preparation. Its ability to interact at the cellular and molecular level, coupled with its magnetic properties, has led to frequent use in the development of multifunctional vesicles for drug loading and imaging in cancer, bacterial, and viral research. Further, Chen et al. conducted a study aimed at optimizing 2D nonlayered Fe_2O_3 engineered Pt-based nanozymes through atomic layer deposition (ALD) to achieve precise control nanometer scale (Chen et al., 2020). The developed sensing approach exhibited enhanced peroxidase activity and a remarkable linear detection range of 20–80 µM with an impressive limit of detection (LOD) of 8.7 µM, demonstrating exceptional sensitivity and precision in glucose sensing. In Liu et al.'s work, TiO_2 nanosheets were utilized as a support matrix to immobilize Hemoglobin (Hb), owing to their ability to create a protective microenvironment that preserves enzymatic stability and activity (Liu et al., 2020). The experimental results highlight the excellent protein stability and bioactivity of the developed structure, making it an ideal choice for Hb immobilization. Additionally, the unique structure of the 2D non-layered material facilitates direct electron transfer of Hb, leading to a high-performance biosensor for detecting H_2O_2. The biosensor showcases an impressive linear range of 7.5–250 µM, an ultra-low detection limit of 3 µM, rapid response, and remarkable stability. Tungsten trioxide (WO_3) is a one of metal oxide semiconductors which exhibits unique electroanalytical properties. In its 2D structure, WO_3 forms nanosheets that possess a high surface area and preferentially exposed facets. These features make it highly attractive for a wide range of applications, including catalysis, sensing, and electrochromic devices (Novak et al., 2021). In line with the exploration of WO_3's potential in the field of biosensing, Sandil et al. (2017) successfully synthesized 2D nanostructured WO_3, capitalizing on its favorable properties for developing a cardiac biomarkers biosensor. To enhance WO_3's functionality, the researchers biofunctionalized the nanosheets with APTES (aminopropyltriethoxysilane) using the electrophoretic deposition technique. Subsequently, they covalently immobilized cTnI (cardiac Troponin I) antibody onto the surface of the APTES-functionalized WO_3 thin films. The

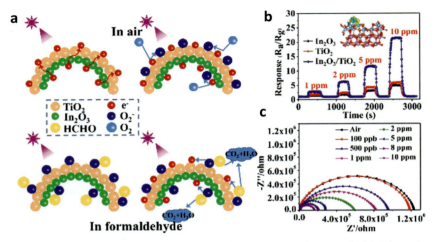

Fig. 2 Enhanced Formaldehyde Sensor based 2D nonlayered materials. (a) Fabrication Process of In$_2$O$_3$@TiO$_2$ structure, (b) Comparative analysis of response curves for In$_2$O$_3$, TiO$_2$, and In$_2$O$_3$/TiO$_2$ nanocomposite sensors, (c) Nyquist plots depicting the response of In$_2$O$_3$/TiO$_2$ nanocomposites to different concentrations of HCHO (Zhang et al., 2023). *Reproduced with permission from Zhang, S., Sun, S., Huang, B., Wang, N., Li, X., 2023. UV-enhanced formaldehyde sensor using hollow In$_2$O$_3$@TiO$_2$ double-layer nanospheres at room temperature. ACS Appl. Mater. Interfaces 15(3), 4329–4342.*

immunosensor was then employed for detecting the presence of cTnI using EIS method. Notably, the immunosensor exhibited a high sensitivity of 26.56 Ω/ng/mL/cm^2 within a linear detection range of 1–250 ng/mL. In another example of combination 2D non-layered materials with biosensors, a double-layer of In$_2$O$_3$ @TiO$_2$ was prepared through a simple water bath method using carbon nanospheres as sacrificial templates. The sensing performance of the chemiresistive-type sensor based on In$_2$O$_3$/TiO$_2$ nanocomposites was investigated for formaldehyde (HCHO) gas detection at room temperature under UV light activation (Fig. 2a). Comparing to sensors based solely on In$_2$O$_3$ or TiO$_2$, the In$_2$O$_3$/TiO$_2$ nanocomposite sensor exhibited significantly improved sensing capabilities for formaldehyde (Fig. 2b). The response of the In$_2$O$_3$/TiO$_2$ nanocomposite-based sensor was analyzed using electrochemical impedance spectroscopy, and the Nyquist plots in Fig. 2c depicted the response to various formaldehyde concentrations ranging from 10 to 100 ppm. The developed biosensor demonstrated excellent sensitivity, with a detectable formaldehyde concentration as low as 0.06 ppm. To further explore the applications of non-layered 2D materials, Table 1 provides additional examples of their utility, showing a diverse range of electrochemical biosensing applications.

Table1 Electrochemical biosensor-based 2D non-layered materials.

Biosensor structure	Target	Method	Linear range	LOD	Refs.
CeO_2/ITO	Hydrogen peroxide	DPV	2–500 μM	0.6 μM	Yagati et al. (2013)
CuS/Cu_2O/CuO	Glucose	DPV	2 μM–4 mM	0.89 μM	Wei et al. (2020)
Gox/TiO_2/C_3N_4-/ITO	Glucose	CV	50 μM–16 mM	10 μM	Liu et al. (2017)
Aptamer/AuNPs/ZnO/ITO	SK-BR-3 breast cancer cells	EIS	1×10^2–1×10^6 cells/mL	58 cells/mL	Liu et al. (2014)
ZnO/MoS_2/GCE	DNA	DPV	1 fM–500 μM	0.66 fM	Yang et al. (2017)
Uricase/CuO/ZnO/ITO	Uric acid	CV	50 μM–1 mM	50 μM	Jindal et al. (2017)
Uricase/ZnS/Au electrode	Uric acid	DPV	5 μM–2 mM	5 μM	Zhang et al. (2006)
NADH/Fe_2O_3/ITO	Formaldehyde	CV	0.01–0.3 mg/L	0.01 mg/L	Kundu et al. (2019)

2.2 Optical biosensors

Optical biosensors have proved to be powerful tools with significant implications for environmental monitoring, food safety and healthcare, leading to their rapid and impressive progression. By incorporating 2D nonlayered materials into the design of optical biosensors, several advantages are achieved. Firstly, the large surface-to-volume ratio of these nanomaterials allows for enhanced interaction with target analytes, leading to improved sensitivity and detection limits (Liu et al., 2022). Additionally, their exceptional optical properties, such as strong light–matter interactions, tunable bandgaps, and high photoluminescence, enable precise and reliable detection of biomolecules or chemical species (Lei and Guo, 2022). Metal oxide nanosheets, such as 2D nonlayered TiO_2, exhibit remarkable enhancement in the optical signal. For instance, the transformation from bulk TiO_2 to 2D nanosheets results in a "blue shift" in the UV–Vis absorption spectrum, increasing their bandgap from approximately 3 eV to 3.65 eV. This structural change leads to thickness limitation but offers advantages such as a higher surface area for easy adsorption of dye molecules and improved electron transport (Kumbhakar et al., 2021). Recently, Lyer et al. designed a fluorescence biosensor by coating glass substrates with a thin film of zinc oxide (Iyer et al., 2014). Subsequently, they immobilized specific DNA sequences on the ZnO thin films to enable detection of dengue virus. ZnO's notable characteristics, such as its chemical inertness, non-toxicity and biocompatibility, as well as its enhanced charge-transfer capability, make it an excellent choice for creating an effective electroactive surface for the advancement of biosensors. Significantly, the authors showcased the capability of the ZnO array platform to simultaneously detect four serotypes of the dengue virus, achieving both qualitative and quantitative detection. Besides, Tereshchenko et al. introduced an enhanced optical biosensor for the specific detection of Grapevine virus A-type (GVA) using a thin film of ZnO deposited through atomic layer deposition (Tereshchenko et al., 2017). Fig. 3a presents the structure of the biosensor and includes an SEM image of the ZnO layer. The ZnO-based films exhibit excellent surface-structural properties, enabling direct immobilization of GVA-specific antibodies. Through this immobilization process, a highly responsive biosensitive layer is formed, which effectively detects and interacts with GVA antigens. Successful immobilization was confirmed by observing changes in the intensity of the primary near band emission (NBE) peak of ZnO, along with the emergence of a prominent

2D Materials in Biosensors, Memristors, and Energy Storage 263

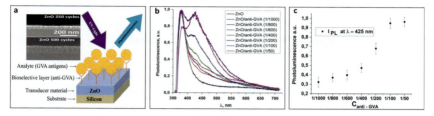

Fig. 3 Photoluminescence immunosensor based 2D ZnO: (a) SEM Image of and schematic design of the biosensor platform, (b) PL Spectra of bare and functionalized ZnO with Anti-GVA (c) Correlation of PL intensity with Anti-GVA concentrations at 425 nm. *Produced with permission from Tereshchenko, A., et al., 2017. ZnO films formed by atomic layer deposition as an optical biosensor platform for the detection of Grapevine virus A-type proteins. Biosens. Bioelectron. 92, 763–769.*

photoluminescence band in the visible range at approximately 425 nm, indicating the presence of immobilized proteins (Fig. 3b). Detection of GVA antigens was achieved by analyzing variations in the behavior and characteristics of this luminescence band. The newly developed biosensor demonstrated exceptional sensitivity to GVA antigens, providing a highly effective label-free detection platform with a detection range spanning from 1 pg/mL to 10 ng/mL, as shown in the calibration curve depicted in Fig. 3c.

Along with it, Kaur et al. utilized the RF sputtering technique to deposit a highly c-axis oriented ZnO thin film on gold-coated glass prisms to create a biosensor for Neisseria meningitidis DNA detection (Kaur et al., 2016). The ZnO thin film's surface was immobilized with single-stranded probe DNA using physical adsorption. The biosensor's linear response within the 10–180 ng/μL concentration range for the target DNA was demonstrated by recording SPR reflectance curves with a laboratory assembled SPR setup and varying the incident angle of a He–Ne laser beam. It exhibited a high sensitivity of around 0.03°/(ng/μL) and a low detection limit of 5 ng/μL. The biosensor showcased notable specificity, extended lifetime, and considerable potential for meningitidis diagnosis. Additionally, ZnO nanostructures possess exceptional photoelectric properties, surpassing conventional bulk or film materials. They exhibit strong optical absorption, light emission, and photoconductive gain, making them promising for revolutionizing the performance of optoelectronic devices. Moreover, these nanostructures have the potential to boost the capabilities of sensors field-effect transistors, solar cells, photodetectors, nanogenerators, and light-emitting diodes, among others (Ding et al., 2018).

2.3 Nano-FET based biosensor

Field-Effect Transistors (FETs) have garnered considerable attention as a promising platform for biosensors, primarily due to their remarkable sensitivity and controllability. Operating on the principle of the field effect, FET-based nanomaterials can modulate electrical conductivity in response to the presence of analytes. These biosensors offer label-free detection and exceptional sensitivity, capitalizing on the integration of advanced 2D nanomaterials at the nanoscale level. What sets them apart is their adaptability and ease of integration into electronic circuits and microfluidic systems, making them highly customizable. In this context, Yang et al. explored this potential impact of FET systems further by conducting a study focused on In_2O_3 as a 2D non layered material (Yang et al., 2021). Specifically, their investigation centered on In_2O_3 electrolyte-gated thin film transistors featuring integrated on-chip gate electrodes. In this innovative design, the In_2O_3 channel acted as a sensitive membrane, while the on-chip gate electrode replaced the conventional reference electrode. The biosensor, with the In_2O_3 channel coated in streptavidin/neutravidin receptors, showcased excellent performance, thus demonstrating the exciting capabilities of this technology.

The developed FET biosensor established precise pH detection capabilities, exhibiting a high sensitivity of approximately $64\,mV/pH$ with only a minor deviation of $10\,mV$. Operating at a low voltage, the streptavidin-modified device showcased its ability to detect minute quantities of biomolecules, reaching an impressive detection limit of $50\,pg/mL$. Additionally, the neutravidin-modified device surpassed expectations with exceptional sensitivity, enabling the detection of even lower concentrations of biotin, achieving an outstanding detection limit of $50\,fg/mL$. Similarly, Liu et al. have reported a powerful approach that enables the fabrication of highly uniform arrays based on In_2O_3 nanoribbons, all without the need for lithography (Liu et al., 2016). Through the use of shadow masks instead of traditional lithography methods, they have achieved low-cost, time-saving, and high-throughput fabrication of In_2O_3 nanoribbon biosensors, eliminating the risk of photoresist contamination. To enhance the detection signal, they have combined these FET structures with enzyme-linked immunosorbent assay. Fig. 4a,b illustrates the fabrication and functionalization process. The resulting In_2O_3 nanoribbon biosensor arrays have been specifically optimized for the early, rapid, and quantitative detection of cardiac biomarkers in diagnosing acute myocardial infarction. In this study, the focus lies on three critical biomarkers associated with heart attacks and heart failure: creatine kinase MB (CK-MB), B-type natriuretic peptide (BNP) and

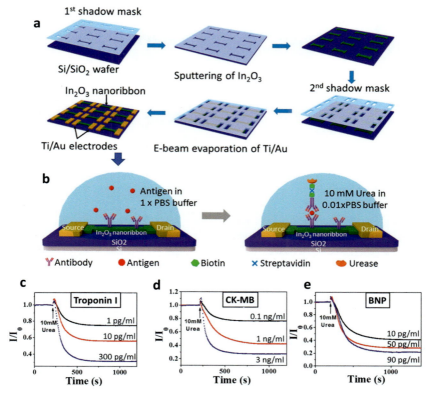

Fig. 4 In$_2$O$_3$ FET-based biosensors through shadow mask fabrication: (a) Lithography-free process for biosensor assembly, (b) Schematic representation of antibody and antigen capture in buffer solution, (c) Real-time sensing results for cTnI antigens at concentrations of 1, 10, and 300 pg/mL, (d) Results for CK-MB Proteins at concentrations of 0.1, 1, and 3 ng/mL, (e) Responses of BNP Proteins at concentrations of 10, 50, and 90 pg/mL. *Reproduced with permission from Liu, Q., et al., 2016. Highly sensitive and quick detection of acute myocardial infarction biomarkers using In$_2$O$_3$ nanoribbon biosensors fabricated using shadow masks. ACS Nano 10(11), 10117–10125.*

cardiac troponin I (cTnI). The real-time responses of three targets are thoughtfully illustrated in Fig. 4c–e. This innovative approach enabled a multi-detection cardiac biomarker without the requirement for labeling, achieving concentrations as low as 0.1 ng/mL for CK-MB, 10 pg/mL for BNP, and 1 pg/mL for cTnI. With their impressive sensitivity, rapid response time, and reusability, the In$_2$O$_3$ nanoribbon biosensors hold substantial potential for clinical tests, enabling early and swift diagnosis of acute myocardial infarction.

Later, the same research group (Liu et al., 2018) have presented a highly sensitive and wearable In$_2$O$_3$-based FET biosensor. The developed platform

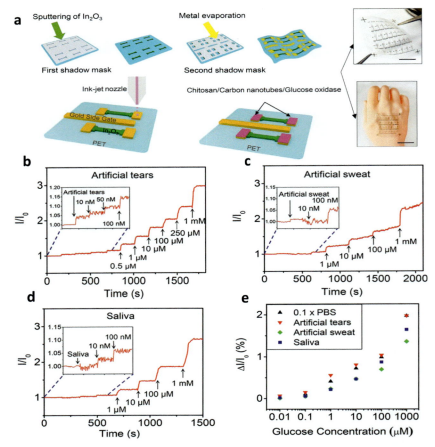

Fig. 5 Flexible and wearable In$_2$O$_3$ FET biosensor: (a) Two-step shadow mask fabrication procedure for In$_2$O$_3$ FETs on a polyethylene terephthalate substrate, (b) Amperometric response of In$_2$O$_3$ FET in artificial tears, (c) Artificial sweat, (d) Saliva, (e) Calibration curve illustrates the correlation between current and glucose concentration, spanning a range from 0.01 to 1000 μM. *Reproduced with permission from Liu, Q., et al., 2018. Highly sensitive and wearable In$_2$O$_3$ nanoribbon transistor biosensors with integrated on-chip gate for glucose monitoring in body fluids. ACS Nano 12(2), 1170–1178.*

was fully integrated on-chip with a gold side gate, enabling detection of glucose in various bodily fluids, including tears sweat and saliva. The fabrication process involves the utilization of two shadow masks. Firstly, a shadow mask was employed to precisely define the sputter-coating of In$_2$O$_3$ nanoribbons, followed by the use of a second shadow mask for the metal deposition of drain, gate and source (Fig. 5a). To enhance the functionality of the biosensor, inkjet printing was utilized to modify the source and drain electrodes single-walled carbon nanotubes (SWCNTs), chitosan polymer and glucose oxidase enzyme (GOx), The integration of these components results in improved performance. Additionally, the gold side-gated In_2O_3 FETs exhibit excellent electrical characteristics, even when implemented on highly flexible substrates. Fig. 5a,b and c display current responses associated with glucose measurement in artificial human tears, artificial human sweat and saliva, respectively. The optimized glucose sensors demonstrate an impressive detection range ranging over five orders of magnitude, offering a versatile detection capability (Fig. 5e). Moreover, these sensors exhibit high sensitivity, with detection limits as low as 10 nM. This breakthrough paves the way for advanced biosensing technologies based 2d non layered material with extensive applications in healthcare and biomedical research (Xie et al., 2018).

3. Memristors based 2D non layered material

Memristors, the "memory resistors," are cutting-edge electronic devices that can store and recall charge, akin to how our brains retain memories. They hold great potential in brain-inspired computing due to their ability to replicate the brain's information processing and storage. By integrating memristors into computing systems, we can emulate the brain's parallelism and distributed processing, enabling efficient and fast data operations. Recently, scientists have explored the use of 2D layered materials, such as MoS_2, and graphene and in memristor devices (Duan et al., 2022; Wang et al., 2018a). These materials enhance performance, enable miniaturization, and offer tunable properties, making them compatible with existing technologies. Hence, the combination of 2D materials and memristors opens up possibilities for novel device architectures and optimized designs, leading to advancements in memory, neuromorphic computing, and analog/digital circuits (Sheykhifar and Mohseni, 2022; Yuan and Lou, 2015). In particular, memristive devices are currently being considered for synaptic devices, as their conductance change closely resembles synaptic weight updates observed in the human brain. On the other hand, 2D-material-based synaptic devices exhibit excellent energy consumption properties compared to bulk materials, owing to their ideal physical attributes. This makes them promising candidates for synaptic devices, highlighting their potential in the development of energy-efficient brain-inspired computing systems (Huh et al., 2020).

However, to date, only a few memristors based on 2D nonlayered materials have been reported. Consequently, there remains a need for thorough exploration of the inherent bionic physics in these devices. In this context Wang et al. successfully fabricated a memristor utilizing 2D Tin selenide (SnSe) films through the pulsed laser deposition technique (Wang et al., 2020b). SnSe, as one of the prominent 2D semiconductors, has become a subject of significant interest in recent research. This is mainly due to its remarkable thermoelectric performance, which is particularly evident in single crystal samples (Wang et al., 2018b; Hou et al., 2018). The material features a layered orthorhombic crystal structure at room temperature, closely resembling the distorted NaCl structure. This specific arrangement of the structure leads to strong Sn-Se bonds within the bc-plane, while the bonds along the a-axis direction are relatively weaker (Zhou et al., 2018; Tian et al., 2018). Notably SnSe exhibits a lattice instability with "ferroelectric-like" characteristics, as recent theoretical calculations have further advanced the understanding of SnSe's behavior, predicting its potential ferroelectricity (Shen et al., 2019; Li et al., 2015b). The fabricated SnSe memristor comprises distinct components: an Au top electrode (representing the Presynaptic region), an intermediate SnSe layer (emulating the Synaptic cleft), and an NSTO bottom electrode (representing the Postsynaptic region), as illustrated in Fig. 6a. The chosen architecture closely resembles biological synapses, which provides a significant advantage in simulating bio-synaptic functions. The atomic structure diagram of the SnSe film, as shown in Fig. 5b, provides a visual representation of how tin (Sn) and selenium (Se) atoms are arranged within the 2D SnSe film. The unique structure of SnSe, consisting of interconnected Sn and Se atoms forming a honeycomb-like pattern, influences the material's electronic and optoelectronic properties. Moreover, the memristor device based on SnSe demonstrates a typical bipolar resistive switching hysteresis loop in the semi-logarithmic current–voltage (I–V) curve, as shown in Fig. 6d. The transition between its high and low resistance states (set and reset) occurs at +2.0 V and −3.5 V, respectively. In Fig. 5e, the current (V_{read}) characteristics of the ON and OFF states for the memristor are depicted in the low-bias region spanning from −1.0 to 1.0 V. Notably, the I–V curve displays a nonlinear and continuous behavior, setting it apart from the typical response seen in other resistive switching devices (Kim et al., 2012). Furthermore, the energy affinity and band gap parameters of NSTO, SnSe, and Au are investigated (Fig. 6f). Without band bending, the energy barrier differences between SnSe/NSTO and SnSe/Au are 0.1 eV and 1.1 eV, respectively. When a positive electric field is applied to the Au top electrode, it induces downward

2D Materials in Biosensors, Memristors, and Energy Storage 269

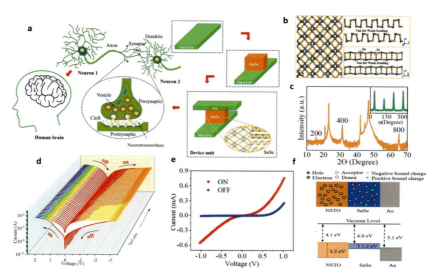

Fig. 6 Au/SnSe/NSTO artificial electronic synapse mimics the functionalities of biological synapses: (a) Preparation process of the artificial synaptic device based on Au/SnSe/NSTO memristors, (b) Top view, side view, and front view of the 2D-SnSe film, (c) XRD pattern measured at various 2θ angles, showcases the characteristics of the SnSe films, (d) Semi-logarithmic current–voltage (*I*–*V*) curve, (e) Measurement of the current (V_{read}) in the low-bias region of the SnSe device in ON or OFF mode, (f) Charge distribution and energy-band diagrams of the Au/SnSe/NSTO system, depicting the spatial arrangement of charges and the energy levels within the device structure. *Reproduced with permission from Wang, H., et al., 2020b. A 2D-SnSe film with ferroelectricity and its bio-realistic synapse application. Nanoscale 12(42), 21913–21922.*

ferroelectric polarization in SnSe. This results in polarized positive charge accumulation at the SnSe/NSTO interface, attracting majority carriers (electrons) from n-type NSTO while not affecting majority carriers (holes) from p-type SnSe. Consequently, the interface's depletion width (W) decreases, and the total potential barrier (Φ) is reduced compared to the initial state.

The SnSe device exhibits a highly reproducible hysteresis loop at room temperature. This suggests that the ferroelectric polarization of the SnSe functional layer plays a significant role in this behavior. This leads to stable and repeatable performance, as observed in the earlier investigation. Such achievements were made possible by the gradual switching polarization of the ferroelectric SnSe layer. The experimental findings demonstrate bio-synaptic functionalities, including spike-timing-dependent plasticity, short-term plasticity, and long-term plasticity, which can be effectively simulated using these two-terminal devices.

In their study, Yin et al. successfully demonstrated the controlled synthesis of 2D CuSe nonlayered nanosheets, which were found to be promising as active layers for non-volatile memristors (Yin et al., 2022). The researchers conducted electrical measurements to uncover the remarkable memristive properties of these 2D copper chalcogenide nanosheets. Notably, these nanosheets exhibited several advantageous characteristics, including a relatively low switching voltage of approximately 0.4 V, rapid switching speed, high uniformity in switching, and a wide operating temperature range from 80 to 420 K. Moreover, the fabricated memristors displayed stable retention and switching behaviors, showcasing their considerable potential for applications in information storage. Importantly, this work represents the first exploration into the intrinsic nonvolatile memristive behaviors of these nanosheets. The observed attributes of the memristors, such as the low switching voltage, fast switching speed, uniformity in switching, extended retention times, and wide working temperature range, make them highly desirable for future investigations and advancements in next-generation electronics.

4. 2D non-layered materials in energy storage

2D materials possess unique characteristics that render them highly attractive for advancing high-performance energy storage devices. A key advantage lies in their slit-shaped ion diffusion channels, which facilitate rapid movement of ions, particularly lithium ions (Pomerantseva and Gogotsi, 2017). This enables swift charging and discharging rates, essential for efficient energy storage. However, to attain optimal energy storage performance, additional factors must be considered, such as electronic conductivity, the availability of intercalation sites, and long-term stability during extended cycling. Ensuring efficient charge transfer through the optimization of electronic conductivity enhances the overall energy storage capabilities of 2D materials. Additionally, maximizing intercalation site density allows for a higher number of ions to be accommodated, effectively increasing the energy storage capacity (Ruan et al., 2021). Moreover, guaranteeing the long-term stability of 2D materials is crucial for sustaining device performance and durability over multiple charge-discharge cycles. To unlock the full potential of 2D materials in energy storage applications, researchers are currently focused on comprehensively addressing these factors. A comprehensive investigation was conducted by Xia et al. to evaluate the Li-ion storage capacity in 2D atomic sheets of non-layered molybdenum dioxide (MoO_2) using both theoretical and experimental approaches (Xia et al., 2018). To fabricate high-quality non-layered MoO_2 sheets, the researchers devised a monomer-assisted reduction process. The ultrathin 2D-MoO_2 electrodes exhibited outstanding reversibility when employed as anodes in lithium-ion batteries. After undergoing 100 cycles, the device's capacity reached an impressive 1516 mAh/g, and remarkably, even after 1050 cycles, it retained a substantial capacity of 489 mAh/g. Notably, these ultrathin 2D sheets demonstrated a different mechanism compared to their bulk form, suggesting a deviation from the typical intercalation-cum-conversion mechanism observed in lithium-ion batteries. Furthermore, the researchers fabricated a micro-supercapacitor based on 2D-MoO_2, which displayed a high areal capacitance of 63.1 mF/cm^2 at 0.1 mA/cm^2, excellent rate performance (81%), and superior cycle stability (86% retention after 10,000 cycles). This pioneering work establishes a novel pathway for creating 2D nanostructures from non-layered compounds, leading to significantly enhanced energy storage capabilities. Another notable example showcasing the potential of 2D non-layered nanostructures in energy storage is presented in a study conducted by Cai et al. (2016). A self-powered glucose biosensor was skillfully fabricated, seamlessly integrating glucose sensing capabilities with both solar energy conversion and electrochemical energy storage. To achieve this, Cai's group used SnO_2 nanosheet arrays to facilitate the generation of photogenerated electron-hole pairs, while rhombus-shaped $NiCo_2O_4$ nanorod arrays were employed for efficient solar energy storage. The biosensor exhibited a stable open circuit voltage of approximately 0.58 V upon full charging, making it well-suited for glucose oxidation. An outstanding feature of this biosensor is its ability to operate in two distinct modes without the need for any external bias voltage. Both modes demonstrated a wide linear range for glucose detection and exhibited exceptional selectivity. Notably, under sunlight, the biosensor exhibited a sensitive decrease in photocurrent upon the addition of varying concentrations of glucose. Similarly, in dark conditions, the charged biosensor responded accordingly to changes in glucose levels, as indicated by alterations in its open circuit voltage. The utilization of 2D non-layered materials in energy storage continues to be an active area of research, with ongoing efforts to understand their fundamental properties, explore new synthesis methods, and develop scalable manufacturing processes. By addressing the challenges and harnessing the unique properties of 2D materials, scientists aim to pave the way for high-performance energy storage devices to meet the ever-growing demands of diverse applications (Wang et al., 2021).

5. Challenges and future perspectives

Two-dimensional non-layered materials offer a wide range of exciting possibilities for advanced applications, such as biosensors, memristors, and energy storage. Their atomic-thin structure and exceptional surface-to-volume ratio contribute to their exceptional sensing capabilities, making them ideal candidates for detecting and analyzing a wide range of analytes and stimuli. The unique electronic, optical, and chemical properties exhibited by these materials enable unprecedented levels of precision and accuracy in sensing applications. Moreover, their compatibility with flexible and wearable devices opens up new opportunities for the development of portable and point-of-care sensing solutions. These materials have the potential to revolutionize the field of diagnostics, enabling rapid and reliable detection of various biomolecules and environmental pollutants. In the realm of energy storage, 2D non-layered materials offer intriguing possibilities. Their large specific surface area and efficient charge transport properties make them excellent candidates for high-performance supercapacitors and batteries. By exploiting the unique interactions between these materials and ions, researchers can push the boundaries of energy storage technology and enhance the efficiency and sustainability of energy storage devices However, to realize their full potential, several significant challenges need to be addressed. One critical challenge is the synthesis and scalability of these materials. Although progress has been made in producing materials like ZnO and CuS, achieving large-scale and reproducible synthesis remains a difficult task. To enable widespread adoption, it is essential to explore novel, cost-effective synthesis techniques, including bottom-up approaches. Another crucial aspect is surface functionalization, which plays a vital role in tailoring material properties. Achieving controlled and uniform functionalization while preserving intrinsic properties is complex. The customization of functional groups and dopants to optimize performance without compromising other desirable attributes is essential. Seamless integration with existing technologies is also crucial for practical implementation. Developing interfaces and adhesion methods that facilitate efficient charge transfer and minimize resistance is vital to ensure compatibility with standard fabrication processes and substrates. Furthermore, long-term stability and environmental impact must be taken into consideration. Understanding degradation mechanisms and evaluating the life-cycle environmental impact are imperative for designing durable and sustainable devices. Improving biosensing performance is another key area of focus. While 2D non layered materials have shown promise in enhancing sensitivity and selectivity, further optimization is required. Novel sensor designs, transduction methods, and hybrid material integration can unlock unprecedented sensing capabilities. The perspectives for addressing these challenges are multifaceted and require a pragmatic approach. Collaborative efforts among material scientists, chemists, engineers, and biologists are not only essential but also likely to yield practical solutions. Establishing effective partnerships and fostering open communication between experts in various fields will be crucial to drive transformative breakthroughs. Embracing emerging synthesis techniques, such as molecular self-assembly and AI-driven computational approaches, holds great promise. However, it is important to acknowledge that these techniques may still require further refinement and validation before widespread adoption in large-scale production. Researchers must continue to invest time and resources to ensure the reliability and scalability of these methods. Environmental considerations must be a top priority in the pursuit of sustainable materials. While green synthesis methods and eco-friendly disposal approaches show potential, the implementation of these practices on a global scale may encounter regulatory, economic, and logistical challenges.

References

Aniu Lincy, S., Allwin Richard, Y., Vinitha, T., Balamurugan, K., Dharuman, V., 2023. Streptavidin Fe_2O_3-gold nanoparticles functionalized theranostic liposome for antibiotic resistant bacteria and biotin sensing. Biosens. Bioelectron. 219, 114849. https://doi.org/10.1016/j.bios.2022.114849.

Cai, B., Mao, W., Ye, Z., Huang, J., 2016. Facile fabrication of all-solid-state SnO_2/$NiCo_2O_4$ biosensor for self-powered glucose detection. Appl. Phys. A 122 (9), 806. https://doi.org/10.1007/s00339-016-0263-9.

Chakraborty, G., Park, I.-H., Medishetty, R., Vittal, J.J., 2021. Two-dimensional metal-organic framework materials: synthesis, structures, properties and applications. Chem. Rev. 121 (7), 3751–3891. https://doi.org/10.1021/acs.chemrev.0c01049.

Chavez-Urbiola, I.R., et al., 2022. Glucose biosensor based on a flexible Au/ZnO film to enhance the glucose oxidase catalytic response. J. Electroanal. Chem. 926, 116941. https://doi.org/10.1016/j.jelechem.2022.116941.

Chen, Y., et al., 2020. Precise engineering of ultra-thin Fe_2O_3 decorated Pt-based nanozymes via atomic layer deposition to switch off undesired activity for enhanced sensing performance. Sens. Actuators B: Chem. 305, 127436. https://doi.org/10.1016/j.snb.2019.127436.

Ding, M., et al., 2018. One-dimensional zinc oxide nanomaterials for application in high-performance advanced optoelectronic devices. Crystals 8 (5), 223. https://doi.org/10.3390/cryst8050223.

Duan, H., et al., 2022. Low-power memristor based on two-dimensional materials. J. Phys. Chem. Lett. 13 (31), 7130–7138. https://doi.org/10.1021/acs.jpclett.2c01962.

Fredj, Z., Sawan, M., 2023. Advanced nanomaterials-based electrochemical biosensors for catecholamines detection: challenges and trends. Biosensors (Basel) 13 (2), 211. https://doi.org/10.3390/bios13020211.

Fredj, Z., Singh, B., Bahri, M., Qin, P., Sawan, M., 2023. Enzymatic electrochemical biosensors for neurotransmitters detection: recent achievements and trends. Chemosensors 11 (7), 388. https://doi.org/10.3390/chemosensors11070388.

Gerbreders, V., et al., 2019. ZnO nanostructure-based electrochemical biosensor for Trichinella DNA detection. Sens. Bio-Sensing Res. 23, 100276. https://doi.org/10.1016/j.sbsr.2019.100276.

Gutiérrez, H.R., 2020. Two-dimensional layered materials offering expanded applications in flatland. ACS Appl. Nano Mater. 3 (7), 6134–6139. https://doi.org/10.1021/acsanm.0c01763.

Hou, S., et al., 2018. The transverse thermoelectric effect in a-axis inclined oriented SnSe thin films. J. Mater. Chem. C 6 (47), 12858–12863. https://doi.org/10.1039/C8TC04633F.

Huh, W., Lee, D., Lee, C.-H., 2020. Memristors based on 2D materials as an artificial synapse for neuromorphic electronics. Adv. Mater. 32 (51), 2002092. https://doi.org/10.1002/adma.202002092.

Iyer, M.A., et al., 2014. Scanning fluorescence-based ultrasensitive detection of dengue viral DNA on ZnO thin films. Sens. Actuators B: Chem. 202, 1338–1348. https://doi.org/10.1016/j.snb.2014.06.005.

Jindal, K., Tomar, M., Gupta, V., 2017. A novel low-powered uric acid biosensor based on arrayed p-n junction heterostructures of ZnO thin film and CuO microclusters. Sens. Actuators B: Chem. 253, 566–575. https://doi.org/10.1016/j.snb.2017.06.146.

Kaur, G., Paliwal, A., Tomar, M., Gupta, V., 2016. Detection of Neisseria meningitidis using surface plasmon resonance based DNA biosensor. Biosens. Bioelectron. 78, 106–110. https://doi.org/10.1016/j.bios.2015.11.025.

Kim, K.-H., et al., 2012. A functional hybrid memristor crossbar-array/CMOS system for data storage and neuromorphic applications. Nano Lett. 12 (1), 389–395. https://doi.org/10.1021/nl203687n.

Krishna, M.S., et al., 2023. A review on 2D-ZnO nanostructure based biosensors: from materials to devices. Mater. Adv. 4 (2), 320–354. https://doi.org/10.1039/D2MA00878E.

Kumbhakar, P., et al., 2021. Emerging 2D metal oxides and their applications. Mater. Today 45, 142–168. https://doi.org/10.1016/j.mattod.2020.11.023.

Kundu, M., Prasad, S., Krishnan, P., Gajjala, S., 2019. A novel electrochemical biosensor based on hematite (α-Fe2O3) flowerlike nanostructures for sensitive determination of formaldehyde adulteration in fruit juices. Food Bioprocess. Technol. 12 (10), 1659–1671. https://doi.org/10.1007/s11947-019-02318-7.

Lei, Z.-L., Guo, B., 2022. 2D material-based optical biosensor: status and prospect. Adv. Sci. 9 (4), 2102924. https://doi.org/10.1002/advs.202102924.

Liu, H., Gao, J., Duan, C., Wu, K., Guo, K., 2020. A novel mediator-free biosensor based on hemoglobin immobilized in the Au nanoparticals and TiO_2 nanosheets co-modified graphene nanocomposite. Mater. Lett. 275, 128142. https://doi.org/10.1016/j.matlet.2020.128142.

Liu, P., Huo, X., Tang, Y., Xu, J., Liu, X., Wong, D.K.Y., 2017. A TiO_2 nanosheet-g-C3N4 composite photoelectrochemical enzyme biosensor excitable by visible irradiation. Anal. Chim. Acta 984, 86–95. https://doi.org/10.1016/j.aca.2017.06.043.

Liu, F., Zhang, Y., Yu, J., Wang, S., Ge, S., Song, X., 2014. Application of ZnO/graphene and S6 aptamers for sensitive photoelectrochemical detection of SK-BR-3 breast cancer cells based on a disposable indium tin oxide device. Biosens. Bioelectron. 51, 413–420. https://doi.org/10.1016/j.bios.2013.07.066.

Liu, W., et al., 2022. Specialty optical fibers and 2D materials for sensitivity enhancement of fiber optic SPR sensors: a review. Opt. Laser Technol. 152, 108167. https://doi.org/10.1016/j.optlastec.2022.108167.

Liu, Q., et al., 2016. Highly sensitive and quick detection of acute myocardial infarction biomarkers using In_2O_3 nanoribbon biosensors fabricated using shadow masks. ACS Nano 10 (11), 10117–10125. https://doi.org/10.1021/acsnano.6b05171.

Liu, Q., et al., 2018. Highly sensitive and wearable In2O3 nanoribbon transistor biosensors with integrated on-chip gate for glucose monitoring in body fluids. ACS Nano 12 (2), 1170–1178. https://doi.org/10.1021/acsnano.7b06823.

Li, P.-P., Cao, Y., Mao, C.-J., Jin, B.-K., Zhu, J.-J., 2019. $TiO_2/g-C_3N_4/CdS$ nanocomposite-based photoelectrochemical biosensor for ultrasensitive evaluation of T4 polynucleotide kinase activity. Anal. Chem. 91 (2), 1563–1570. https://doi.org/10.1021/acs.analchem.8b04823.

Li, X., Zhao, C., Liu, X., 2015a. A paper-based microfluidic biosensor integrating zinc oxide nanowires for electrochemical glucose detection. Microsyst. Nanoeng. 1 (1), 15014. https://doi.org/10.1038/micronano.2015.14.

Li, C.W., et al., 2015b. Orbitally driven giant phonon anharmonicity in SnSe. Nat. Phys. 11 (12), 1063–1069. https://doi.org/10.1038/nphys3492.

Novak, T.G., Kim, J., DeSario, P.A., Jeon, S., 2021. Synthesis and applications of WO_3 nanosheets: the importance of phase, stoichiometry, and aspect ratio. Nanoscale Adv. 3 (18), 5166–5182. https://doi.org/10.1039/D1NA00384D.

Pomerantseva, E., Gogotsi, Y., 2017. Two-dimensional heterostructures for energy storage. Nat. Energy 2 (7), 17089. https://doi.org/10.1038/nenergy.2017.89.

Ruan, S., et al., 2021. Synthesis and functionalization of 2D nanomaterials for application in lithium-based energy storage systems. Energy Storage Mater. 38, 200–230. https://doi.org/10.1016/j.ensm.2021.03.001.

Sandil, D., et al., 2017. Biofunctionalized nanostructured tungsten trioxide based sensor for cardiac biomarker detection. Mater. Lett. 186, 202–205. https://doi.org/10.1016/j.matlet.2016.09.107.

Shen, X.-W., Fang, Y.-W., Tian, B.-B., Duan, C.-G., 2019. Two-dimensional ferroelectric tunnel junction: the case of monolayer In:SnSe/SnSe/Sb:SnSe homostructure. ACS Appl. Electron. Mater. 1 (7), 1133–1140. https://doi.org/10.1021/acsaelm.9b00146.

Shetti, N.P., Bukkitgar, S.D., Reddy, K.R., Reddy, Ch.V., Aminabhavi, T.M., 2019. ZnO-based nanostructured electrodes for electrochemical sensors and biosensors in biomedical applications. Biosens. Bioelectron. 141, 111417. https://doi.org/10.1016/j.bios.2019.111417.

Sheykhifar, Z., Mohseni, S.M., 2022. Highly light-tunable memristors in solution-processed 2D materials/metal composites. Sci. Rep. 12 (1), 18771. https://doi.org/10.1038/s41598-022-23404-5.

Tereshchenko, A., et al., 2017. ZnO films formed by atomic layer deposition as an optical biosensor platform for the detection of Grapevine virus A-type proteins. Biosens. Bioelectron. 92, 763–769. https://doi.org/10.1016/j.bios.2016.09.071.

Tian, H., et al., 2018. A hardware Markov chain algorithm realized in a single device for machine learning. Nat. Commun. 9 (1), 4305. https://doi.org/10.1038/s41467-018-06644-w.

Wang, J., Malgras, V., Sugahara, Y., Yamauchi, Y., 2021. Electrochemical energy storage performance of 2D nanoarchitectured hybrid materials. Nat. Commun. 12 (1), 3563. https://doi.org/10.1038/s41467-021-23819-0.

Wang, Y., Zhang, Z., Mao, Y., Wang, X., 2020a. Two-dimensional nonlayered materials for electrocatalysis. Energy Environ. Sci. 13 (11), 3993–4016. https://doi.org/10.1039/D0EE01714K.

Wang, H., et al., 2020b. A 2D-SnSe film with ferroelectricity and its bio-realistic synapse application. Nanoscale 12 (42), 21913–21922. https://doi.org/10.1039/D0NR03724A.

Wang, M., et al., 2018a. Robust memristors based on layered two-dimensional materials. Nat. Electron. 1 (2), 130–136. https://doi.org/10.1038/s41928-018-0021-4.

Wang, Z., et al., 2018b. Defects controlled hole doping and multivalley transport in SnSe single crystals. Nat. Commun. 9, 47. https://doi.org/10.1038/s41467-017-02566-1.

Wei, C., Zou, X., Liu, Q., Li, S., Kang, C., Xiang, W., 2020. A highly sensitive non-enzymatic glucose sensor based on CuS nanosheets modified Cu_2O/CuO nanowire arrays. Electrochim. Acta 334, 135630. https://doi.org/10.1016/j.electacta.2020.135630.

Xia, C., et al., 2018. Anomalous Li storage capability in atomically thin two-dimensional sheets of nonlayered MoO_2. Nano Lett. 18 (2), 1506–1515. https://doi.org/10.1021/acs.nanolett.7b05298.

Xie, Z., et al., 2018. Ultrathin 2D nonlayered tellurium nanosheets: facile liquid-phase exfoliation, characterization, and photoresponse with high performance and enhanced stability. Adv. Funct. Mater. 28 (16), 1705833. https://doi.org/10.1002/adfm.201705833.

Yagati, A.K., Lee, T., Min, J., Choi, J.-W., 2013. An enzymatic biosensor for hydrogen peroxide based on CeO_2 nanostructure electrodeposited on ITO surface. Biosens. Bioelectron. 47, 385–390. https://doi.org/10.1016/j.bios.2013.03.035.

Yang, Z.-X., et al., 2012. Controllable synthesis, characterization and photoluminescence properties of morphology-tunable CdS nanomaterials generated in thermal evaporation processes. Appl. Surf. Sci. 258 (19), 7343–7347. https://doi.org/10.1016/j.apsusc.2012.04.010.

Yang, T., Chen, M., Kong, Q., Luo, X., Jiao, K., 2017. Toward DNA electrochemical sensing by free-standing ZnO nanosheets grown on 2D thin-layered MoS_2. Biosens. Bioelectron. 89, 538–544. https://doi.org/10.1016/j.bios.2016.03.025.

Yang, P., Rong, H., Wu, Z., Pei, Y., 2021. Biosensor based on In_2O_2 electrolyte gated thin film transistor with integrated on-chip gate electrode. IEEE Trans. NanoBioscience 20 (3), 287–290. https://doi.org/10.1109/TNB.2021.3065725.

Yin, L., et al., 2022. High-performance memristors based on ultrathin 2D copper chalcogenides. Adv. Mater. 34 (9), 2108313. https://doi.org/10.1002/adma.202108313.

Yuan, J., Lou, J., 2015. Memristor goes two-dimensional. Nat. Nanotechnol. 10 (5), 389–390. https://doi.org/10.1038/nnano.2015.94.

Zhang, F., et al., 2006. ZnS quantum dots derived a reagentless uric acid biosensor. Talanta 68 (4), 1353–1358. https://doi.org/10.1016/j.talanta.2005.07.051.

Zhang, S., Sun, S., Huang, B., Wang, N., Li, X., 2023. UV-enhanced formaldehyde sensor using hollow In_2O_3@TiO_2 double-layer nanospheres at room temperature. ACS Appl. Mater. Interfaces 15 (3), 4329–4342. https://doi.org/10.1021/acsami.2c19722.

Zheng, Z., Yao, J., Li, J., Yang, G., 2020. Non-layered 2D materials toward advanced photoelectric devices: progress and prospects. Mater. Horiz. 7 (9), 2185–2207. https://doi.org/10.1039/D0MH00599A.

Zhou, W., Guo, Y., Liu, J., Wang, F.Q., Li, X., Wang, Q., 2018. 2D SnSe-based vdW heterojunctions: tuning the Schottky barrier by reducing Fermi level pinning. Nanoscale 10 (28), 13767–13772. https://doi.org/10.1039/c8nr02843e.

Zhou, N., Yang, R., Zhai, T., 2019. Two-dimensional non-layered materials. Mater. Today Nano 8, 100051. https://doi.org/10.1016/j.mtnano.2019.100051.

CHAPTER TEN

Environment applications of non-layered 2D materials

Mohamed Bahri[a,b,*] and Peiwu Qin[a,b,*]

[a]Center of Precision Medicine and Healthcare, Tsinghua-Berkeley Shenzhen Institute, Shenzhen, Guangdong Province, P.R. China
[b]Institute of Biopharmaceutical and Health Engineering, Tsinghua Shenzhen International Graduate School, Tsinghua University, Shenzhen, P.R. China
*Corresponding authors. e-mail address: mohamedbahri@sz.tsinghua.edu.cn; pwqin@sz.tsinghua.edu.cn

Contents

1. Introduction	278
2. Assessing environmental quality and remediation applications	279
2.1 NH_3 and H_2S quantification	279
2.2 Heavy metals monitoring	282
2.3 pH sensing	283
3. Catalysis	285
3.1 Catalysis of water splitting	285
3.2 Catalysis of CO conversion and organic reactions	287
4. Summary and outlooks	289
Acknowledgments	291
References	292

Abstract

Non-layered two-dimensional (2D) materials, including oxides (WO_3, Co_3O_4, TiO_2, and ZnO), chalcogenides (Cds, PbSe, In_2S_3, etc.), Nitrides (GaN, TiN, BN, etc.), and perovskites like $CH_3NH_3PbX_3$ (X = Cl, I, Br) have recently garnered significant attention due to their unique properties over conventional materials like graphene or transition metal dichalcogenides (TMCDs). Such non-layered 2D materials display unique properties, including large surface area, high mechanical strength, enhanced reactivity, tunable properties, and are environmentally friendly, making them auspicious candidates for high-performance environmental applications. Therefore, this chapter highlights non-layered 2D materials' promising potential for environmental remediation applications by overviewing their distinctive properties suitable for photocatalytic hydrogen generation and CO conversion and assessing environmental quality. Moreover, the underlying working principle and performance evaluation metrics within specific case studies are also discussed here. Finally, the chapter concludes with future perspectives and emerging trends in the field, emphasizing the need for continued research to unleash the full potential of non-layered 2D materials for advanced environment remediation technologies.

1. Introduction

Since the more than half-century-old discovery of 2D materials, extensive progress has been achieved in the study and development of 2D layered materials-based applications, leading to significant breakthroughs (Kong et al., 2021; Bahri et al., 2023a, 2022, 2023b, 2019). However, the range of available species and quantities of these materials is still limited. In contrast, non-layered materials, including noble metals (Rh, Au, and Pd), metal oxides (In_2O_3, CeO_2, TiO_2, Fe_2O_3, WO_3, SnO_2, etc.) and metal chalcogenides (SnSe, PbS, ZnS, CuS, CdSe, CuSe, ZnSe, etc.), with their three-dimensional chemical bonding crystal structure, present attractive properties and functionalities in optoelectronics, biomedical, and energy harvesting and storage (Wang et al., 2017; Zhou et al., 2019). Compared to layered materials, the surfaces of non-layered materials are filled with dangling bonds, combining the benefits of atomically thin structures with highly active surfaces, manifesting numerous intriguing properties. This enhanced surface reactivity equips them with superior capabilities for various catalysis and sensing applications.

The need for innovative materials and technologies becomes paramount in an ever-evolving world where environmental concerns and sustainable practices take center stage. Among the myriad solutions, 2D non-layered materials have emerged as versatile candidates for addressing environmental challenges. With their unique properties and applications, these attractive materials hold great promise for monitoring various environmental parameters and advancing catalytic processes. Therefore, this chapter delves into the fascinating realm of 2D non-layered materials and unveils their diverse potential in environmental science and engineering. Our focus revolves around monitoring critical parameters such as NH_3, H_2S, pH, heavy metals and their potential in catalytic applications, including hydrogen generation and carbon monoxide conversion, as displayed with representative examples in Scheme 1. Through an interdisciplinary lens, we aim to lighten the latest advancements, cutting-edge techniques, and emerging trends in the field, with a strong emphasis on practical applications and sustainable solutions. Ultimately, we believe this chapter will serve as a valuable resource for researchers, scientists, engineers, and enthusiasts alike, inspiring further exploration, innovation, and collaboration in the realm of 2D non-layered materials for environmental monitoring and catalysis.

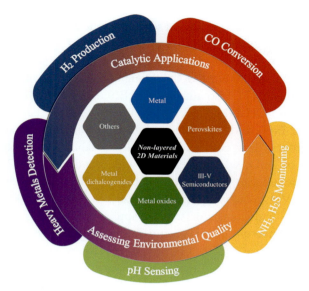

Scheme 1 The main research hotspots of non-layered 2D materials in the environment monitoring and catalytic fields.

2. Assessing environmental quality and remediation applications

2.1 NH₃ and H₂S quantification

Ammonia (NH₃) and hydrogen sulfide (H₂S) are major harmful gases that can irritate the human respiratory system, eyes, and skin, possibly leading to pneumonedema, headaches, vomiting, and in severe cases, even death (Wiener et al., 2006; Khan et al., 2017). Such gases were also recorded as a consequence of bacterial activity during food spoilage (Preethichandra et al., 2023). Therefore, it is vital to promptly detect the presence of NH₃ and H₂S in food stocks to safeguard human health and effectively manage valuable food resources. Moreover, NH₃ and H₂S are classified as corrosive gases produced by human and industrial activities. Consequently, their presence in polluted wastewater, including seas, rivers, brackish, and sewage waters, represents a significant environmental concern (Wiener et al., 2006). Therefore, understanding the properties and effects of both gases is vital for effectively addressing and optimizing their use in various environmental processes, from soil fertilization to wastewater treatment and refrigeration.

Non-layered 2D materials-based sensing application presents a potential candidate for quantifying and monitoring harmful gases. Sensors-based oxides like WO_3 (Lee et al., 2014; Srivastava and Jain, 2008), ZnO (Badadhe and Mulla, 2009), Co_3O_4 (Deng et al., 2015) and TiO_2 (Gopala Krishnan et al., 2017), also nitride like BN (Ahmed et al., 2023), and perovskites like $(CH_3NH_3)PbI_3$ (Maity et al., 2019), has been widely performed for NH_3 and H_2S due to their large surface area, simple manufacturing process, low cost, and regional adaptability advantages. Giving the example of ZnO, which presents a wide bandgap of 3.37 eV, and its thin film structure can be synthesized using various methods, including sputtering, CVD, reactive deposition, and sol–gel-method making it attractive for a wide range of sensing applications (Shetti et al., 2019). However, the detection of NH_3 using ZnO is usually performed at a high operating temperature (>250 °C) to accelerate the reaction of NH_3 on the ZnO surface as well as its adsorption and desorption governing (Wei et al., 2011). Li et al. (2014) followed the sol–gel method to develop ZnO thin film with good crystallinity using zinc acetate, monoethanolamine, and 2-methoxyethanol as precursors. The developed NH_3 gas sensor displays an enhanced response of 18% and 57% as going from 50 to 600 ppm within a recovery time of 660 s and 160 s as response time at a moderate operating temperature of 150 °C. Despite the achieved enhanced metrological parameters, the suggested detection strategy is hindered by the operating temperature. Further, Maity and Ghosh (2018) developed a room-temperature, cheap, and disposable with a rapid response ammonia paper sensor made of perovskite halide $CH_3NH_3PbI_3$ (MAPI) film with a lower LOD down to 10 ppm and a response time of 10 s. The easily manufactured sensor paper displays a color variation behavior from black to yellow in the presence of the toxic NH_3 (Fig. 1a–c). Whereas the selectivity test was carried out in the presence of nitrous oxide (N_2O), methane (CH_4), and carbon dioxide (CO_2) at a concentration value of up to 500 ppm of each interfering gas. The conversion mechanism of MAPI to PbI_2 has been well investigated through a sequence of characterization techniques, including XRD, EDX, UV, and PL techniques.

WO_3 displays a wide bandgap of 2.4–2.8 eV, a high physical and chemical stability, along with low cost and non-toxicity properties, which uplift it to be an attractive sensing material for versatile toxic gases including CH_3COCH_3, NH_3, NO_2, and H_2S (Shi et al., 2016). Shi et al. (2016) used reduced graphene oxide (RGO) coated 2D hexagonal WO_3 nanosheet (h-WO_3) to form 3D hybrid nanocomposites of rGO/h-WO_3

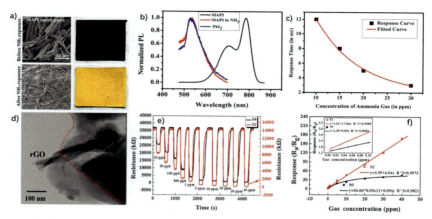

Fig. 1 NH₃ and H₂S gas sensors characterization and performances. (a) Color and morphology variation of the MAPI-coated paper before and after NH₃ gas exposure. (b) PL spectra of pristine MAPI, PbI₂ film, and exposed MAPI to NH₃, confirming the transformation of MAPI to PbI₂ under the exposure of NH₃. (c) Time response of the MAPI-based NH₃ gas sensor to various NH₃ gas concentrations. (d) TEM image of rGO/h-WO₃ composites, (e) Dynamic response-recovery of pure WO₃ (S0), and rGO/h-WO₃ composites (S2) under various H₂S concentrations ranging from 10 to 40 ppm. (f) The corresponding response value of S0, and S2 to various H₂S concentrations ranging from 10 to 40 ppm. *(a-c) Reproduced with permission Maity, A., Ghosh, B., 2018. Fast response paper based visual color change gas sensor for efficient ammonia detection at room temperature. Sci. Rep. 8, 16851. Copyright Springer Nature (2018). (d-f) Reproduced with permission Shi, J., Cheng, Z., Gao, L., Zhang, Y., Xu, J., Zhao, H., 2016. Facile synthesis of reduced graphene oxide/hexagonal WO₃ nanosheets composites with enhanced H2S sensing properties Sens. Actuators B Chem. 230, 736–745. Copyright Elsevier (2016).*

for H₂S sensing application. The synthesized hybrid nanocomposite was comprehensively characterized using TEM, HRTEM, and FESEM, showing porous structure-like channels favorable for gas diffusion (Fig. 1d). The developed H₂S sensor showed a sensitivity of 168.58 when exposed to 40 ppm, a response time of 7 s at 10 ppm, and a low LOD of 10 ppb. A comparative study of the used rGO/h-WO₃ nanocomposites efficiency with pure WO₃ was performed, recording an enhanced sensitivity of 3.7 times attributed to the mixing between the good electrical properties of rGO and the efficient gas transport channels of the suggested heterojunction (Fig. 1e,f). On the other hand, boron nitride (BN) was described as a potential sensing material for toxic gas sensing applications with interesting adsorption coefficients and sensitivity. Recently, Ahmed et al. (2023) conducted an analysis on manganese (Mn) and cobalt (Co) doped boron nitride (BN) nanosheets for multiple toxic gases sensing, including NH_3 and H_2S using density functional theory calculation (DFT). The conducted simulation indicates a strong interaction between the Mn/Co-doped B.N. nanosheet and the hazardous gases with a high absorption coefficient of 10^5 cm^{-1} order leading to high adsorption energy with a subtle variation in bond length. The dopped BN nanostructure with Co and Mn reveals a higher sensitivity than pristine BN nanosheet toward toxic gases with a high recovery time. However, a subtle arises in the band gap was recorded for the metallic Co–BN after gas exposure, proposing enhancing its suitability for gas sensing applications.

2.2 Heavy metals monitoring

Extensive discharge of industrial waste and manufacturing have resulted in significant release of heavy metals into the water and soil, leading to severe environmental pollution (Unnikrishnan et al., 2021). Among these pollutants, the contamination of water with mercury (Hg), cadmium (Cd), metalloid arsenic (As), and lead (Pb) is particularly becoming today's concern (Rahman and Singh, 2019). Such heavy metals have been classified as potent carcinogens, prompting the US Environmental Protection Agency (EPA) to establish a permissible range of heavy metals in drinking water ranging from 2 to 15 μg L^{-1} (Lv et al., 2018). Therefore, developing a sensitive, rapid, and stable sensing tool for quantifying heavy metals is becoming of great significance for environmental pollution prevention and human health protection.

Co_3O_4 is a highly fascinating p-type semiconductor with extensive utilization across a diverse array of heavy metals monitoring with a wide range of morphologies (Zou et al., 2021). As a microsheets structure, liu et al. (Liu et al., 2013) electrochemically investigated the adsorption efficiency of Pb(II) into porous and layered Co_3O_4 microsheets. The porous Co_3O_4/nafion modified glassy carbon electrode (GCE) was exposed to Pb (II) at a concentration of 0.05–0.275 μM, leading to a sensitivity of y, 71.57 μA μM^{-1} and limit of detection (LOD) of 0.018 μM, with a maximum adsorption capacity of 26.69 mg g^{-1}. Whereas lower adsorption capacities of 6.56 mg g^{-1} with reduced sensitivity and LOD of 28.26 μA μM^{-1}, 0.052 μM, respectively, using layered Co_3O_4/Nafion electrode. The high sensing efficiency of porous Co_3O_4 microsheets was mainly attributed to their unique structure with well-arranged microstructures and the high number of nanopores distributed throughout the microsheets. Altogether

2.3 pH sensing

The pH value can significantly influence the physiological, biological, and medical state of an individual well-being. From skin disease to cellular processes and enzymatic reactions, pH value presents a crucial indicator of the disease type and underlying causes (Kohn et al., 2002; Fredj et al., 2021, 2023). Similarly, in food processing, the pH value holds significance in various aspects, including microbial growth, gel formation, and protein denaturation (Gibson et al., 1988; Kress-Rogers, 1991). In addition, the pH value serves as a reliable indicator for monitoring both industrial wastewater and water pollution conditions, where a stable pH level ranging from 6.5 to 8.5 needs to be maintained to ensure optimal water quality (Kress-Rogers, 1991).

Nowadays, recent advancements in pH sensing technology have been primarily focused on improving sensitivity, biocompatibility, rapid response, cost-effectiveness, and operational lifetime across a broader pH range through the utilization of proper sensitive materials. Various 2D non-layered materials have been recently explored to advance pH sensors in food quality assessment and pollution monitoring (Manjakkal et al., 2020). Giving the example of Ghafouri et al. work, in which the author suggested using ZnO nanosheets coupled ethylene vinyl acetate (EVA)-made millifluidic channels for pH sensing (Ghafouri and Manavizadeh, 2023). The developed pH sensor exhibits excellent linearity and repeatability across a broad pH range of 2.5–12.5 due to the high crystallinity and both hydrophilic and amphoteric properties of the chemical bath deposition ZnO nanosheets. The synthesized structure was employed for quasi-single-electrode and quasi-contact separation; two operational modes of triboelectric nanogenerator (TENG) technologies. Quasi-single-electrode mode displayed higher sensitivity of $-480 \, mV/pH$. Whereas lower sensitivity of $-244.8 \, mV/pH$ was recorded when using quasi-contact-separation mode, but with superior stability, reproducibility, and higher linearity (Fig. 2a–c). On the other hand, TiO_2, with its stability, distinctive chemical characteristics, and good electrical properties, grants a close Nernstian sensitivity. Zulkefle et al. (2016) investigated the pH sensitivity of various TiO_2 film thickness-issued sol–gel deposition technique at various spin speeds and times. Accordingly, the highest sensitivity of $68 \, mV/pH$ with good linearity of 0.9943 was obtained at an optimum thickness of $20.73 \, nm$ of TiO_2 film elaborated at a spinning speed of $3000 \, rpm$ at a fixed spinning time of $75 \, s$.

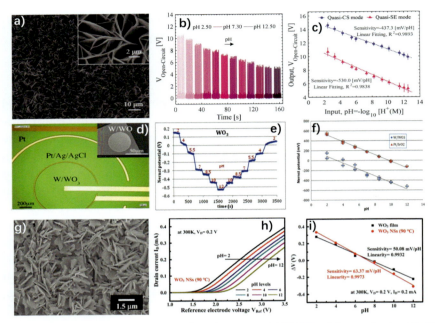

Fig. 2 pH sensors characterization and performances. (a) FE-SEM of the synthesized ZnO nanosheet, (b) Proficiency of the developed ZnO-based quasi-single-electrode pH sensor under different HCl and NH₃ solutions concentration. (c) Linearity index comparison between the two operation modes (quasi-CS and quasi-SE). (d) Optical and SEM images of W/WO₃-Pt-Ag/AgCl electrochemical microcell chips. (e) Time response of the developed W/WO₃ microelectrode at various pH values from 2 to 12. (f) pH calibration curves of the developed W/WO₃ and Pt/IrO₂ microelectrodes. (g) SEM image of the synthesized WO₃ nanosheets, (h) Current–Potential characteristics of the WO₃ nanosheets based pH sensor electrode at various pH values ranging from 2 to 12, (i) Comparison of ΔV–pH characteristics of the developed WO₃ and WO₃ nanosheets-based pH sensing electrode. *(a-c) Reproduced with permission from Ghafouri, T., Manavizadeh, N., 2023. A 3D-printed millifluidic device for triboelectricity-driven pH sensing based on ZnO nanosheets with super-Nernstian response. Anal. Chim. Acta 1267, 341342. Copyright Elsevier (2023). (d-f) Reproduced with permission from Lale, A., Tsopela, A., Civélas, A., Salvagnac, L., Launay, J., Temple-Boyer, P., 2015. Integration of tungsten layers for the mass fabrication of WO₃-based pH-sensitive potentiometric microsensors Sens. Actuators B Chem. 206 152–158. Copyright Elsevier (2015). (g-i) Reproduced with permission Kuo, C.Y., Wang, S.J., Ko, R.M., Tseng, H.H., 2018. Super-Nernstian pH sensors based on WO₃ nanosheets. Jpn. J. Appl. Phys. 57. Copyright IOP Publishing 2018.*

Tungsten oxide (WOx)-based materials, specifically WO₃, have garnered significant attention in pH sensing applications due to its potential in pH monitoring across a broad pH range (pH 2–12) within a high sensitivity very close to the Nernstian value. A sensitivity value of 55 mV/pH was reached by Lale et al. using magnetron sputtered WO_3-based microelectrode (Fig. 2d–f) (Lale et al., 2015). Whereas lower pH sensitivity of 45–56 mV/pH was recorded using sputter deposition of amorphous WO_3 thin film integrated ion-sensitive field-effect transistor (ISFET) system (Chiang et al., 2001). On the other hand, a higher sensitivity of 59 mV/pH, a very close to the Nernstian value was attaint by Salazar et al. using DC magnetron sputtering of amorphous nanocolumnar porous WO_3 thin-film electrode (Woias et al., 1998). Surface morphology effect on pH sensing efficiency between hydrothermally grown WO_3 nanosheets and sputtering WO_3 film was investigated by Kuo et al. (Kuo et al., 2018). Interestingly, advanced outcomes were reached using WO_3 nanosheets structure in terms of super-Nernstian sensitivity (63.37 mV/pH), and linearity (0.9973) over WO_3 film structure (50.08 mV/pH and 0.9932) (Fig. 2g–i). The enhanced pH performances displayed by using WO_3 nanosheets structure was mainly attributed to the increase in surface ion adsorption sites and the occurrence of local electric field improvement following the corners and sharp edges of the nanosheets.

3. Catalysis

Catalytic reactions primarily occur on or near the nanomaterial surfaces. Therefore, 2D materials hold significant promise in surface-active catalysis due to their exceptionally large specific surface area, enhanced structural stability, and short electron diffusion path (You et al., 2019). Moreover, in the case of 2D non-layered materials, unsaturated dangling bonds on their surfaces provide abundant active sites for catalysis, further augmenting their catalytic capabilities and expanding their potential applications. Recent studies have demonstrated that ultrathin 2D non-layered materials exhibit highly efficient catalytic performance across various catalytic processes, such as water splitting, CO conversion, and other organic reactions.

3.1 Catalysis of water splitting

The photocatalytic water splitting process presents an environmentally sustainable approach to obtaining clean and reusable fuel by utilizing solar energy, giving rise to the promising prospect of green hydrogen hope of powering our planet. Essentially, photocatalysis involves converting light energy into chemical energy and consists of three fundamental steps: flux of photon absorption, photogeneration of charge carrier separation, and photogeneration of charge transfer to carry out the water photocatalysis. Among these phases, charge separation plays a key role in high-energy charge carriers' generation that control the oxygen/hydrogen-evolution reactions. Non-layered 2D materials show great potential in driving photocatalysis due to their distinct characteristics, including the high number of exposed active atoms featuring unsaturated coordination, which is unheeded in their bulk counterparts.

Since the revolutionary work conducted by Fyjishima and Honda in 1972, which showcased the electrochemical photolysis potential of TiO_2 semiconductor in water splitting (Green et al., 1979), numerous other materials have been investigated for water-splitting photocatalysis. For instance, Lei et al. (2014) developed 5-atom-thick porous non-layered sheets of In_2O_3 with varying levels of O-vacancy defects through the fast heating of cubic $In(OH)_3$ sheets. A remarkable photocurrent value of $1.73\,mA\,cm^{-2}$ was recorded using rich vacancy defects of In_2O_3 sheets, surpassing the poor vacancy defects sheets and bulk In_2O_3 structures by 2.5 and 15 times, respectively. The improved catalytic activity was mainly attributed to the introduction of O-vacancy defects, which narrow bandgaps and improve carrier concentration in In_2O_3 sheets, thus enhancing the efficiency of visible light absorption and water splitting. Sun et al. (2012) reached a higher photocurrent density of $2.14\,mA\,cm^{-2}$ with a notable photon-to-current conversion efficiency of 42.5% using four atomic thick freestanding non-layered ZnSe, surpassing the performance of both bulk ZnSe and ZnSe/n-propylamine composite layers. The enhanced photostability and efficiency in solar water splitting were referred to as the good electronic structure and distinct stability induced by surface distortion of the obtained ZnSe layers. Similarly, Xu et al. (2013) reported the usage of 4 nm CdS nanosheets as visible-light-driven water-splitting photocatalysts for hydrogen generation. The synthesized CdS nanosheets following the ultrasonication-induced aqueous exfoliation approach demonstrated excellent stability, referred to as the presence of L-cysteine. Under the exposition to visible light, the obtained CdS nanosheets displayed highly efficient performance in the photocatalytic H_2 evolution reaction, with no noticeable decrease detected even in a 12 h test. Compared with CdS nanosheet-based aggregates, and CdSe diethylenetriamine hybrid nanosheets, the well-dispersed CdS nanosheets display enhanced performance with approximately 6.1 times and 5.5 times higher, respectively (Fig. 3a–c). Recently, Sendeku et al. (2021) reported for the first time the CVD

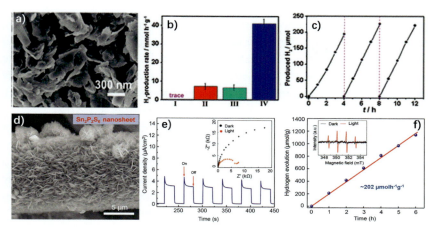

Fig. 3 Cds and Sn$_2$P$_2$S$_6$ based water splitting applications characterization and performances. (a) SEM image of the CdS- diethylenetriamine nanosheet. (b) Comparison of the hydrogen production rate of different Cds-based photocatalysts: from (I) to (IV) refer to CdS nanoparticles, CdS–diethylenetriamine nanosheets, CdS nanosheet-based aggregates, and ultrathin CdS nanosheet dispersion, respectively. (c) Time evolution of H$_2$ production over CdS nanosheet structure. (d) SEM image of CVD-grown Sn$_2$P$_2$S$_6$. (e) Photocurrent response of Sn$_2$P$_2$S$_6$ electrode at OCP under 100 mW cm^{-2} solar light stimulation. Inset is the electrochemical impedance spectroscopy response under dark and irradiation conditions. (f) H$_2$ production time evolution in pure water under solar light stimulation of A.M 1.5 G, 100 mW cm^{-2}. Inset is the ESR response under dark and solar light irradiation for comparison. *(a-c) Reproduced with permission Xu, Y., Zhao, W., Xu, R., Shi, Y., Zhang, B., 2013. Synthesis of ultrathin CdS nanosheets as efficient visible-light-driven water splitting photocatalysts for hydrogen evolution. Chem. Commun. 49, 9803–9805. Copyright Royal Society of Chemistry (2013). (d-f) Reproduced with permission from Sendeku, M.G., Wang, F., Cheng, Z., Yu, P., Gao, N., Zhan, X., et al., 2021. Nonlayered tin thiohypodiphosphate nanosheets: controllable growth and solar-light-driven water splitting. ACS Appl. Mater. Interfaces 13, 13392–13399. Copyright American Chemical Society (2021).*

synthesis of non-layered tin thiohypodiphosphate (Sn$_2$P$_2$S$_6$) nanosheets and unveiled its potential activity for H$_2$ generation in pure water. The as-obtained Sn$_2$P$_2$S$_6$ nanosheets down to 10 nm thickness display an indirect bandgap of 2.25 eV, making it a promising catalyst for water splitting. Furthermore, the CVD-grown non-layered Sn$_2$P$_2$S$_6$ catalyst demonstrated an auspicious photocatalytic hydrogen gas production at a ratio of 202 μmol h^{-1} g^{-1} (Fig. 3e–f).

3.2 Catalysis of CO conversion and organic reactions

Catalytic oxidation of carbon monoxide (CO) presents a promising solution for environmental pollution concerns, and it is extensively applied in various applications, including automotive exhaust treatment, indoor air purification, and breathing apparatus (Green et al., 1979). In addition to the water splitting application, non-layered 2D materials are widely performed for catalytic oxidation of CO due to their attractive abundance structure with highly reactive lattice oxygen atoms and low energy requirement for oxygen vacancy formation and low toxicity (Sasikala et al., 2001; Zeng et al., 2012).

Due to the numerous catalytically active sites, the tetragonal structure of single-layered tin(IV) dioxide (SnO_2) finds its attractive path toward superior catalytic performance for CO. conversion. A thin SnO_2 nanosheet with a thickness of 0.66 nm and 40% surface atom occupancy exhibited significantly high CO catalytic performance, marked by a reduced apparent activation energy of 59.2 kJ mol^{-1} and lower full-conversion temperature (Sun et al., 2013a). Compared to 1.9 nm SnO_2 sheets, bulk SnO_2, and SnO_2 nano-particles, the 0.66 nm thickness of SnO_2 nanosheets recorded superior CO conversion capability without deactivation of the catalytic reaction during 54 h at 250 °C. Meanwhile, Sun et al. (2013b) suggested pits confined ultrathin CeO_2 nanosheets for the evaluation of carbon monoxide catalytic oxidation. Compared to bulk CeO_2, the surface-confined pits of the CeO_2 nanosheet display twice lower apparent activation energy (61.7 kJ mol^{-1}), leading to improved catalytic ability for carbon monoxide at a reduced conversion temperature (Fig. 4a–c). On the other hand, Hou et al. (2013) reported ultrathin Rh nanosheets supported on CeO_2 with high performance in CO catalyzing oxidation. The temperature required for achieving complete con-version using Rh-loaded CeO_2 catalysts is significantly lower when compared to the blank CeO_2. Specifically, the ultrathin Rh nanosheets achieve 100% conversion at 110 °C, while the Rh nanoparticles require 120 °C. The improved catalytic activity of the suggested catalysts was attributed to the presence of Rh catalysts with a clean surface, which facilitates the formation of oxygen species at lower temperatures and promotes the creation of oxygen vacancies. Similarly, Duan et al. (2014) affirmed the remarkable catalytic activity of PVP-capped Rh nanosheets in hydrogenation and hydroformyla-tion reactions. Through a comparison with PVP-capped Rh nanoparticles and commercial Rh, the suggested PVP-capped Rh nanosheets demonstrated higher efficient catalysts for the conversion of hydroformylation of 1–octene and phenol, attributed to the complete exposure of Rh atoms with unsaturated coordination. Following the suggested PVP-capped Rh nanosheets, the hydrogenation of phenol achieved an impressive conversion rate of 99.9% within four-hour at room temperature (30 °C) under a hydrogen pressure of 1.0 MPa (Fig. 4d–g). Table 1 summarizes the synthesis technique and recorded

Environment applications of non-layered 2D materials

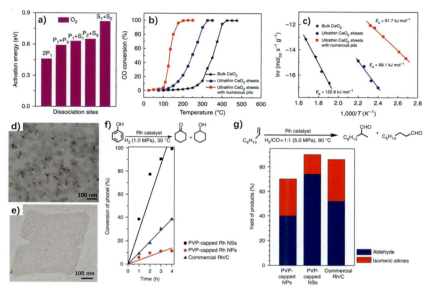

Fig. 4 CeO$_2$ and Rh catalyst for CO conversion application. (a) Activation energies for the dissociation of O$_2$ with CeO$_2$ sheets under various pits: P1, P2, P3, S1, S2, and S3 refer to 5,4,6,6,6 and 6. (b) Catalytic activity versus temperatures for CeO$_2$ bulk, CeO$_2$ sheet, and CeO$_2$ ultrathin sheet with pits. (c) The Arrhenius plot and activation energies of CeO$_2$ bulk, CeO$_2$ sheet, and CeO$_2$ ultrathin sheet with pits. (d) Low and (e) high magnification TEM image of the synthesized PVP-capped Rh nanosheets. (f) Phenol hydrogenation and (g) 1-octene hydroformylation using PVP-capped Rh nanosheets, PVP-capped Rh nanoparticles, and commercial Rh. *(a–c) Reproduced with permission from Sun, Y., Liu, Q., Gao, S., Cheng, H., Lei, F., Sun, Z., et al., 2013b. Pits confined in ultrathin cerium(IV) oxide for studying catalytic centers in carbon monoxide oxidation. Nat. Commun. 4, 2899. Copyright Springer Nature (2013). (d–g) Reproduced with permission from Duan, H., Yan, N., Yu, R., Chang, C.R., Zhou, G., Hu, H.S., et al., 2014. Ultrathin rhodium nanosheets. Nat. Commun. 5, 3093. Copyright Springer Nature (2014).*

performances of the previous detailed works and others related to non-layered 2D materials-based CO conversion and water-splitting catalysis.

4. Summary and outlooks

In summary, this chapter aims to explore the fascinating properties of 2D non-layered materials and unveil their wide-ranging applications in various fields, namely NH$_3$ and H$_2$S monitoring, pH sensing, heavy metals detection, and catalytic applications for H$_2$ production and CO conversion. The recent advances in the fabrication of sensing platforms using 2D

Table 1 Synthesis technique and performance summary of non-layered 2D materials-based CO conversion and water splitting catalysis applications.

Materials	Synthesis method	Application	Findings	Ref
Rh	Solvothermal	Hydrogenation of phenol and hydroformylation of 1-octene	>99.9% conversion within 4 h at 30 °C under (1.0 MPa) H_2 pressure	Duan et al. (2014)
CeO_2	Hydrothermal	CO conversion	~50% conversion at 131 °C, with activation energy of 61.7 kJ mol^{-1}	Sun et al. (2013b)
Ru	Hydrothermal	CO conversion	~38% selectivity at 200 °C	Yin et al. (2012)
In_2O_3	Hydrothermal	Water splitting	Visible light photocurrent 1.73 mA cm^{-2}	Lei et al. (2014)
Cds	CSS & exfoliation	Water splitting	H_2-generation rate of 7.5 mmol h^{-1} g^{-1} with 1.38% **QE***	Xu et al. (2013)
ZnSe	Exfoliation	Water splitting	**IPCE*** of 42.5%, and photocurrent density of 2.14 mA cm^{-2}	Sun et al. (2012)
Co	Ligand confined	CO_2 reduction	10.59 mA cm^{-2}, 90.1% selectivity	Gao et al. (2016)
SnO_2	Solvothermal	CO conversion	~50% conversion at 165 °C	Sun et al. (2013a)
$Sn_2P_2S_6$	CVD	Water splitting	H_2-generation rate of 202 μmol h^{-1} g^{-1}	Sendeku et al. (2021)

QE*: Quantum efficiency; **IPCE***: photon-to-current conversion efficiency.

non-layered materials and their performance in target detection were overviewed, highlighting their improved performance and underlying their sensing mechanisms for real-world applications in environmental monitoring and industrial processes. Meanwhile, the current research trends in 2D non-layered materials-based catalytic applications for H_2 production and CO conversion were also highlighted, emphasizing their enhanced activity and potential in advancing the fields of renewable energy and carbon capture.

While significant progress has been achieved in the related fields, there are still plenty of challenges and opportunities for further investigations. The future outlook for 2D non-layered materials in sensing applications involves the usage of flexible and wearable substrates, enhancing their selectivity towards specificity, and developing large-scale fabrication methods for commercialization. Further research is needed to optimize the stability and explore their applications in food safety and environmental monitoring. On the other hand, future studies should focus on developing a multifunctional sensing platform capable of simultaneously detecting multiple targets within a low detection limit and high sensitivity. Additionally, efforts should be directed toward understanding the long-term stability and the practical application of these sensing platforms in assessing environmental quality and remediation applications. Despite the continuous demonstration of the remarkable catalytic potential, future research should aim to enhance the catalytic activity of 2D non-layered materials, explore their potential in other important reactions, and develop scalable and cost-effective manufacturing processes for catalyst synthesis.

Overall, the chapter highlights the tremendous potential of 2D non-layered materials in various environment applications and sets the stage for further research and technological advancements in this rapidly evolving field. Therefore, the authors are convinced that acquiring a comprehensive background in 2D non-layered materials-based environment applications will be of interest to pave the way for a sustainable and prosperous cleaner future for all.

Acknowledgments

We acknowledge the support from the National Natural Science Foundation of China (31970752); Science, Technology, Innovation Commission of Shenzhen Municipality (JCYJ-20190809180003689, JSGG20200225150707332, JCYJ20220530143014032, ZDSYS20200-820165400003, WDZC20200820173710001, JSGG20191129110812708); Shenzhen Bay Laboratory Open Funding (SZBL2020090501004).

References

Kong, X., Bahri, M., Djebbi, K., Shi, B., Zhou, D., Tlili, C., et al., 2021. Graphene-based liquid gated field-effect transistor for label-free detection of DNA hybridization. In: 18th IEEE International Multi-Conference on Systems, Signals and Devices, SSD. pp. 387–391. ⟨https://doi.org/10.1109/SSD52085.2021.9429414⟩.

Bahri, M., Amin Elaguech, M., Nasraoui, S., Djebbi, K., Kanoun, O., Qin, P., et al., 2023a. Laser-Induced graphene electrodes for highly sensitive detection of DNA hybridization via consecutive cytosines (polyC)-DNA-based electrochemical biosensors. Microchem. J. 185, 108208. https://doi.org/10.1016/J.MICROC.2022.108208.

Bahri, M., Shi, B., Elaguech, M.A., Djebbi, K., Zhou, D., Liang, L., et al., 2022. Tungsten disulfide nanosheet-based field-effect transistor biosensor for DNA hybridization detection. ACS Appl. Nano Mater. 5, 5035–5044. https://doi.org/10.1021/ACSANM.2C00067/SUPPL_FILE/AN2C00067_SI_001.PDF.

Bahri, M., Gebre, S.H., Elaguech, M.A., Dajan, F.T., Sendeku, M.G., Tlili, C., et al., 2023b. Recent advances in chemical vapour deposition techniques for graphene-based nanoarchitectures: from synthesis to contemporary applications. Coord. Chem. Rev. 475, 214910. https://doi.org/10.1016/J.CCR.2022.214910.

Bahri, M., Baraket, A., Zine, N., Ben Ali, M., Bausells, J., Errachid, A., 2019. Capacitance electrochemical biosensor based on silicon nitride transducer for TNF-α cytokine detection in artificial human saliva: heart failure (HF). Talanta 209, 120501. https://doi.org/10.1016/j.talanta.2019.120501.

Wang, F., Wang, Z., Shifa, T.A., Wen, Y., Wang, F., Zhan, X., et al., 2017. Two-dimensional non-layered materials: synthesis, properties and applications. Adv. Funct. Mater. 27, 1603254. https://doi.org/10.1002/adfm.201603254.

Zhou, N., Yang, R., Zhai, T., 2019. Two-dimensional non-layered materials. Mater. Today Nano 8, 100051. https://doi.org/10.1016/j.mtnano.2019.100051.

Wiener, M.S., Salas, B.V., Quintero-Núñez, M., Zlatev, R., 2006. Effect of H2S on corrosion in polluted waters: a review. Corros. Eng. Sci. Technol. 41, 221–227. https://doi.org/10.1179/174327806×132204.

Khan, M.A., Qazi, F., Hussain, Z., Idrees, M.U., Soomro, S., Soomro, S., 2017. Recent trends in electrochemical detection of NH3, H2S and NOx gases. Int. J. Electrochem. Sci. 12, 1711–1733. https://doi.org/10.20964/2017.03.76.

Preethichandra, D.M.G., Gholami, M.D., Izake, E.L., O'Mullane, A.P., Sonar, P., 2023. Conducting polymer based ammonia and hydrogen sulfide chemical sensors and their suitability for detecting food spoilage. Adv. Mater. Technol. 8, 2200841. https://doi.org/10.1002/admt.202200841.

Lee, I., Choi, S.J., Park, K.M., Lee, S.S., Choi, S., Kim, I.D., et al., 2014. The stability, sensitivity and response transients of ZnO, SnO2 and WO3 sensors under acetone, toluene and H2S environments. Sens. Actuators B Chem. 197, 300–307. https://doi.org/10.1016/j.snb.2014.02.043.

Srivastava, V., Jain, K., 2008. Highly sensitive NH3 sensor using Pt catalyzed silica coating over WO3 thick films. Sens. Actuators B Chem. 133, 46–52. https://doi.org/10.1016/j.snb.2008.01.066.

Badadhe, S.S., Mulla, I.S., 2009. H2S gas sensitive indium-doped ZnO thin films: preparation and characterization. Sens. Actuators B Chem. 143, 164–170. https://doi.org/10.1016/j.snb.2009.08.056.

Deng, J., Zhang, R., Wang, L., Lou, Z., Zhang, T., 2015. Enhanced sensing performance of the Co3O4 hierarchical nanorods to NH3 gas. Sens. Actuators B Chem. 209, 449–455. https://doi.org/10.1016/j.snb.2014.11.141.

Gopala Krishnan, V., Purushothaman, A., Elango, P., 2017. A study of the physical properties and gas-sensing performance of TiO2 nanofilms: automated nebulizer spray pyrolysis method (ANSP). Phys. Status Solidi (A) Appl. Mater. Sci. 214, 1700020. https://doi.org/10.1002/pssa.201700020.

Ahmed, M.T., Hasan, S., Islam, S., Ahmed, F., 2023. First principles investigations of Cobalt and Manganese doped boron nitride nanosheet for gas sensing application. Appl. Surf. Sci. 623, 157083. https://doi.org/10.1016/j.apsusc.2023.157083.

Maity, A., Raychaudhuri, A.K., Ghosh, B., 2019. High sensitivity NH3 gas sensor with electrical readout made on paper with perovskite halide as sensor material. Sci. Rep. 9, 7777. https://doi.org/10.1038/s41598-019-43961-6.

Shetti, N.P., Bukkitgar, S.D., Reddy, K.R., Reddy, C.V., Aminabhavi, T.M., 2019. ZnO-based nanostructured electrodes for electrochemical sensors and biosensors in biomedical applications. Biosens. Bioelectron. 141, 111417. https://doi.org/10.1016/j.bios.2019.111417.

Wei, A., Wang, Z., Pan, L.H., Li, W.W., Xiong, L., Dong, X.C., et al., 2011. Room-temperature NH3 gas sensor based on hydrothermally grown ZnO nanorods. Chin. Phys. Lett. 28, 080702. https://doi.org/10.1088/0256-307X/28/8/080702.

Li, C.F., Hsu, C.Y., Li, Y.Y., 2014. NH3 sensing properties of ZnO thin films prepared via sol-gel method. J. Alloy. Compd. 606, 27–31. https://doi.org/10.1016/j.jallcom.2014.03.120.

Maity, A., Ghosh, B., 2018. Fast response paper based visual color change gas sensor for efficient ammonia detection at room temperature. Sci. Rep. 8, 16851. https://doi.org/10.1038/s41598-018-33365-3.

Shi, J., Cheng, Z., Gao, L., Zhang, Y., Xu, J., Zhao, H., 2016. Facile synthesis of reduced graphene oxide/hexagonal WO3 nanosheets composites with enhanced H2S sensing properties. Sens. Actuators B Chem. 230, 736–745. https://doi.org/10.1016/j.snb.2016.02.134.

Unnikrishnan, B., Lien, C.W., Chu, H.W., Huang, C.C., 2021. A review on metal nanozyme-based sensing of heavy metal ions: challenges and future perspectives. J. Hazard. Mater. 401, 123397. https://doi.org/10.1016/j.jhazmat.2020.123397.

Rahman, Z., Singh, V.P., 2019. The relative impact of toxic heavy metals (THMs) (arsenic (As), cadmium (Cd), chromium (Cr)(VI), mercury (Hg), and lead (Pb)) on the total environment: an overview. Environ. Monit. Assess. 191, 419. https://doi.org/10.1007/s10661-019-7528-7.

Lv, M., Liu, Y., Geng, J., Kou, X., Xin, Z., Yang, D., 2018. Engineering nanomaterials-based biosensors for food safety detection. Biosens. Bioelectron. 106, 122–128. https://doi.org/10.1016/j.bios.2018.01.049.

Zou, W., Tang, Y., Zeng, H., Wang, C., Wu, Y., 2021. Porous Co3O4 nanodisks as robust peroxidase mimetics in an ultrasensitive colorimetric sensor for the rapid detection of multiple heavy metal residues in environmental water samples. J. Hazard. Mater. 417, 125994. https://doi.org/10.1016/j.jhazmat.2021.125994.

Liu, Z.G., Chen, X., Liu, J.H., Huang, X.J., 2013. Well-arranged porous Co3O4 microsheets for electrochemistry of Pb(II) revealed by stripping voltammetry. Electrochem. Commun. 30, 59–62. https://doi.org/10.1016/j.elecom.2013.02.002.

Kohn, D.H., Sarmadi, M., Helman, J.I., Krebsbach, P.H., 2002. Effects of pH on human bone marrow stromal cells in vitro: implications for tissue engineering of bone. J. Biomed. Mater. Res. 60, 292–299. https://doi.org/10.1002/jbm.10050.

Fredj, Z., Baraket, A., Ben Ali, M., Zine, N., Zabala, M., Bausells, J., et al., 2021. Capacitance electrochemical pH sensor based on different hafnium dioxide (HfO2) thicknesses. Chemosensors 9, 1–13. https://doi.org/10.3390/chemosensors9010013.

Fredj, Z., Singh, B., Bahri, M., Sawan, M., Montreal, P., Qin, P., 2023. Enzymatic Electrochemical Biosensors for Neurotransmitters Detection: Recent Achievements and Trends. ⟨https://doi.org/10.3390/chemosensors11070388⟩.

Gibson, A.M., Bratchell, N., Roberts, T.A., 1988. Predicting microbial growth: growth responses of salmonellae in a laboratory medium as affected by pH, sodium chloride and storage temperature. Int. J. Food Microbiol. 6, 155–178. https://doi.org/10.1016/0168-1605(88)90051-7.

Kress-Rogers, E., 1991. Solid-state pH sensors for food applications. Trends Food Sci. Technol. 2, 320–324. https://doi.org/10.1016/0924-2244(91)90735-2.

Manjakkal, L., Szwagierczak, D., Dahiya, R., 2020. Metal oxides based electrochemical pH sensors: current progress and future perspectives. Prog. Mater. Sci. 109, 100635. https://doi.org/10.1016/j.pmatsci.2019.100635.

Ghafouri, T., Manavizadeh, N., 2023. A 3D-printed millifluidic device for triboelectricity-driven pH sensing based on ZnO nanosheets with super-Nernstian response. Anal. Chim. Acta 1267, 341342. https://doi.org/10.1016/j.aca.2023.341342.

Zulkefle, M.A.H., Abdul Rahman, R., Yusof, K.A., Abdullah, W.F.H., Rusop, M., Herman, S.H., 2016. Spin speed and duration dependence of TiO2 thin films pH sensing behavior. J. Sens. 2016, 9746156. https://doi.org/10.1155/2016/9746156.

Lale, A., Tsopela, A., Civélas, A., Salvagnac, L., Launay, J., Temple-Boyer, P., 2015. Integration of tungsten layers for the mass fabrication of WO3-based pH-sensitive potentiometric microsensors. Sens. Actuators B Chem. 206, 152–158. https://doi.org/10.1016/j.snb.2014.09.054.

Chiang, J.L., Jan, S.S., Chou, J.C., et al., 2001. Study on the temperature effect, hysteresis and drift of pH-ISFET devices based on amorphous tungsten oxide. Sens. Actuators B Chem. 76, 624–628. https://doi.org/10.1016/S0925-4005(01)00657-8.

Woias, P., Meixner, L., Fröstl, P., 1998. Slow pH response effects of silicon nitride ISFET sensors. Sens. Actuators B Chem. 48, 501–504. https://doi.org/10.1016/S0925-4005(98)00032-X.

Kuo, C.Y., Wang, S.J., Ko, R.M., Tseng, H.H., 2018. Super-Nernstian pH sensors based on WO3 nanosheets. Jpn. J. Appl. Phys. 57, 04FM09. https://doi.org/10.7567/JJAP.57.04FM09.

You, H., Zhuo, Z., Lu, X., Liu, Y., Guo, Y., Wang, W., et al., 2019. 1t'-mote2-based on-chip electrocatalytic microdevice: a platform to unravel oxidation- dependent electrocatalysis. CCS Chem. 1, 396–406. https://doi.org/10.31635/ccschem.019.20190022.

Green, I.X., Tang, W., Neurock, M., Yates, J.T., 1979. Spectroscopic observation of dual catalytic sites during oxidation of CO on a Au/TiO2 catalyst. Science 333 (2011), 736–739. https://doi.org/10.1126/science.1207272.

Lei, F., Sun, Y., Liu, K., Gao, S., Liang, L., Pan, B., et al., 2014. Oxygen vacancies confined in ultrathin indium oxide porous sheets for promoted visible-light water splitting. J. Am. Chem. Soc. 136, 6826–6829. https://doi.org/10.1021/ja501866r.

Sun, Y., Sun, Z., Gao, S., Cheng, H., Liu, Q., Piao, J., et al., 2012. Fabrication of flexible and freestanding zinc chalcogenide single layers. Nat. Commun. 3, 1057. https://doi.org/10.1038/ncomms2066.

Xu, Y., Zhao, W., Xu, R., Shi, Y., Zhang, B., 2013. Synthesis of ultrathin CdS nanosheets as efficient visible-light-driven water splitting photocatalysts for hydrogen evolution. Chem. Commun. 49, 9803–9805. https://doi.org/10.1039/c3cc46342g.

Sendeku, M.G., Wang, F., Cheng, Z., Yu, P., Gao, N., Zhan, X., et al., 2021. Nonlayered tin thiohypodiphosphate nanosheets: controllable growth and solar-light-driven water splitting. ACS Appl. Mater. Interfaces 13, 13392–13399. https://doi.org/10.1021/acsami.1c00038.

Sasikala, R., Gupta, N.M., Kulshreshtha, S.K., 2001. Temperature-programmed reduction and CO oxidation studies over Ce-Sn mixed oxides. Catal. Lett. 71, 69–73. https://doi.org/10.1023/A:1016656408728.

Zeng, X., Zhang, R., Xu, X., Wang, X., 2012. Study on ceria-modified SnO2 for CO and CH4 oxidation. J. Rare Earths 30, 1013–1019. https://doi.org/10.1016/S1002-0721(12)60171-9.

Sun, Y., Lei, F., Gao, S., Pan, B., Zhou, J., Xie, Y., 2013a. Atomically thin tin dioxide sheets for efficient catalytic oxidation of carbon monoxide. Angew. Chem. - Int. Ed. 52, 10569–10572. https://doi.org/10.1002/anie.201305530.

Sun, Y., Liu, Q., Gao, S., Cheng, H., Lei, F., Sun, Z., et al., 2013b. Pits confined in ultrathin cerium(IV) oxide for studying catalytic centers in carbon monoxide oxidation. Nat. Commun. 4, 2899. https://doi.org/10.1038/ncomms3899.

Hou, C., Zhu, J., Liu, C., Wang, X., Kuang, Q., Zheng, L., 2013. Formaldehyde-assisted synthesis of ultrathin Rh nanosheets for applications in CO oxidation. CrystEngComm 15, 6127–6130. https://doi.org/10.1039/c3ce40837j.

Duan, H., Yan, N., Yu, R., Chang, C.R., Zhou, G., Hu, H.S., et al., 2014. Ultrathin rhodium nanosheets. Nat. Commun. 5, 3093. https://doi.org/10.1038/ncomms4093.

Yin, A.X., Liu, W.C., Ke, J., Zhu, W., Gu, J., Zhang, Y.W., et al., 2012. Ru nanocrystals with shape-dependent surface-enhanced raman spectra and catalytic properties: controlled synthesis and DFT calculations. J. Am. Chem. Soc. 134, 20479–20489. https://doi.org/10.1021/ja3090934.

Gao, S., Lin, Y., Jiao, X., Sun, Y., Luo, Q., Zhang, W., et al., 2016. Partially oxidized atomic cobalt layers for carbon dioxide electroreduction to liquid fuel. Nature 529, 68–71. https://doi.org/10.1038/nature16455.

CHAPTER ELEVEN

Biomedical applications of non-layered 2DMs

Seyedeh Nooshin Banitaba[a,b], Abeer Ahmed Qaed Ahmed[c], Mohammad-Reza Norouzi[d], and Sanaz Khademolqorani[b,d,*]

[a]Department of Textile Engineering, Amirkabir University of Technology, Tehran, Iran
[b]Emerald Experts Laboratory, Isfahan Science and Technology Town, Isfahan, Iran
[c]Department of Environmental Sciences, School of Agriculture and Environmental Sciences, University of South Africa, Florida, Johannesburg, South Africa
[d]Department of Textile Engineering, Isfahan University of Technology, Isfahan, Iran
*Corresponding author. e-mail address: s.khademolqorani@alumni.iut.ac.ir

Contents

1. Introduction   298
2. The recent advancement in the synthesis and structure of NL2DMs   299
3. Biosensors devices based on 2D non-layered composition   304
4. Antibacterial activity of the NL2DMs   307
5. The application of non-layered 2D arrays in controlled drug delivery systems   311
6. The use of NL2DMs 2D non-layered structures in tissue engineering   315
7. Current challenges   319
References   320

Abstract

The emergence of novel nonlayered 2D materials (NL2DMs) has been considered a key strategy to render a bright prospect in versatile biomedical applications in both fundamental research and applied technologies. Metal oxides, sulfides, metal dichalcogenides, and perovskite are the most known non- or few-layered compositions explored for biomedical usages. Atomically thin structures, as well as outstanding physiochemical characteristics have implied their critical roles in various biomedical aspects. Their unique optical and electrical properties have made them highly attractive for bioimaging and biosensing. In addition, the high surface-to-volume ratio of NL2DMs offers ample functionalization and surface modification opportunities, allowing for enhanced drug-loading capacities and effective interactions with biological systems, such as cells and tissues. Moreover, the unique structure of NL2DMs enables the precise control of drug release kinetics, in which tailorable drug release could be attained by altering the material composition and structure. Also, the integrated and modified NL2DMs empower the attachment of targeting moieties, such as ligands or antibodies, which can selectively bind to specific cells or tissues, improving the drug delivery efficiency while minimizing off-target effects. Furthermore, many NL2DMs exhibit excellent biocompatibility, meaning they can be used in contact with biological systems without causing adverse effects. In this chapter, specific attention has been

Semiconductors and Semimetals, Volume 113
ISSN 0080-8784, https://doi.org/10.1016/bs.semsem.2023.09.013

Copyright © 2023 Elsevier Inc.
All rights reserved.

297

addressed to recent advancements in biomedical applications of 2D non-layered constituents. Eventually, the key challenges facing further development toward progressing of the biomedical research era are remarked, followed by several strategies capable of addressing the obstacles. Overall, the fascinating role of nonlayered 2D materials implied the high potential of these structures to bridge the translational gap in biomedical apparatus.

1. Introduction

Two-dimensional materials (2DMs) have depicted worthy prospects attributed to their outstanding features of atomic thin crystals, including nanoscale lateral size, flat structure, and low thickness of atom layer (<5 nm). The emergence of graphene as a crystalline carbon layer opened a new window in the field of ultrathin 2DMs and their applications. 2DMs could be classified into two main categories, containing layered and non-layered structures. As the most well-known few layered 2DMs (L2DMs), transition metal dichalcogenides (TMDCs) have revealed strong chemical interaction, in which each layer is connected via van der Waals bonding. Contrarily, non-layered 2DMs (NL2DMs) are created by three-dimensional chemical bonding. Stack up against L2DMs, NL2DMs presented unsaturated surface atoms, which could provide an additional degree of freedom to form a heterostructure, as well as an adjustable configuration.

So far, tremendous merits have been declared for 2DMs, including flexibility, optical transparency, and mechanical strength provided by their covalent bonding and ultrathin thickness; proper electronic carrier with electrical characteristics as a consequence of electrons confinement in 2D; and great surface activity and tunable structural characteristics, owing to the high proportion of surface atoms and broad-spectrum response. It is noteworthy that 2DMs paved the way for expanding the biosensor and bioimaging applications because of their intelligent sensitivity, as well as their photothermal and electrical properties. Compared with graphene-based nanomaterials, 2DMs have shown lower cytotoxicity due to the absence of sharp corners which makes them a great choice for tissue regeneration fields. According to their physic- and photochemical properties, the antimicrobial properties were expected by damaging the bacteria membranes and oxidation stress. Finally, NIR absorbance, photothermal property, high porosity, and surface volume ratio of 2DMs facilitate the conditions for successful drug delivery strategies, tissue engineering, and tumor therapy. Fig. 1 schematically illustrates the main merits and biomedical applications of the NL2DMs configurations.

In the perspective view of the well-promising properties of 2DMs for medical and sensors fields, in this chapter, we focused on 2DMs, especially on the nonlayered and ultra-thin few-layered structures, including metal oxides, dichalcogenides, sulfides, oxides, as well as organic-inorganic perovskite in biomedical and healthcare applications.

2. The recent advancement in the synthesis and structure of NL2DMs

Recently, specific attention has been dedicated to the synthesis NL2DMs applicable in various fields, due to their fascinating properties (Ma et al., 2022). So far, a diverse range of non-layered structures has been effectively transformed into 2D structures, including metals, metal chalcogenides, metal oxides, organic-inorganic hybrid perovskites, and so forth (Wang et al., 2017a). Achieving accurate managing over synthesis is decisive for comprehensively investigating their fundamental properties and identifying new applications. Thus, it is imperative to explore advanced synthetic methods that can produce NL2DMs with specific morphology and characteristics (Tao et al., 2019).

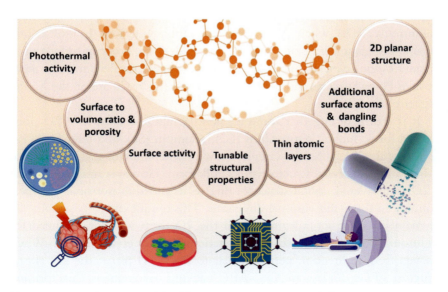

Fig. 1 Lucrative features and healthcare opportunities revealed by 2DMs.

The distinct physical and chemical properties of metallic materials enable them to create NL2DMs, possessing robust metallic bonds between their atoms that promote the creation of well-ordered structures. Additionally, metallic materials often have high melting points and can resist corrosion, rendering them appropriate for use in challenging environments with high temperatures (Zhou et al., 2019). Non-layered structures can be formed by metal oxides, such as In_2O_3, Gd_2O_3, Co_3O_4, TiO_2, etc. As an example, In_2O_3 has a non-layered configuration where In and O layers are settled in a cubic crystal phase. This crystal contains O-vacancy defects due to the release of lattice oxygen into O_2 during the transformation of $In(OH)_3$ to In_2O_3. The O-vacancy defects in the In_2O_3 sheet generate a new intermediate energy level, which results in a narrower band gap, representing the energy difference between the conduction and valence links, and improving carrier separation efficiency. This phenomenon leads to enhanced of light energy absorption and generation of electron-hole pairs, which are critical for photocatalytic reactions (Fareza et al., 2022; Banitaba et al., 2021, 2019).

TMDCs have also received considerable attention as a class of NL2DMs. These materials exhibit different crystal structures depending on the stacking sequence. For instance, CdX (X represents S, Se, and Te) has two crystal structures: a cubic phase zinc-blende structure consisting ABCABC repeating units for the close-packed (111) planes, and a hexagonal phase wurtzite structure comprising ABAB repeating units (Lai et al., 2021). Perovskites have the formula ABX_3, where A, B, and X are attributed to ions. In most hybrid organic–inorganic perovskites studied today, the A-site is occupied by a small organic cation, typically methylammonium, while a divalent metallic atom (such as Pb or Sn) is on the B-site, and halogens (I, Br, Cl) are on the X-sites (Egger et al., 2016).

It is worth noting that some elements could also exist in a 2D non-layered form, such as germanium (Ge), selenium (Se), and tellurium (Te) (Zhou et al., 2019). As an example, Ge is a semiconductor material that has a diamond-like crystal structure in its bulk form. However, it can also form a 2D non-layered crystal structure due to its intrinsic isotropic chemical bonding in three-dimensional directions. This unique property enables Ge to maintain its structural integrity in the 2D form without the need for additional bonds or interlayer interactions (Yuan et al., 2022). When viewed from a direction perpendicular to the (111) plane, the 2D crystal structure of Ge displays an ABC-ABC stacking arrangement, thus forming a non-layered crystal configuration (Hu et al., 2018). In addition to energy storage and conversion, these materials were investigated for their possible use in electronic and optoelectronic devices (Ji et al., 2017). The ability of these substances to create NL2DMs expands the range of potential materials available for use in a variety of fields, and ongoing research in this area continues to reveal new insights and potential applications. According to the literature, the NL2DMs could be successfully synthesized through wet chemical synthesis or chemical vapor deposition techniques.

Wet chemical synthesis is a class of methods widely used to create materials through chemical reactions that occur in a liquid phase, often with the assistance of surfactants or polymers (Nikam et al., 2018). These techniques are exceedingly advantageous due to their high yield, low cost, versatility, and controllability, making them a promising route for producing a variety of NL2DMs with controlled thickness (Tan and Zhang, 2015). Additionally, the resulting 2DMs are easily dispersed in organic solvents or water, making them ideal for use in liquid-phase usages (Li et al., 2022). This makes wet chemical synthesis an attractive option for researchers and engineers seeking to develop new materials for various fields and demands. This method can be classified into several types based on different mechanisms, including surface-energy-controlled, confined space, template-directed, colloidal, and other synthesizing methods (Wang et al., 2017a).

In the class of wet chemical synthesis strategies, a surface-energy-controlled technique could be employed to produce metals, such as Pd, Ag, Ru, and Rh. In this process, surfactant agents like Poly (vinylpyrrolidone) and cetyltrimethylammonium bromide are commonly employed to bind to the low-index surfaces of NL2DMs (Duan et al., 2014). Accordingly, the growth rate of these surfaces is reduced, which facilitates 2D growth. For instance, Yin et al. (2014) reported the formation and mechanical properties of multilayered Pd nanosheets. The study identified an anisotropic assembly of Pd stacked vertically with rotational disparities, causing unique diffractions and Moiré patterns. In another study, Duan et al. (2014) reported the fabrication of single-layered Rh nanosheets, with a thickness lower than 4 Å and a planar single-atom-layered sheet structure. Moreover, they revealed that the prepared Rh nanosheet encompasses a δ-bonding network that alleviates the structure, along with the poly(vinylpyrrolidone) ligands.

Confined space synthesis (CSS) is another method in this era, that creates a restricted environment for a particular reaction to occur. This technique confines the reactive precursor ions to a two-dimensional space, forcing them to align themselves in a crystallographic face to minimize their surface energy (Li et al., 2020). This alignment ultimately leads to the formation of 2D non-layered nanosheets. The CSS technique is useful for creating nanosheets with unique properties that cannot be achieved through traditional synthesis methods, and it has potential applications in electronics, catalysis, and energy storage fields (Wang et al., 2016). Using this strategy, Wang et al. (2016) demonstrated the employment of surfactant monolayers as soft templates for the bottom-up preparation of 2D nanomaterials beyond the limitations of van der Waals solids. The researchers successfully synthesized ZnO nanosheets using CSS, which exhibited distinctive characteristics such as improved mechanical strength, customized optical response, and enhanced electrical conductivity.

Unlike the restricted growth approach, templated-directed synthesis (TDS) employs 2D templates to facilitate 2D growth. Graphene oxide has been traditionally utilized as a 2D template to steer the growth of diverse non-layered materials, including Au, SnO_2, and Al_2O_3 (Huang et al., 2016). Ding et al. (2011) developed a novel hydrothermal technique, which enables the direct growth of SnO_2 nanosheets on a graphene oxide substrate that is later transformed into graphene. The resulting hybrid structure of SnO_2/graphene is exceptional in terms of its superior lithium storage capabilities, featuring high reversible capacities, and exhibiting good cycling performance. In contrast to TDS, oleic acid, oleyl amine, and Octylamine surfactants act as colloidal templates that direct crystal growth (Ithurria et al., 2011). These surfactants, which feature long-chain ligands based on oleyl amine or oleic acid, enable precursor assembly along a single face (Zhang and Yan, 2012). Following oriented assembly, organic shells are eliminated, leaving behind 2DMs. Pelleteir et al. reported a successful 2D-oriented attachment of lead sulfide (PbS) nanocrystals into ultrathin, micrometer-scale single-crystal sheets. The process was initiated by co-solvents and driven through the dense packing of oleic acid ligands on facets of PbS (Schliehe et al., 2010).

Apart from wet chemical methods, CVD is the most common strategy for producing 2DMs with high crystal quality, clean surfaces, customizing morphology, and adjustable arrangement (Shi et al., 2015). Meanwhile, conventional CVD techniques are not suitable for synthesizing 2D non-layered crystals due to their intrinsic isotropic chemical bonding, leading to the emergence of optimized methods, such as van der Waals epitaxy (vdWE) growth, space-confined CVD, and self-limited epitaxial growth (Samad et al., 2016).

The vdWE growth method has the potential to achieve extremely anisotropic NL2DMs growth by regulating growth factors and presenting appropriate epitaxial substrates (Cheng et al., 2013). Unlike the conventional method of epitaxial growth, lattice matching is not strictly obligatory between 2DMs and substrates, and the weak interaction between them facilitates the migration of reactive molecules and anisotropic growth of large, high-crystallinity 2DMs. This method could also be employed to create 2D van der Waals heterojunctions with meticulous crystal orientation and stacking style when L2DMs are used as growth substrates (Liu and Xue, 2021). A space-confined CVD approach involves using two stacked mica substrates to generate a confined reaction space, which limits the growth rate of the material by controlling the diffusion rate of the precursor or the reaction rate of adsorbed species (Chen et al., 2021). This is effective in hindering thermodynamic growth and introducing kinetic growth to obtain ultrathin NL2DMs. The mica substrates with atomic smoothness accelerate horizontal growth and suppress vertical one, resulting in the formation of ultrathin NL2DMs with regular shape and thickness (Zhou et al., 2018). Compared to other methods, the space-confined CVD is relatively easy to use and offers greater control over the material's morphology and crystallinity (Wang et al., 2017a).

The self-limited epitaxial growth technique is another modified CVD approach to synthesizing ultrathin NL2DMs (Jin et al., 2018a). In this method, foreign elements like Indium or halide are introduced into the reaction system to restrict the certain crystal planes growth, enabling sustained synthesis of the desired material. The technique necessitates a substrate with a chemically inert surface to facilitate epitaxial growth and foreign elements to inhibit growth in the vertical direction (Hu et al., 2018). This approach offers an avenue forthe anisotropic growth of ultrathin NL2DMs. Studies have successfully synthesized CdS and Ge nanoplates using this strategy through employing CdS powder and layered mica as precursors and epitaxy as substrate loaded with In_2S_3 powder to restrict the specific crystal faces growth (Wang et al., 2017b).

In conclusion, the synthesis of NL2DMs presents both exciting opportunities and significant challenges. While wet chemical synthesis and chemical vapor deposition techniques have shown a new pave to approach precise control over the synthesis process remains a major obstacle. Moreover, the scalability and cost-effectiveness of these methods for large-scale production require further development. Nonetheless, ongoing research in this area holds great promise for the development of new materials with unique properties and potential opportunities in a variety of fields.

3. Biosensors devices based on 2D non-layered composition

Biosensors have become an essential tool in several arenas, including medical diagnostics, food safety, and environmental monitoring, while traditional biosensors have limitations in terms of sensitivity, selectivity, and cost (Peng et al., 2017; Lam et al., 2021). The appearance of NL2DMs has opened up new possibilities for biosensing applications, offering exceptional physicochemical properties that enable efficient detection of biological and chemical analytes through revealing high surface area, proper electron mobility, and the ability to be integrated with other functional materials and structures (He and Tian, 2016; Banitaba et al., 2023).

Metal oxide-based biosensors are highly targeted and accurate, rendering them valuable tools for detecting various biological and chemical analytes, including gases, liquids, and biological molecules, suitable for a wide range of usages (Sedki et al., 2020). These biosensors have widespread applications in disease diagnosis, drug discovery, and monitoring of physiological parameters (Solanki et al., 2011). The field of biosensing is continually advancing, with researchers developing and optimizing new metal oxide-based materials for specific applications. As an example, Zhu et al. (2015) designed an electrochemical biosensor using a novel composite made of polyaniline and TiO_2 nanotubes. They selected TiO_2 nanoparticles as a precursor due to their unique physicochemical properties, comprising large surface-to-volume ratio and excellent biocompatibility. The polyaniline/TiO_2 nanotube composite demonstrated high conductivity and biocompatibility, leading to improved biosensor performance. The animation of the polyaniline/TiO_2 nanotubes fabrication steps is displayed in Fig. 2A. The prepared biosensor showed efficient direct electron transfer of the enzyme glucose oxidase, low background charging current, and a high signal-to-noise ratio. Aiming to develop an ultrasensitive biosensor for microRNA (miRNA)−21 detection in cancer cells, Zhang et al. (2019) used luminol-functionalized Au and ZnO nanomaterials. The ZnO nanostars were employed because of their exceptional catalytic performance in enhancing electrochemiluminescence luminous efficiency. These material choices were critical for the success of the biosensor in cancer biomarker monitoring and clinical diagnosis.

Likewise, MDCs-based biosensors have figured out remarkable stability in harsh environments, resulting from their outstanding electronic and optical features, which made them ideal for environmental monitoring and

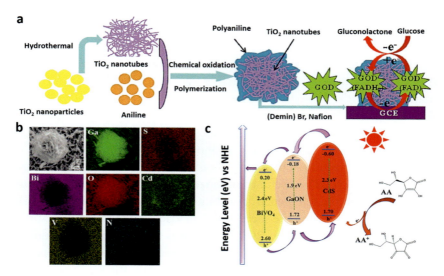

Fig. 2 (A) Schematic of the polyaniline/TiO$_2$ nanotubes fabrication procedure, (B) elemental distribution of the BiVO$_4$/GaON/CdS immunosensor and (C) the possible light-induced electron transfer mechanism in the developed structure. *(A) Reproduced from reference Zhu, J., et al., 2015. Sen. Actuators B: Chem. 221, 450, with Elsevier Copyright. (B and C) Reproduced from reference Li, S., et al., 2021a. Sens. Actuators B: Chem. 329, 129244, with permission from Elsevier.*

food safety usages (Hu et al., 2017). MDCs have a bandgap that is sensitive to changes in the local environment, making them highly responsive to the presence of biological and chemical analytes. When analytes bind to the MDC surface, they can alter the bandgap, resulting in changes in the material's optical and electronic properties that can be detected by a biosensor (Wang et al., 2017c). Additionally, MDCs have high electron mobility, which enables the fast and efficient transfer of charge carriers, leading to suitable structures for use in field-effect transistor (FET) biosensors. In FET biosensors, the binding of analytes to the MDC surface can change the conductance of the material, causing a measurable change in the FET output that can be used to detect the presence of analytes (Presutti et al., 2022). In fact, their large surface area allows for a greater number of analyte molecules to bind to the surface, ensuring a stronger signal that is easier to detect (Tajik et al., 2022). Detecting miRNA is crucial for disease diagnosis and can act as a biomarker. Researchers have focused on creating biosensors using MDC nanosheets for miRNA detection. Furthermore, the integration of these materials with other functional structures, such as nanoparticles and microfluidic channels, can enhance the sensitivity and selectivity of biosensors (Yola et al., 2014; Limbut et al., 2023; Zhang et al., 2022).

By enabling rapid and precise identification of various biomarkers associated with specific illnesses, such as cancer, cardiac disease, or infectious diseases, the biosensor devices comprising 2D nanomaterials could also be a game changer in the diagnosis of diseases (Hu et al., 2019). For example, Li et al. (2021a) grew the CdS structure on the $BiVO_4$/GaON composite structure to generate a photoelectrochemical biosensor for procalcitonin detection. The strong photoelectric effect of CdS compound on visible light could be useful to sensitize wide band gap semiconductors. Fig. 2B shows the elemental distribution in the synthesized $BiVO_4$/GaON/CdS, confirming the successful CdS loading due to the high surface area of $BiVO_4$/GaON. Fig. 2C also exhibits the possible light-induced electron transfer mechanism, corroborating the quick transference of electrons in CdS to GaON semiconductors and then $BiVO_4$. Notably, the ITO glass collected the remaining electrons, and an effective energy band level could be obtained, leading to a remarkable amplified response. The designed immunosensor displayed a detection limit of 0.03 pg.mL^{-1}. The advancement of portable and wearable biosensors based on NL2DMs also is an exciting area of research (Yu et al., 2022). These devices may be able to enable real-time monitoring of biological parameters, e.g. glucose, lactate, or pH, making them suitable for a wide variety of applications, including sports and fitness tracking, personalized medical care, and the management of diseases (Hu et al., 2017; Yu et al., 2022). In a study, Gao et al. (2014) developed a wearable and sensitive film based on PbS quantum dots integrated with multiwalled carbon nanotubes to detect the heart-rate. The synergetic effect of great light sensitivity provided by PbS, as well as high conductivity and outstanding mechanical features of the multiwalled carbon nanotubes led to the excellent responsivity 583 mA.W^{-1} and high stability of the structure after 10,000 cycles.

Perovskite-based electrochemical biosensors are also candidates for various bioanalysis tools due to their excellent electronic and optical properties. They offer high sensitivity, selectivity, and stability, as well as low cost, easy fabrication and miniaturizing, and successful detection of various biomolecules. These biosensors have potential for use in clinical diagnosis, environmental monitoring, and food safety testing, and could contribute to the development of next-generation point-of-care testing devices (Durai and Badhulika, 2022; Jia et al., 2022). Dai et al. (2018). deployed a third-generation glucose biosensor utilized perovskite-type

SrTiO$_3$ nanoparticles to modify glassy carbon electrodes (GCE) for the immobilization of glucose oxidase (GOx). The researchers found that the designed nanoparticles provided a larger specific surface area and a beneficial microenvironment that promoted the direct electron transfer of GOx. The GOx retained its biological activity and high affinity, showing a quasi-reversible surface-controlled process with two-proton and two–electron transfer with a wide linear range, acceptable reproducibility, and proper stability.

Despite outstanding advantages of NL2DMs, several downsides have remained unsolved, requiring more evaluations in the future, which include ensuring their biocompatibility, approaching higher sensitivity and selectivity, evaluating their long-term stability in complex biological environments, achieving reproducibility and scalability in their production, providing more cost-efficient synthesis and fabrication routes, and certifying their compatibility with existing sensor technologies, signal readout systems, and data analysis methods.

4. Antibacterial activity of the NL2DMs

Infections are assumed as seriously hard to treat health problems, leading to the emergence of even more challenging antibiotic resistance bacteria. This has caused an increasing need for alternative methods of fighting bacterial infections. Discovering versatile NL2DMs suitable antibacterial activity has been considered a potential strategy during recent decades (Zheng et al., 2018; Li et al., 2018).

In this field, metal oxides could be potential agents for combating bacterial infections, owing to their unique physical and chemical properties. Among a wide range of substances, the metal oxides based on silver, zinc, copper, titanium, and iron have received great attention (Menazea and Ahmed, 2020; Besinis et al., 2014). These nanoparticles are capable of damaging bacterial cell membranes and disrupting their cell walls. Moreover, they can generate ROS that can cause oxidative stress, resulting in bacterial cell death (de Dicastillo et al., 2020). Additionally, some metal oxide nanoparticles can attach themselves to bacterial DNA and impede the replication of bacteria (Rehman et al., 2020). Tayel et al. (2011) conducted a study to evaluate the antibacterial effectiveness of ZnO nanoparticles compared to conventional ZnO powder against nine bacterial strains. The results showed that ZnO nanoparticles were more effective than the powder in inhibiting bacterial growth, especially against gram-positive bacteria. Salmonella typhimurium and Staphylococcus aureus were eliminated within 8 and 4 h, respectively, when exposed to relevant minimal inhibitory concentrations of ZnO nanoparticles. The study suggests that ZnO nanoparticles could be an effective antibacterial agent against foodborne pathogens.

TMDCs have also attracted attention as a potential solution to antibacterial resistance due to their unique physicochemical characteristics. TMDCs have a large surface area and a distinctive electronic structure, resulting in strong and specific interactions with bacterial cells. Moreover, TMDCs have shown excellent biocompatibility and low toxicity toward mammalian cells, which make them a promising candidate for antibacterial agents (Mondal and De, 2022). As an example, Zhu et al. (2018). evaluated the antimicrobial activity and biological performance of In_2Se_3 nanosheets. Morphological structure of the developed In_2Se_3 structure represented the formation of an amorphous morphology with lateral size of about 300 nm. Based on the solution obtained from the dissolving of In_2Se_3 in the NMP solvent, sufficient nanostructure exfoliation with proper dispersion was identified. Mapping images of the designed structure exhibited homogenous distribution of the In and Se elements. The bacterial viability of In_2Se_3 with various concentrations confirmed an apparent reduction in the viable bacteria through an enhancement in the In_2Se_3 concentration. Additionally, all bacteria could be eliminated after incubation with In_2Se_3, resulting from the heat production by In_2Se_3. Moreover, no antimicrobial effect was observed without laser irradiation or incubation, while partial wrinkling and cracking were identified in the *E. coli* cells after laser irradiation, confirming the photothermal behavior of In_2Se_3 for antimicrobial applications.

Correspondingly, as a potential solution to antibiotic resistance, metal sulfides have also garnered interest. Non-layered 2D sulfides belong to a class of materials that possess a large surface area and a unique electronic structure, which enables them to interact strongly and specifically with bacterial cells (Argueta-Figueroa et al., 2017). Furthermore, metal sulfides are highly biocompatible and have low toxicity toward mammalian cells, making them an attractive option for use as antibacterial agents (Roy et al., 2022). Recent research has revealed that metal sulfides, such as CuS, FeS, and CoS, have substantial antibacterial activity against various bacterial strains, including multidrug-resistant ones (Argueta-Figueroa et al., 2017). The antibacterial mechanism of sulfides involves disrupting the bacterial cell membrane and inducing oxidative stress, leading to bacterial death (Roy et al., 2022). As an example, Ahmed and Anbazhagan (2017) synthesized CuS nanoparticles using N-lauryltyramine as a stabilizer and evaluated their antibacterial activity. Fig. 3A–D illustrate the SEM micrograph and fluorescence microscopic images of E. coli for treated and untreated samples, in which live and dead cells showed green and red colors, respectively. CuS nanoparticles produced ROS and disturbed the bacterial membrane, exhibiting excellent antibacterial activity against both gram-positive and gram-negative bacteria (Fig. 3E). Bacterial colonies viability was observed in Fig. 3F, in which by adding CuS nanoparticles, the number of viable bacteria was reduced. Accordingly, CuS nanoparticles

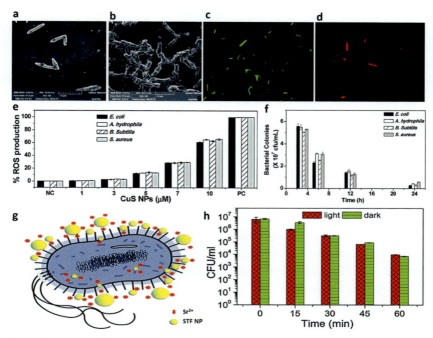

Fig. 3 Copper sulfide nanoparticles and their antibacterial activity; (A and B) SEM images of E. coli, (C and D) fluorescence microscopy images of E. coli, (E) forming of ROS by CuS nanoparticle against both gram-negative and gram-positive, and (F) viable bacterial colonies, bactericidal effect of perovskite strontium titanate ferrite metal oxide; (G) Antibacterial process of STFx and (H) bactericidal effect of STF$_{0.8}$ in light and dark condition against E. coli. (A–F) Reproduced from reference Ahmed, K. B. A., and Anbazhagan, V., 2017. RSC Adv. 7(58), 36644, with Roya Society of Chemistry Copyright. (G and H) Reproduced from reference Zhang, L., et al., 2014. Colloids Surf. A: Physicochem. Eng. Asp. 456, 169, with permission from Elsevier.

could be suggested for preventing waterborne diseases. They are also impactful in treating bacterial infections in zebrafish and are compatible with human red blood cells. The study found no toxicity in the liver and brain of zebrafish, even at a concentration eight times higher than their minimum inhibitory concentration.

Likewise, perovskite nanomaterials have revealed strong antibacterial activity against a range of bacterial strains, including both gram-positive and gram-negative bacteria (Abdel-Khalek et al., 2022). Moreover, perovskite nanomaterials have several advantages over conventional antibiotics. For example, they can be synthesized using cost-effective and scalable methods, and they can be easily tuned to optimize their antibacterial properties. In addition, perovskite nanomaterials have shown an enhancement in the efficacy of conventional antibiotics, potentially reducing the likelihood of antibiotic resistance (Abirami et al., 2020). In this era, Akbari et al. (2019) compared the antibacterial properties of $CsPbBr_3$ nanoparticles as a perovskite with that of the ZnO nanoparticles. Their results declared superior antibacterial activity of $CsPbBr_3$ against gram-negative rod-shaped E. coli compared with that of the ZnO. Zhang et al. (2014) investigated the potential of utilizing strontium titanate ferrite (STF_x) metal oxide as a bactericidal agent (see Fig. 3G). The study synthesized perovskite STF_x nanoparticles with varying levels of iron and evaluated their antibacterial efficacy on E. coli. $STF_{0.8}$ nanoparticles demonstrated excellent bactericidal activity, eliminating all E. coli within 15 min, regardless of light exposure. Fig. 3H represents the comparison of the antibacterial behavior of $STF_{0.8}$ in light and dark conditions against E. coli loading. This effect was attributed to the combination of positive surface charge, high pH environment, Sr^{2+} dissociation, and the nano-size of $STF_{0.8}$ metal oxide, indicating the potential of STF_x as a promising material for water purification and microorganism destruction.

The use of 2DMs as antibacterial agents is a capable approach, but it also raises concerns about potential toxicity to human cells, bacterial resistance, and environmental impacts. Although 2DMs have been found to exhibit low toxicity to mammalian cells, long-term effects are not fully understood. Bacteria may eventually develop mechanisms to overcome their antibacterial activity, and the production and disposal of 2DMs could have environmental impacts. Researchers need to conduct rigorous safety and toxicity testing of these materials, monitor the development of bacterial resistance, and explore ways to mitigate their environmental impacts (Naikoo et al., 2022; Tu et al., 2018).

5. The application of non-layered 2D arrays in controlled drug delivery systems

In recent decades, controlled therapeutic delivery has received great interest in the nanomedicine field to compensate for the shortcomings of conventional drug administration, including general drug distribution in the body, inadequate local drug concentration, and lack of ability to target specific tissues, cells, or organs in the body. Targeted delivery minimizes the exposure of healthy tissues to the drug while maximizing the concentration at the desired site, which can improve therapeutic outcomes and reduce systemic side effects. In addition, drug delivery systems allow for the controlled release of drugs over a specific period. This ensures a sustained therapeutic effect, avoiding the need for frequent dosing. The controlled release can also help maintain drug concentrations within the therapeutic range, leading to improved efficacy and patient compliance. The bioavailability of drugs could be enhanced, which refers to the fraction of the drug that reaches the systemic circulation and is available for action. By protecting the drug from degradation or clearance mechanisms, drug delivery systems can improve the absorption and utilization of the drug, leading to better therapeutic outcomes. Moreover, drug delivery systems can protect drugs from degradation, enzymatic activity, or harsh physiological conditions, ensuring the stability of the drug during storage and delivery, maintaining its efficacy, and extending its shelf life (Khademolqorani and Banitaba, 2022).

In this era, NL2DMs have prompted the drug delivery progress by reducing the deleterious side effects, along with attesting to more effective therapy. For example, in the oxides category, magnetic nanoparticles have been extensively applied as carriers to release drugs controllably by an external magnetic field. Kayal and Ramanujan (2010) synthesized PVA-coated iron oxide nanoparticles to carry doxorubicin drug, showing its great potential to be used in magnetically targeted drug delivery. Additionally, ZnO compositions have been broadly employed since they are cost-effective and environmentally friendly semiconductors (Anjum et al., 2021). Accordingly, Yuan et al. (2010) figured out that the addition of ZnO quantum dots into the chitosan carriers could sustain the doxorubicin delivery, possibly due to the binding affinity of this drug to the deployed nanocarriers, influencing drug release response. Porous TiO_2 was also synthesized in a study to sustainably deliver the doxorubicin drug. Based on the investigations, the hydroxyl, carboxyl, and carbonyl groups of the used drug could interact with Ti-OH groups via hydrogen bonding. In the releasing profile, the drug was delivered within two stages of 3 and 21 h, resulting from the quick dissolving of the weakly adsorbed drug molecules in the fast stage, and slowly releasing of the strongly bonded drug molecules in the TiO_2 pore structures (Cui et al., 2021). Meng et al. (2016) also depicted that the initial release rate in the SiO_2/Al_2O_3 carrier composition could be altered by manipulating the pore sizes, in which higher release could be obtained via creating larger pores due to an increase in the loaded drug.

The TMDCs nano architectures are also easily adjusted as targeted drug carrier platforms to provide superior properties in nanomedicine, based on their biocompatibility, high surface area, cell uptake, and surface modification. Among various well-known structures, the metal sulfide nanostructures composed of metal ions and sulfur elements have been introduced for intelligent drug release, due to customizing structure, metalloid features, high conductivity, and valence states diversity. With the aim of designing an effective curcumin delivery system, Kheirandish et al. (2018) synthesized a chitosan substrate embedded with ruthenium sulfide nanoparticles (RuS_2-NPs) and compared with the chitosan loaded with ruthenium oxides (RuO_2). Accordingly, the lower electronegativity of sulfur than that of oxygen causes MxSy-NPs to be naturally versatile in comparison to the highly exploited metal oxide family (Shetty et al., 2023). In this case, the modified chitosan with RuO_2 and RuS_2 was stable in the intracellular environment because of the RuO_2 and RuS_2 luminescent properties. The electrostatic interaction between curcumin and Ru-based nanoparticles was obtained through hydrogen bond and π-π bond of curcumin and functional groups of nanoparticles, providing a proper condition for curcumin loading. As a result, the combination of chitosan via RuS_2 displayed more than 91% curcumin drug loading efficiency in 90 min and high drug release in buffer solution compared with RuO_2 in 15 min treatment.

As a TMDCs structure, FeS_2 has shown a great potential for theranostic applications. In a study by Jin et al. (2018b), synthesized FeS_2 nanodots through a protein-templated biomineralization procedure. Schematic illustration of the synthesis procedure is depicted in Fig. 4A, showing the conjugation of FeS_2 nanodots with chlorin e6 (Ce6), which is a typical PS molecule. The synthesized nanodots illustrated the average size of about 7 nm. According to their magnetic features, a concentration-dependent behavior was observed, in which the desirable r2 relaxivity of $85.36 \, mM.S^{-1}$ was obtained (see Fig. 4B). The designed nanodots were then subjected to PDT treating of breast cancer cells, confirming no cell toxicity without light irradiation, while a significant death rate was observed in the cell numbers after 24 h incubation with the

Biomedical applications of non-layered 2DMs 313

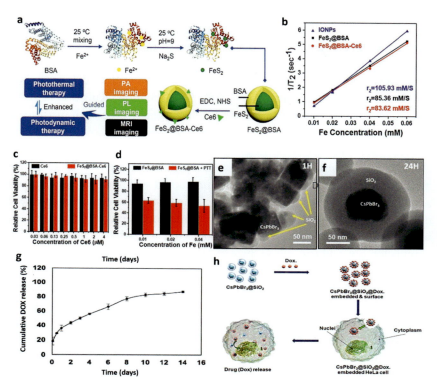

Fig. 4 Properties of FeS2@BSA-Ce6 nanodots for theranostic applications; (A) schematic illustration of the synthesis procedure, (B) magnetic properties, and (C and D) cell viability data attained from the MTT assay. CsPbBr$_3$ @SiO$_2$ core-shell structure for simultaneous bioimaging and drug delivery; TEM images of the synthesized core-shell structure after (E) 1 h and (F) 24 h reaction times, (G) the release profile of doxorubicin drug molecules from the designed nanocarriers, and (H) schematic illustration of the drug loading mechanism in the structure. *(A–D) Reproduced from reference Jin, Q., et al., 2018. ACS Appl. Mater. Interfaces 10(1), 332, with permission from American Chemical Society. (E–H) Reproduced from reference Kumar, P., et al., 2020, with permission from the Royal Society of Chemistry.*

light presence (see Fig. 4C and D). In a similar approach, She et al. (2020) demonstrated significant inhabitation of 4T1 breast tumor growth without any side effects. It was declared that the provided structure could show remarkable inhabitation of the cell growth in the presence of H$_2$O$_2$, as a result of the creation of OH radical with highly toxicity. Additionally, a greater inhibition cell rate was observed after treating with 1064 nm laser irradiation, compared with that of the 808 nm irradiation, which could be resulting from the superior photothermal efficiency of 1064 nm laser.

It is noteworthy that perovskite non-layered structures have triggered numerous studies for developing photodetector and bioimaging devices, resulting from fascinating characteristics of electronic and photophysical features. Meanwhile, few attempts have employed them in drug delivery systems (Lotfi et al., 2018; Soleymani et al., 2015). As an example, Kumar et al. (2020) physically attached doxorubicin to the $CsPbBr_3@SiO_2$ core-shell structure for simultaneous bioimaging and drug delivery. Fig. 4E and F show the formation of a SiO_2 shell around the $CsPbBr_3$ perovskite structure with the thickness of about 10 and 50 nm after 1 and 24 h reaction times, respectively. The increase in the shell thickness with the reaction time could be linked to enhancing the adsorption of oligomers around the core. The best mesoporous structure with the highest water stability and photo luminance emission intensity was observed for the core-shell structure subjected to 24 h reaction time. The cumulative doxorubicin release profile of this sample is illustrated in Fig. 4G, showing the release of 18% of the drug within 2 h followed by slow release of the remained drug over time. The first rapid drug release could be assigned to the weak bonded drug around the nanocarrier's outer surface. Fig. 4H illustrates the schematic mechanism of doxorubicin loading in the designed core-shell framework, exhibiting an electrostatic reaction between the negatively charge nanocrystals and positively charge drug molecules, as well as the adsorption of drug molecules in the SiO_2 pores. Due to the presence of $CsPbBr_3$ perovskite in the core of the nanocarriers, the suggested structure also figured out a great efficiency for in vitro bioimaging.

Despite the progress observed in designing smart drug delivery systems using NL2DMs, there are several concerns, which should be considered to ensure their safety and efficiency. One of the primary concerns is the biocompatibility of these structures, as some of them may trigger an immune response or cause toxicity in the body. Additionally, these substances can undergo chemical reactions or transformations in certain environments, affecting their stability and performance as drug carriers. Various factors, such as particle size, surface properties, and drug loading capacity require to be prudently optimized to achieve controlled release kinetics. Moreover, achieving high specificity and targeting efficiency could be challenging, needing the development of strategies, including surface modifications or functionalization. The loading capacity of these drug carriers is another challenge, since some of them may have limited drug loading capacity, which can affect the overall efficiency of the drug delivery system. Finally, the synthesis and manufacturing processes need to be optimized to ensure consistent quality and reproducibility on a larger scale. Additionally, the cost-effectiveness of the manufacturing process is also a consideration.

6. The use of NL2DMs 2D non-layered structures in tissue engineering

Artificial scaffolds are apprehended as porous architectures, implanted temporarily in the body to boost the formation of new tissues by supporting the cells in the whole regeneration procedure. Cell adhesion and their interpenetration in the located scaffold are key factors, enabling the cells to colonize properly in the entire network. Additionally, the cells should pursue proliferation furtherly followed by providing a suitable framework with the potential to transport different growth factors to the cells (Feng et al., 2023). Accordingly, the scaffold characteristics, including surface-to-volume ratio, topography, pore features, porosity ratio, biocompatibility, and biodegradability directly influence the entire biological phenomena. Additionally, the engineered scaffolds must possess appropriate mechanical strength since they should maintain their original porous structure in the tissue regeneration procedure, and degrade based on a customized time frame (Khademolqorani et al., 2021a, 2021b).

Oxides have recently caught great interest in tissue engineering, in particular to dope polymeric and non-polymeric components. According to the literature, many oxide materials (e.g., TiO_2, Al_2O_3, and ZnO) have excellent biocompatibility, meaning that they are well-tolerated by living tissues and do not elicit adverse immune responses or toxic effects when in contact with biological systems. Also, oxides could be engineered to have specific surface features, such as controlled roughness and enhanced hydrophilicity, promoting cell adhesion, proliferation, and differentiation (Bhushan et al., 2023; Hakimi et al., 2023). Moreover, oxides have shown desirable mechanical properties, which are significant for load-bearing applications in tissue engineering (Wang et al., 2023). Antibacterial characteristics have also been observed in some oxides, making them potential to inhibit bacterial infections and promote successful tissue integration. Their inherent electrical conductivity could also be advantageous for nerve regeneration or cardiac tissue engineering, where electrical stimulation plays a crucial role in tissue function (Soorani, 2023; Acuña et al., 2023).

In a research study, Kaushik et al. (2019) synthesized ZnO nanoparticles in various annealing temperatures from 300 to 900 °C, to investigate the effect of particle size on their bioactivity behavior. According to the results, the proliferation of fibroblast cells was enhanced by increasing the annealing temperature, resulting in the formation of larger particles and a lower agglomeration rate. In contrast, smaller particles showed lower proliferation and migration rates, which could be attributed to an increase in their cellular uptake by the cells. In another study, Li et al. (2021b) fabricated PCL/collagen nanofibers incorporated with 0.25, 0.50, and 0.75 $\mu g.mL^{-1}$ ZnO quantum dots to evaluate the biological wound healing mechanism. Accordingly, the antibacterial efficiency of the scaffolds was increased against E. coli and S. aureus via a rise in the ZnO concentration, corroborating the antibacterial performance of the loaded ZnO quantum dots. The proposed structure also exhibited an enhanced wound repair rate, assigned to the release of Zn^{2+} during the ZnO quantum dot decomposition.

The role of Al_2O_3 nanorods in synthesizing a versatile PHB/chitosan-based cartilage scaffold was also evaluated by Toloue et al. (2019). Based upon the results obtained, the addition of Al_2O_3 into the composite could enhance the hydrophilicity nature, which is very beneficial in the scaffold cellular behavior. Also, the high intrinsic strength nature of Al_2O_3 led to an increase in the nanocomposite's mechanical strength. Regarding the effect of iron oxides on the property enhancement of soft and hard tissues, Fallahiarezoudar et al. (2022) declared that the addition of 1 wt% Υ-Fe_2O_3 into a TPU-based 3D printed scaffold could result in the enhancement of Young's modulus from 35.72 to 112.28 MPa and slowing down the degradation rate from 4.60% to 2.12% due to the strong Fe-O bonds with the polymer chains. Also, the osteoinductive effect of the loaded nanoparticles, along with the formation of numerous tiny magnetic fields in the network caused accelerated the proliferation rate. Improving the bioactivity behavior through incorporating other oxide-based particles, such as magnesium (Armenise et al., 2020), cerium (Hosseini and Mozafari, 2020), silicon (Fielding et al., 2012), copper (Sahmani et al., 2019), and zirconium (Bhowmick et al., 2016) have also been widely reported.

TMDCs have gained attention for advancing the field of tissue due to their unrivaled properties, such as biocompatibility, surface modifiability, mechanical strength, photothermal effects, as well as electrical features. The TMDCs structures could mimic the extracellular matrix of native tissues as a consequence of the high surface area to volume ratio, porous structure, and proper topography for cell adhesion and proliferation. For example, Kang et al. (2023) designed a 3D scaffold based on PCL and FeS_2 for bone regeneration. A self-customized cell recruitment model was utilized for the bone formation simulation. The fluorescence images of the proposed membrane in days 1, 4, and 10, are represented in Fig. 5A, showing the cell migration in the developed scaffolds. All membranes exhibited an increasing trend in the number of cells (see Fig. 5B). Compared with the

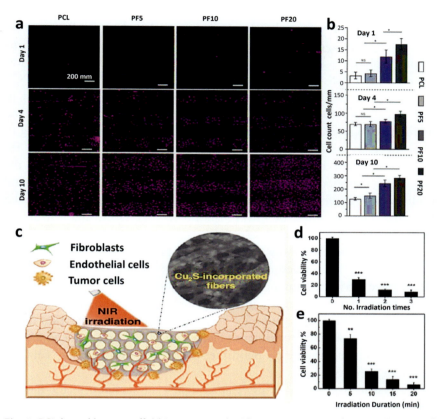

Fig. 5 PCL-based bone scaffold incorporated with 0%, 5%, 10%, and 20% FeS$_2$, named as PCL, PF5, PF10, and PF20; (A) fluorescence illustrations of stained scaffolds and (B) number of cell growth on the scaffolds, features of the electrospun PDLLA/PCL nanocomposites integrated with Cu$_2$S nanoparticles; (C) schematic of tumor therapy and skin regeneration process through irradiating of NIR and (D and E) photothermal therapeutic efficiency of the membrane against times and duration of irradiation. *(A and B) Reproduced from reference Kang, D., et al., 2023. Int. J. Bioprint. 9(1), with ACCSCIENCE Copyright. (C–E) Reproduced from reference Wang, X., et al., 2017. ACS Nano 11(11), 11337, with permission from the American Chemical Society.*

control, the scaffolds containing FeS$_2$ revealed greater number of cells. Additionally, the incorporation of 20% FeS$_2$ into the structure led to approaching more desirable features.

Cu$_x$S as semiconductor cuprous sulfide nanomaterials possesses an excellent photothermal capability due to its NIR region absorption, making it a successful additive for tissue engineering and tumor therapy.

With the aim of wound healing and skin tumor therapy, Wang et al. (2017d) synthesized electrospun micropatterned nanocomposites, containing poly(D, L-lactic acid)/poly(ε-caprolactone) (PDLLA/PCL) integrated with Cu_2S Nanoflowers. The composition of the PDLLA/PCL solution provided both stability and electrospinnability. As is shown in Fig. 5C, NIR was irradiated on the injured surfaces and the designed membranes figured out controllable photothermal performance, leading to high mortality of skin tumor cells in mice model (see Fig. 5D and E). Additionally, the homogenous encapsulated Cu_2S nanoparticles in the polymeric membranes could support skin cell adhesion, migration, and proliferation, along with playing a vital role in healing tumor-induced wounds. The presence of Cu_2S nanoflowers in electrospun structure provides Cu^{2+} ions, which could enhance the expression of main ECM molecules and accelerate the wound healing process (Salvo and Sandoval, 2022). Briefly, nanostructured TMDCs materials and the metal sulfides family have offered exciting opportunities for advanced tumor therapy and tissue regeneration strategies, including neural tissue regeneration (Shah et al., 2015), bone regeneration (Wu et al., 2018), skin regeneration (Xiao et al., 2020), and cancer treatment (Tian et al., 2011), as a consequence of their electrical conductivity, photothermal activity, outstanding stability, and high porosity.

Perovskite materials have also gained significant attention in the field of tissue engineering due to their unique properties and capabilities, including biocompatibility, optoelectronic properties, electrical stimulation, tunable chemical and structural properties, and formation versatilities. Optoelectronic features enable the Perovskite materials great candidates for photothermal therapy or photodynamic therapy to selectively target and treat diseased tissues. Their excellent electrical conductivity also is advantageous for tissue engineering applications that require electrical stimulation, such as nerve regeneration or cardiac tissue engineering. Moreover, the material properties could be tailored by altering the chemical composition or doping elements into the perovskite lattice to meet specific tissue engineering requirements, which enables the optimization of mechanical strength, degradation rates, and other important characteristics for various tissue types. Furthermore, these materials could be processed into various forms, providing a three-dimensional network for supporting cell growth, tissue formation, and nutrient transport, mimicking the extracellular matrix, and attesting mechanical sustenance to cells during tissue regeneration (Sanaullah et al., 2023). For example, $Ag_{0.3}Na_{1.7}La_2Ti_3O_{10}$ Perovskite was embedded in a PLGA nanofibrous structure by Gora et al. (2020) to enhance the antimicrobial properties of the designed scaffold. Accordingly, compatible mechanical strength with the human skin was obtained for the loaded skin tissue regeneration with perovskite material. In addition, the antimicrobial properties were enhanced against P. aeruginosa and S. saprophyticus in low filler concentration compared with the bare PLGA nanofibers, due to the presence of silver in the Perovskite structure. Moreover, the membrane-embedded with a high concentration of filler illustrated antimicrobial behavior against K. pneumonia, P. aeruginosa, and S. saprophyticus. In another case study, Baghchi et al. (2014) evaluated the bioactivity behavior of PCL electrospun layers loaded by three different perovskite nanoparticles, including calcium titanate (CT), strontium titanate (ST), and barium titanate (BT) for bone tissue regeneration. According to the data obtained, the tendency of sedimentation was increased from adding CT to BT, resulting from increasing in density. The membrane crystallinity was also enhanced through the addition of CT as it acts as a heteronucleating agent, while ST and BT had a lower heteronucleating impact, mighty due to their agglomeration and reduced interfacial interaction with PCL. Eventually, the osteogenic differentiation could be promoted through adding the perovskite nanoparticles, although the best results were revealed for CT nanofillers. The observed difference between the biological responses could be attributed to the soluble Ca^{+2} and Sr^{+2} released from the applied nanoparticles.

Overall, the usage of non-layered 2D frameworks could be leveraged to design high-efficient scaffolds. Meanwhile, it is vital to note that there are still several challenges to overcome, including scalability, long-term stability, evaluating the long-term effect on living tissues, and so forth. It is expected that with ongoing research and advancements in NL2DMs, exciting developments will occur in the future of tissue engineering.

7. Current challenges

Ongoing research in NL2DMs has confirmed diverse strategies to synthesize tunable structures with potential biomedical applications concomitant to designing versatile biosensors, antibacterial agents, smart drug delivery systems, and tissue scaffolds. Meanwhile, the progress has remained sluggish yet due to several downsides declared for these compounds. First, ensuring the biocompatibility of NL2DMs is crucial for their successful integration into biomedical applications, as some induce cytotoxicity or immune response when in contact with living cells or tissues. In addition, scaling up the production of such materials with consistent quality and properties is a challenge, requiring further optimization. Moreover, NL2DMs can be susceptible to degradation in biological environments. Factors such as exposure to enzymes, pH changes, and oxidation can impact their stability and performance over time. Developing strategies to enhance the stability and durability of these materials in physiological conditions is essential. Also, NL2DMs often require surface functionalization or engineering to improve their biocompatibility, stability, and functionality. However, achieving precise control over the surface modification and maintaining the desired properties can be challenging. NL2DMs need to be efficiently cleared from the body after their biomedical application. Understanding their clearance mechanisms and potential long-term accumulation in organs is critical for their safe use. Additionally, developing biodegradable NL2DMs that can be metabolized or eliminated from the body is an area of active investigation. Standardizing the synthesis, characterization, and testing protocols for NL2DMs in biomedical applications is essential. Establishing guidelines and regulatory frameworks for their safe and effective use will promote their translation from the lab to clinical settings.

References

Abdel-Khalek, E., et al., 2022. Phys. B: Condens. Matter 628, 413573.
Abirami, R., et al., 2020. J. Solid. State Chem. 281, 121019.
Acuña, D., et al., 2023. Mater. Lett. 337, 133997.
Ahmed, K.B.A., Anbazhagan, V., 2017. RSC Adv. 7 (58), 36644.
Akbari, A., et al., 2019. Int. Nano Lett. 9, 349.
Anjum, S., et al., 2021. Cancers 13 (18), 4570.
Argueta-Figueroa, L., et al., 2017. Mater. Sci. Eng. C 76, 1305.
Armenise, V., et al., 2020. Coatings 10 (4), 356.
Bagchi, A., et al., 2014. Nanotechnology 25 (48), 485101.
Banitaba, S.N., et al., 2021. Mater. Perform. Charact. 10 (1), 819.
Banitaba, S.N., et al., 2019. Mater. Res. Express 6 (8), 0850d6.
Banitaba, S.N., et al., 2023. Mater. Today Electron. 5, 100055.
Besinis, A., et al., 2014. Nanotoxicology 8 (1), 1.
Bhowmick, A., et al., 2016. Carbohydr. Polym. 151, 879.
Bhushan, S., et al., 2023. Int. J. Biol. Macromolecules 236, 123813.
Chen, F., et al., 2021. CrystEngComm 23 (6), 1345.
Cheng, Y., et al., 2013. RSC Adv. 3 (38), 17287.
Cui, X.-g, et al., 2021. Front. Mater. 8, 649237.
Dai, H., et al., 2018. J. Electroanal. Chem. 810, 95.
de Dicastillo, C.L., et al., 2020. Antimicrobial Resistance-A One Health Perspective.
Ding, S., et al., 2011. Chem. Commun. 47 (25), 7155.
Duan, H., et al., 2014. Nat. Commun. 5 (1), 3093.

Durai, L., Badhulika, S., 2022. ACS Omega 7 (44), 39491.
Egger, D.A., et al., 2016. Acc. Chem. Res. 49 (3), 573.
Fallahiarezoudar, E., et al., 2022. Polymers 14 (13), 2561.
Fareza, A.R., et al., 2022. J. Mater. Chem. A 10 (16), 8656.
Feng, Y., et al., 2023. Processes 11 (7), 2105.
Fielding, G.A., et al., 2012. Dental Mater. 28 (2), 113.
Gao, L., et al., 2014. Appl. Phys. Lett. 105, 15.
Góra, A., et al., 2020. Nanomaterials 10 (6), 1127.
Hakimi, F., et al., 2023. Int. J. Biol. Macromol. 233, 123453.
He, X.P., Tian, H., 2016. Small 12 (2), 144.
Hosseini, M., Mozafari, M., 2020. Materials 13 (14), 3072.
Hu, T., et al., 2019. Sci. Bull. 64 (22), 1707.
Hu, X., et al., 2018. J. Am. Chem. Soc. 140 (40), 12909.
Hu, Y., et al., 2017. Mater. Chem. Front. 1 (1), 24.
Huang, X.-L., et al., 2016. RSC Adv. 6 (91), 87945.
Ithurria, S., et al., 2011. Nat. Mater. 10 (12), 936.
Ji, X., et al., 2017. Adv. Optical Mater. 5 (1), 1600592.
Jia, D., et al., 2022. Biosensors 12 (9), 754.
Jin, B., et al., 2018a. Adv. Funct. Mater. 28 (20), 1800181.
Jin, Q., et al., 2018b. ACS Appl. Mater. Interfaces 10 (1), 332.
Kang, D., et al., 2023. Int. J. Bioprinting 9, 1.
Kaushik, M., et al., 2019. Appl. Surf. Sci. 479, 1169.
Kayal, S., Ramanujan, R.V., 2010. Mater. Sci. Eng. C 30 (3), 484.
Khademolqorani, S., Banitaba, S.N., 2022. J. Appl. Sci. Nanotechnol. 2 (2), 1.
Khademolqorani, S., et al., 2021a. Mater. Sci. Eng. C 122, 111867.
Khademolqorani, S., et al., 2021b. Polym. Adv. Technol. 32 (6), 2367.
Kheirandish, S., et al., 2018. Appl. Organomet. Chem. 32 (2), e4035.
Kumar, P., et al., 2020. J. Mater. Chem. B, 8, 10337–10345.
Lai, Z., et al., 2021. Nat. Mater. 20 (8), 1113.
Lam, C.Y.K., et al., 2021. J. Compos. Sci. 5 (7), 190.
Li, P., et al., 2021b. J. Bionic Eng. 18 (6), 1378.
Li, S., et al., 2021a. Sens. Actuators B: Chem. 329, 129244.
Li, Y.-F., et al., 2018. Theranostics 8 (20), 5713.
Li, Z., et al., 2020. Adv. Energy Mater. 10 (11), 1900486.
Li, Z., et al., 2022. Natl Sci. Rev. 9 (5) nwab142.
Limbut, W., et al., 2023. Emerging functional materials for microfluidic biosensors. Microfluidic Biosensors. Elsevier, pp. 195.
Liu, H., Xue, Y., 2021. Adv. Mater. 33 (17), 2008456.
Lotfi, S., et al., 2018. J. Superconductivity Nov. Magnetism 31, 2187.
Ma, H., et al., 2022. Chin. Phys. B.
Menazea, A., Ahmed, M., 2020. J. Mol. Struct. 1218, 128536.
Meng, L.-Y., et al., 2016. J. Ind. Eng. Chem. 37, 14.
Mondal, A., De, M., 2022. Tungsten 1.
Naikoo, G.A., et al., 2022. Chemico-Biological Interact. 110081.
Nikam, A., et al., 2018. CrystEngComm 20 (35), 5091.
Peng, J., et al., 2017. Small 13 (15), 1603589.
Presutti, D., et al., 2022. Materials 15 (1), 337.
Rehman, S., et al., 2020. Biomolecules 10 (4), 622.
Roy, S., et al., 2022. Crit. Rev. Food Sci. Nutr. 1.
Sahmani, S., et al., 2019. Eur. Phys. J. Plus 134, 1.
Salvo, J., Sandoval, C., 2022. Burn. Trauma. 10 tkab047.
Samad, L., et al., 2016. ACS Nano 10 (7), 7039.

Sanaullah, I., et al., 2023. J. Mech. Behav. Biomed. Mater. 138, 105635.

Schliehe, C., et al., 2010. Science 329 (5991), 550.

Sedki, M., et al., 2020. Sensors 20 (17), 4811.

Shah, P., et al., 2015. Nanotechnology 26 (31), 315102.

She, D., et al., 2020. Chem. Eng. J. 400, 125933.

Shetty, A., et al., 2023. Molecules 28 (6), 2553.

Shi, Y., et al., 2015. Chem. Soc. Rev. 44 (9), 2744.

Solanki, P.R., et al., 2011. NPG Asia Mater. 3 (1), 17.

Soleymani, M., et al., 2015. Polym. J. 47 (12), 797.

Soorani, M., 2023. Using Molecular Dynamic Simulation to Study the Structural Role of Cu^+ and Cu^{++} Oxides in Bioactive Glasses for Design and Fabrication of Scaffolds for Tissue Engineering. Loughborough University.

Tajik, S., et al., 2022. Biosens. Bioelectron. 114674.

Tan, C., Zhang, H., 2015. Nat. Commun. 6 (1), 7873.

Tao, P., et al., 2019. J. Mater. Chem. A 7 (41), 23512.

Tayel, A.A., et al., 2011. J. Food Saf. 31 (2), 211.

Tian, Q., et al., 2011. ACS Nano 5 (12), 9761.

Toloue, E.B., et al., 2019. J. Med. Signals Sens. 9 (2), 111.

Tu, Z., et al., 2018. Adv. Mater. 30 (33), 1706709.

Wang, F., et al., 2017a. Adv. Funct. Mater. 27 (19), 1603254.

Wang, F., et al., 2016. Nat. Commun. 7 (1), 10444.

Wang, Q., et al., 2023. Int. J. Mol. Sci. 24 (3), 2691.

Wang, X., et al., 2017d. ACS Nano 11 (11), 11337.

Wang, Y., et al., 2017b. Sci. Bull. 62 (24), 1654.

Wang, Y.-H., et al., 2017c. Biosens. Bioelectron. 97, 305.

Wu, S., et al., 2018. ACS Appl. Nano Mater. 1, 337.

Xiao, Y., et al., 2020. Theranostics 10 (4), 1500.

Yin, X., et al., 2014. Nano Lett. 14 (12), 7188.

Yola, M.L., et al., 2014. Electrochim. Acta 125, 38.

Yu, W., et al., 2022. Small 18 (14), 2105383.

Yuan, J., et al., 2022. Optical Mater. 125, 112115.

Yuan, Q., et al., 2010. Acta Biomater. 6 (7), 2732.

Zhang, J., et al., 2022. Biosensors 12 (4), 254.

Zhang, L., et al., 2014. Colloids Surf. A: Physicochem. Eng. Asp. 456, 169.

Zhang, Q., Yan, B., 2012. Chem. Eur. J. 18 (17), 5150.

Zhang, X., et al., 2019. Biosens. Bioelectron. 135, 8.

Zheng, H., et al., 2018. Sci. Bull. 63 (2), 133.

Zhou, N., et al., 2019. Mater. Today Nano 8, 100051.

Zhou, S., et al., 2018. Nano Res. 11, 2909.

Zhu, C., et al., 2018. Chem. Eur. J. 24 (71), 19060.

Zhu, J., et al., 2015. Sens. Actuators B: Chem. 221, 450.

CHAPTER TWELVE

Thermoelectric applications of non-layered 2-D materials

Ajay Kumar Verma[a,b,d,*], Rahul Mitra[a,c,d], Bhasker Gahtori[b,d], and Sumeet Walia[a]

[a]School of Engineering, RMIT University, Melbourne, Victoria, Australia
[b]CSIR–National Physical Laboratory, Dr. K.S. Krishnan Marg, New Delhi, India
[c]Materials Chemistry Department, CSIR–Institute of Minerals and Materials Technology, Sachivalaya marg, Bhubaneswar, India
[d]Academy of Scientific and Innovative Research (AcSIR), Ghaziabad, India
*Corresponding author. e-mail address: s3922285@student.rmit.edu.au

Contents

1. Introduction	324
1.1 Thermoelectric effect	324
1.2 Thermoelectric devices and materials	325
2. Non-layered 2-D materials	327
2.1 Metalloids	328
2.2 Metal oxides	329
2.3 Metal chalcogenides	330
2.4 III–V Semiconductors	331
3. Thermoelectric applications	332
3.1 Power generation	332
3.2 Cooling applications	333
3.3 Space technologies	333
3.4 Wearable medical applications	333
3.5 Self-powered sensors	334
3.6 Other applications	334
References	335

Abstract

In the quest for innovative solutions to harness waste heat and promote sustainable energy technologies, thermoelectric materials have emerged as pivotal components. Currently, in the field, several new compounds are being explored for the development of thermoelectric materials that exhibit high efficiency. Here, this book chapter focuses on the intriguing domain of non-layered 2-D materials, which could have great potential as viable options for thermoelectric applications. The unique properties of these materials, such as anisotropic electronic structures, exceptional electrical conductivities, and tunable thermal properties, make them favourable for efficient thermoelectric conversion. This chapter thoroughly examines the most prospective non-layered 2-D materials, including metalloids, metal oxides, metal

---

Semiconductors and Semimetals, Volume 113
ISSN 0080-8784, https://doi.org/10.1016/bs.semsem.2023.10.002

Copyright © 2023 Elsevier Inc.
All rights reserved.

323

chalcogenides, III–V Semiconductors, etc. Here we will discuss the practicality of these materials in various temperature regimes and potential applications, from portable energy harvesting devices to industrial waste heat recovery systems. The insights provided herein will pave the way for exciting innovations in energy conversion and sustainable power generation.

1. Introduction

In the modern era, humans are aspiring to live a standard life by developing numerous technologies, which have a need for large amounts of electricity. For that, we are highly dependent on conventional energy sources, which are limited and costly (Jaziri et al., 2020). Some studies on the environment suggest that the burning of fossil fuels is a major cause of global warming and climate change. Due to the limited amount of this fossil fuel, it is a crucial demand of the current scenario, we have to slow down the consumption of fossil fuel and move towards the development of renewable energy sources. In the way of renewable energy sources, thermoelectric technology can play a significant role in the utilization of waste heat in the form of usable electricity.

1.1 Thermoelectric effect

Thermoelectric effect had been started from a most fundamental & early founded Phenomenon known as the Seebeck effect. The Seebeck effect, coined after the renowned physicist Thomas Johann Seebeck, refers to the phenomenon wherein an electric voltage is produced in the presence of a temperature gradient across a conductive substance. The phenomenon arises as a result of the migration of charge carriers, either electrons or holes, from the region of higher temperature to the lower temperature side within the material. The ratio of voltage developed ΔV to temperature gradient ΔT is an intrinsic property of materials, known as the Seebeck coefficient (α). The magnitude of the Seebeck coefficient, which serves as an indicator of the strength of the impact, depends on the material. Thermoelectric devices that exhibit high efficiency are characterized by the presence of materials possessing a high Seebeck coefficient. Seebeck coefficient remains low for metals and high for semiconductors. A thermoelectric generator works on the Seebeck effect (Fig. 1a), and could be utilized in various applications. After some time, Peltier discovered the reverse effect of Seebeck, known as the Peltier effect. He showed that when an electric current passes through the two dissimilar metals then one

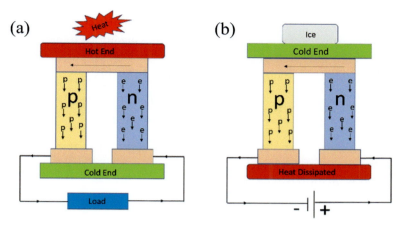

Fig. 1 The schematic diagram of a single unit of (a) a thermoelectric generator (based on the Seebeck effect), and (b) a Peltier cooler (based on the Peltier effect).

junction will cool down and the other one will be hot due to the movement of charge carriers. These charge carriers move under the electric force with taking some amount of heat and dissipating this heat at another end. The Peltier coefficient is described by the ratio of the rate of heat absorbed or rejected at the junction to the current flowing through the junction, that is $\pi = Q/I$ (Beretta et al., 2019). And both these effects are interrelated to each other- $\pi = \alpha T$. Peltier cooler works on this effect (Fig. 1b) and nowadays Peltier coolers are utilizing in a lot of applications because they are in solid state form, environmentally green, and can be made in desired compact size. Here the beneficial site is that a single module can be run as a thermoelectric generator as well as a thermoelectric cooler.

1.2 Thermoelectric devices and materials

Thermoelectric devices facilitate the efficient conversion of thermal energy into electrical energy, as well as the reverse process, so presenting viable options for both power generation and the recapture of waste heat. A single thermoelectric device can be used for both the purpose of power generation as well as for cooling applications. Thermoelectric generators (TEG) (Snyder and Toberer, 2011) have their own incessant advantages over other renewable technologies such as frictionless, high reliability, desired shape and size, eco-friendly, and free of noise pollution. However, its practical applications are limited owing to some hindrances such as low efficiency, high cost of materials as well as toxicity of materials. Therefore, in this direction, several research groups are making efforts to develop cost-effective, high-efficiency thermoelectric materials by employing advanced approaches. The efficiency of a TEG can be defined η,

$$\eta = \frac{T_h - T_c}{T_h} \frac{\sqrt{1 + ZT} - 1}{\sqrt{1 + ZT} + T_c/T_h}$$

Where, T_h & T_c are the hot & cold side temperatures respectively and ZT is the figure of merit.

which is given by;

$$ZT = \frac{\alpha^2 \sigma T}{\kappa_e + \kappa_l}$$

Where, α, σ, T, κ_e & κ_l are Seebeck coefficient, electrical conductivity, average temperature of both ends, electronic thermal conductivity & lattice thermal conductivity, respectively. High ZT leads to high efficiency therefore ZT is an important parameter for the choice of TE material, which is purely material's property. An assembly of TEG consists of several n- and p-type legs, which are connected electrically in series and thermally in parallel (Fig. 2). According to Wiedemann–Franz law κ_e is directly proportional to electrical conductivity, so κ_e can't be reduced to increase ZT, only κ_l can be decrease employing several strategies. Due to the difference in the mean free path of phonons and electrons there is a window to reduce the lattice thermal conductivity. Currently in worldwide many research groups are working on the development of high-efficiency TE materials and continuously making their effort to increase the power factor

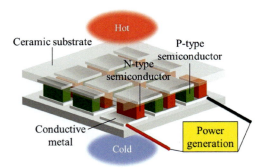

Fig. 2 The schematic of a thermoelectric device, consists of several p- and n-type counterparts (Byon and Jeong, 2020).

and decrease the thermal conductivity. Recently commercialized thermoelectric materials are PbTe (Fu et al., 2016), BiTe (Liu et al., 2016), and SiGe (Yu et al., 2012) but issues with them are toxicity, low abundance, and expansiveness. So, the research is continuously growing to develop other thermoelectric materials like; Chalcogenides (Shi et al., 2019; Gautam et al., 2022), zintl (Zhu et al., 2017), Skutterudites (Shi et al., 2011; Bhardwaj et al., 2022), Silicides (Zhou et al., 2010), Clathrates (Saramat et al., 2006), Oxides (Shin and Murayama, 2000) and half Heusler (Fu et al., 2014; Verma et al., 2023a, 2023b) alloy. The issues in device fabrication with thermoelectric materials are thermal expansion, thermal and chemical stability, repeatability, reliability, mechanical stability, and cost. As the current focus is turned towards the flexible and low–dimensional materials for better utilization of wide heat sources, the research is continuing to find new low–dimensional materials. The 2–D materials could be a viable option for enhancing the sector of flexible thermoelectric materials and can be utilized in several applications.

2. Non-layered 2-D materials

The utilization of two–dimensional (2–D) thermoelectric materials is a highly promising approach for enhancing the efficiency of thermal energy conversion (Li et al., 2020). Layered 2–D thermoelectric materials have recently gained significant attention as a highly intriguing category of materials that exhibit outstanding potential for the development of efficient thermoelectric devices. These materials consist of layers that are only a few atoms thick and possess distinctive electrical and thermal characteristics (Liu et al., 2018). As a result, they are highly desirable for applications employing recovery of wasted heat, the creation of portable power, and the facilitation of energy-efficient cooling. Extensive research has been conducted on thermoelectric applications of layered 2–D materials such as transition metal dichalcogenides (TMDCs), graphene, and black phosphorus. These materials exhibit distinctive electrical and thermal characteristics that can be utilized for thermoelectric applications. Apart from the layered 2–D materials, non-layered 2–D materials are also being explored to find high–performance thermoelectric materials. Both categories of materials have demonstrated potential for use in thermoelectric applications, and the appropriateness of a particular material is contingent upon several criteria, such as its electronic structure, thermal properties, and the precise conditions of the intended application. Scientists are currently engaged in ongoing research and refinement efforts aimed at improving the thermoelectric performance of both types of materials. The tunability of electronic properties of non-layered 2-D materials is generally enhanced, allowing for greater freedom in customization. The thermoelectric performance of non-layered 2-D materials by employing several approaches like as strain engineering, chemical doping, and heterostructure synthesis could be enhanced and these methods allow for the precise adjustment of the band structures and electrical properties of these materials. The concept of reduced dimensionality is applicable to both layered and non-layered 2-D materials. However, non-layered materials have the potential to provide greater control over dimensionality, which can result in the emergence of unique electrical and thermal properties (Li et al., 2020). Here we will discuss some non-layered 2-D thermoelectric materials and possible thermoelectric applications.

2.1 Metalloids

The emergence of non-layered 2-D metalloids has introduced a new category of materials that display fascinating characteristics in terms of electrical and thermal transport. Ongoing research in this field has focused on the investigation of various non-layered 2-D metalloids, with the aim of exploring their uses in thermoelectrics. Here are some metalloids with effective transport properties for thermoelectric applications. Germanene is a two-dimensional material that is composed of a single layer of germanium atoms (Fig. 3). It possesses the desirable characteristic of a modifiable bandgap, rendering it an appealing material for utilization in electronic devices. The electrical characteristics of the material can be altered through the application of strain engineering and functionalization techniques. Germanene has anisotropic thermal conductivity and shows notably reduced thermal conductivity along the armchair direction. The aforementioned characteristic is highly sought after in the context of thermoelectric applications (Ali et al., 2017). Antimonene possesses a direct bandgap, and its electrical characteristics can be modulated through the application of external strain or the formation of heterostructures. Antimonene demonstrates favourable thermoelectric characteristics because of its decreased thermal conductivity (Kripalani et al., 2018; Ares et al., 2018). Another example is Arsenene which is similar to other metalloids in the

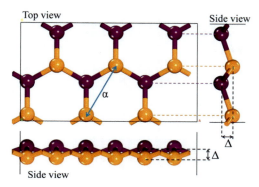

Fig. 3 Atomic structure of germanene can be defined by the hexagonal lattice constant (a) and the buckling high (Δ). Here, in order to facilitate the visualization of these buckled sheets, atoms with the same out-of-plane atomic positions are colored in either orange or purple (Mortazavi et al., 2016).

non-layered two-dimensional category, demonstrates anisotropic heat conductivity. The thermal conductivity is greater in the zigzag direction and lower in the armchair direction. The utilization of this anisotropy has the potential to be advantageous in thermoelectric applications, as it facilitates the reduction of heat conduction (Zeraati et al., 2016).

2.2 Metal oxides

Non-layered 2-dimensional metal oxides are considered a novel and very promising group of materials with potential uses in numerous fields, such as thermoelectricity. Some non-layered 2-D metal oxide materials show favourable electronic and thermal transport properties for thermoelectric applications. TiO_2 is one of the metal oxides (Fig. 4) which has been extensively investigated, revealing their very high thermal stability and enormously low thermal conductivity. Nevertheless, the comparatively modest electrical conductivity of these materials has spurred extensive investigations aimed at enhancing their performance through techniques such as doping and nanostructuring (Lu et al., 2006). Another example is VO_2 which also has potential for utilisation in thermoelectric applications (Wu et al., 2011). Next in line, WO_3 also could be utilized in thermoelectric applications owing to their advantageous electrical characteristics and reduced thermal conductivity (Kieslich et al., 2014). Thus, the non-layered 2-D metal oxides could be explored to extend the capability of thermoelectric devices.

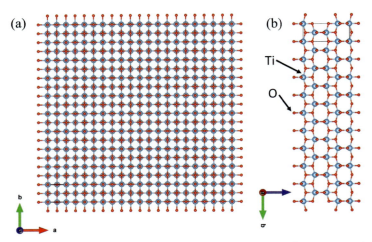

Fig. 4 Crystal structure of TiO$_2$, (a) single layer array, (b) small array at a different orientation.

2.3 Metal chalcogenides

Non-layered 2-dimensional metal chalcogenides is another category of materials that can play a crucial role in assessing their appropriateness for thermoelectric purposes. Non-layered 2-D metal chalcogenides have the potential for modulating electronic band topologies. The band structures can be altered by researchers through the implementation of doping or strain engineering techniques in order to enhance the Seebeck coefficient, resulting in boots the thermoelectric applications. The well-known non-layered 2-D metal chalcogenides are PbS, Pb$_{1-x}$Sn$_x$Se, Pb$_{1-x}$Sn$_x$Te, CdS, CdSe, and CdTe (Rogacheva et al., 2001; Wang et al., 2017). The thermoelectric properties of non-layered 2-dimensional PbS are characterized by notable features, such as elevated electrical conductivity and reduced thermal conductivity, which could be promising for the field of thermoelectric generators. 2-D Pb$_{1-x}$Sn$_x$Se and Pb$_{1-x}$Sn$_x$Te materials, which are non-layered in structure (as shown for PbSe Fig. 5), belong to the family of Lead–Tin Chalcogenides. The thermoelectric properties of these non-layered two-dimensional materials, which are composed of solid solutions of PbSe and PbTe with Sn substitution, exhibit the ability to be tuned with properties. The selection of the variable x enables customization of the bandgap, rendering them adaptable for a wide range of thermoelectric applications. These materials could be particularly employed in space

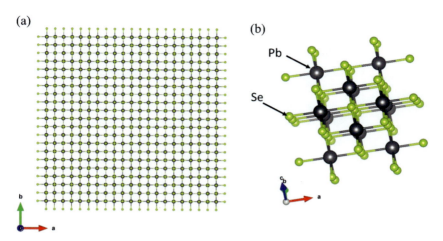

Fig. 5 Crystal structure of PbSe, (a) single layer array, (b) small array at a different orientation.

missions, remote power generation, and conditions characterized by elevated temperatures.

2.4 III–V Semiconductors

The structural characteristics of non-layered 2-D III–V semiconductors involve the composition of elements from Group III (such as Ga and In) and Group V (such as N and P) as shown in Fig. 6. They exhibit a diverse range of crystals formations and electronic bandgap characteristics (Vashishtha et al., 2023). The non-layered 2-D III–V semiconductors have gained significant interest due to their potential utilization in thermoelectric applications (Theja et al., 2022). The approach of band engineering involves the manipulation of electronic band structures in non-layered 2-D III–V semiconductors by techniques such as doping and alloying and providing a high degree of freedom in their design (Geller et al., 2001). This allows for the optimization of the Seebeck coefficient (α) and electrical conductivity (σ). High electron mobility is observed in certain III–V materials, such as GaN, resulting in enhanced electrical conductivity. Non-layered 2-D III–V semiconductors possess the ability to attain low thermal conductivity because of phonon scattering. The comprehension of thermal stability is of greatest importance in the context of high-temperature thermoelectric applications for these materials. Thus, the investigation of non-layered two-dimensional III–V semiconductors is currently a subject

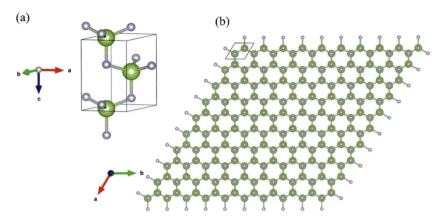

Fig. 6 Crystal structure of GaN, (a) unit cell, (b) single layer array at a different orientation.

of intense research due to its considerable promise in the field of thermoelectric applications.

3. Thermoelectric applications

Due to the low dimension of these non-layered 2-D materials they have high flexibility and low weight, thus they could be utilized in more applications compared to bulk thermoelectric materials, as well as their properties easily can be tuned for achieving high thermoelectric performance. Some of the important applications of these materials can be:

3.1 Power generation

The utilization of non-layered two-dimensional materials in thermoelectric devices for power generation is currently a subject of extensive investigation. These materials have distinctive electrical and thermal characteristics, declaring them highly attractive contenders for effective energy conversion. In the most efficient thermoelectric generators, the generated electrical signals are considerably smaller in magnitude when compared to the daily electricity demand. Additionally, the power supply capacity of TEGs falls significantly short of meeting current requirements. Nevertheless, certain thermoelectric generators have the potential to operate effectively across several compact micro power devices (Kanahashi et al., 2020). After the advancement, thermoelectric generators could be utilized in several areas such as micro-power generators for terrestrial areas, and in automobile vehicles to enhance their efficiency, and energy harvesting from power plants.

3.2 Cooling applications

It is anticipated that thermoelectric devices could become a viable option for powering many cooling applications. It is believed that the flexible and desired shape's thermoelectric device made by non-layered 2-D materials can solve the issue of refrigeration. Current cooling sources are very expensive, and less durable, thus a solid-state cooler would be the first choice of people, which could be utilized to cool the electronics, in automobile vehicle sheets to control the temperature, and in refrigerators, etc (Hou et al., 2018). Furthermore, it is worth noting that chloro-fluorocarbons have been prohibited under environmental conventions. Thus, thermoelectric devices would facilitate the exploration of novel market opportunities.

3.3 Space technologies

Nowadays every country is in a race for Space exploration, and for that, they are need a reliable and durable energy source for space missions. Like NASA is using Radioisotopes thermoelectric generators as an energy source in interstellar space (Bennett et al., 1984; Ambrosi et al., 2019) missions such as Voyager-1, Voyager-2, Cassini, etc. As some of the non-layered 2-D materials have very high thermal stability, lightweight design, high dependability, and an extended operational lifespan of many years, they could be a viable option for interstellar space missions.

3.4 Wearable medical applications

Wearable thermoelectric devices have a significant impact on the acqui-sition, transformation, and retention of energy in the context of human movement, as well as in the detection of health-related indicators (Zhang et al., 2019). Certain implantable electronic devices have the capacity to not only facilitate real-time health monitoring, but also offer electrical energy to stimulate certain organs within patients (such as cardiac pace-makers or atrial defibrillators). Furthermore, these devices possess the capability to forecast certain diseases, as well as aid in the treatment of cardiovascular and nervous system disorders. The first wearable electro-cardiogram (ECG) detecting device was invented that was powered by a wearable thermoelectric generator (Kim et al., 2018). By harnessing the temperature disparity between the human body and the surrounding environment, this breakthrough has significantly advanced the progress and commercial viability of wearable medical electronic devices. These medical sensors can be implanted at some body parts, where they can recover the required power from human body heat. The non-layered 2-D materials would be suitable for these wearable medical devices owing to their low dimensionality, which makes them more flexible and efficient.

3.5 Self-powered sensors

A thermoelectric generator (TEG) can be used in many areas as a self-powered sensor. A TEG could be utilized as self-powered temperature sensor device (Yang et al., 2012). A stretchable TEG is capable of functioning as a self-powered strain sensor. It maintains a constant temperature difference between its two sides and exhibits a steady thermoelectric current even in the absence of any strain. When a tensile strain is applied to the thermoelectric generator, it results in a rise in resistance, which subsequently leads to a drop in the associated thermoelectric (TE) current. Therefore, the TEG can be utilized as a strain sensor to monitor tensile strain by analyzing the relationship between strain and current (Taroni et al., 2018). The simultaneous observation of temperature and pressure holds significant importance in various domains such as artificial intelligence, E-skin technology, bionic robotics, and related topics. Historically, it has been documented that the majority of sensors were designed to monitor a single parameter exclusively, posing a challenge for E-skin or intelligent components to simultaneously detect both temperature and pressure. Zhu et. al have been made a flexible dual-parameter self-powered temperature–pressure sensors based on organic TE materials (Zhang et al., 2015). Non-layered 2-D thermoelectric materials can also be utilized for self-powered sensors with improved sensitivity.

3.6 Other applications

These non-layered 2-D thermoelectric materials would not be limited to the above-mentioned areas, they could be used everywhere the flexible energy generator, flexible sensors, and flexible cooler are required as shown in Fig. 7. They could also be utilized for temperature-controlled clothes, Internet of things based (IOT-based) applications, and supportive energy source in many applications. Thus, these non-layered 2-D thermoelectric materials will pave the way for the advancement of the modern era of science.

Thermoelectric applications of non-layered 2-D materials

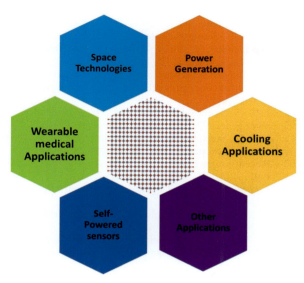

Fig. 7 Different thermoelectric applications of non-layered 2-D materials.

References

Ali, M., et al., 2017. Electronic and magnetic properties of graphene, silicene and germanene with varying vacancy concentration. AIP Adv. 7 (4), 045308.

Ambrosi, R.M., et al., 2019. European radioisotope thermoelectric generators (RTGs) and radioisotope heater units (RHUs) for space science and exploration. Space Sci. Rev. 215, 1–41.

Ares, P., et al., 2018. Recent progress on antimonene: a new bidimensional material. Adv. Mater. 30 (2), 1703771.

Bennett, G., Lombardo, J., Rock, B., 1984. US radioisotope thermoelectric generators in space. Nucl. Eng. 25, 49–58.

Beretta, D., et al., 2019. Thermoelectrics: from history, a window to the future. Mater. Sci. Eng.: R Rep. 138, 100501.

Bhardwaj, R., et al., 2022. CoSb3 based thermoelectric elements pre-requisite for device fabrication. Solid. State Sci. 129, 106900.

Byon, Y.-S., Jeong, J.-W., 2020. Phase change material-integrated thermoelectric energy harvesting block as an independent power source for sensors in buildings. Renew. Sustain. Energy Rev. 128, 109921.

Fu, C., et al., 2014. High band degeneracy contributes to high thermoelectric performance in p-type half-Heusler compounds. Adv. Energy Mater. 4 (18), 1400600.

Fu, T., et al., 2016. Enhanced thermoelectric performance of PbTe bulk materials with figure of merit $zT > 2$ by multi-functional alloying. J. Materiomics 2 (2), 141–149.

Gautam, S., et al., 2022. Structural, electronic and thermoelectric properties of Bi2Se3 thin films deposited by RF magnetron sputtering. J. Electron. Mater. 51 (5), 2500–2509.

Geller, C.B., et al., 2001. Computational band-structure engineering of III–V semiconductor alloys. Appl. Phys. Lett. 79 (3), 368–370.

Hou, W., et al., 2018. Fabrication and excellent performances of Bi0. 5Sb1. 5Te3/epoxy flexible thermoelectric cooling devices. Nano Energy 50, 766–776.

Jaziri, N., et al., 2020. A comprehensive review of thermoelectric generators: technologies and common applications. Energy Rep. 6, 264–287.

Kanahashi, K., Pu, J., Takenobu, T., 2020. 2D materials for large-area flexible thermoelectric devices. Adv. Energy Mater. 10 (11), 1902842.

Kieslich, G., et al., 2014. Enhanced thermoelectric properties of the n-type Magnéli phase WO 2.90: reduced thermal conductivity through microstructure engineering. J. Mater. Chem. A 2 (33), 13492–13497.

Kim, C.S., et al., 2018. Self-powered wearable electrocardiography using a wearable thermoelectric power generator. ACS Energy Lett. 3 (3), 501–507.

Kripalani, D.R., et al., 2018. Strain engineering of antimonene by a first-principles study: mechanical and electronic properties. Phys. Rev. B 98 (8), 085410.

Li, D., et al., 2020. Recent progress of two-dimensional thermoelectric materials. Nano-Micro Lett. 12, 1–40.

Liu, Y., et al., 2018. Recent advances of layered thermoelectric materials. Adv. Sustain. Syst. 2 (8–9), 1800046.

Liu, Y., Zhou, M., He, J., 2016. Towards higher thermoelectric performance of Bi2Te3 via defect engineering. Scr. Mater. 111, 39–43.

Lu, Y., Hirohashi, M., Sato, K., 2006. Thermoelectric properties of non-stoichiometric titanium dioxide TiO2−x fabricated by reduction treatment using carbon powder. Mater. Trans. 47 (6), 1449–1452.

Mortazavi, B., et al., 2016. Application of silicene, germanene and stanene for Na or Li ion storage: a theoretical investigation. Electrochim. Acta 213, 865–870.

Rogacheva, E.I., et al., 2001. Temperature and thickness dependences of thermoelectric properties of PbS/EuS bilayers. MRS Online Proc. Library 691.

Saramat, A., et al., 2006. Large thermoelectric figure of merit at high temperature in Czochralski-grown clathrate Ba 8 Ga 16 Ge 30. J. Appl. Phys. 99 (2), 023708.

Shi, X., et al., 2011. Multiple-filled skutterudites: high thermoelectric figure of merit through separately optimizing electrical and thermal transports. J. Am. Chem. Soc. 133 (20), 7837–7846.

Shi, Y., Sturm, C., Kleinke, H., 2019. Chalcogenides as thermoelectric materials. J. Solid State Chem. 270, 273–279.

Shin, W., Murayama, N., 2000. High performance p-type thermoelectric oxide based on NiO. Mater. Lett. 45 (6), 302–306.

Snyder, G.J. and E.S. Toberer, 2011. Complex thermoelectric materials. In: Materials for Sustainable Energy: A Collection of Peer-reviewed Research and Review Articles from Nature Publishing Group. World Scientific, pp. 101–110.

Taroni, P.J., et al., 2018. Toward stretchable self-powered sensors based on the thermoelectric response of PEDOT: PSS/polyurethane blends. Adv. Funct. Mater. 28 (15), 1704285.

Theja, V.C., et al., 2022. Thermoelectric properties of sulfide and selenide-based materials. Sulfide and Selenide Based Materials for Emerging Applications. Elsevier, pp. 293–328.

Vashishtha, P., et al., 2023. GaN-djoser pyramidal self powered UV photodetector for optical signal detection in rugged environments. J. Alloy. Compd. 930, 167267.

Verma, A.K., et al., 2023b. Coupling of electronic transport and defect engineering substantially enhances the thermoelectric performance of p-type TiCoSb HH alloy. J. Alloy. Compd. 947, 169416.

Verma, A.K., et al., 2023a. Realization of band convergence in p-type TiCoSb half-Heusler alloys significantly enhances the thermoelectric performance. ACS Appl. Mater. Interfaces 15 (1), 942–952.

Wang, F., et al., 2017. Two-dimensional non-layered materials: synthesis, properties and applications. Adv. Funct. Mater. 27 (19), 1603254.

Wu, C., et al., 2011. Hydrogen-incorporation stabilization of metallic VO2 (R) phase to room temperature, displaying promising low-temperature thermoelectric effect. J. Am. Chem. Soc. 133 (35), 13798–13801.

Yang, Y., et al., 2012. Nanowire-composite based flexible thermoelectric nanogenerators and self-powered temperature sensors. Nano Res. 5, 888–895.

Yu, B., et al., 2012. Enhancement of thermoelectric properties by modulation-doping in silicon germanium alloy nanocomposites. Nano Lett. 12 (4), 2077–2082.

Zeraati, M., et al., 2016. Highly anisotropic thermal conductivity of arsenene: an ab initio study. Phys. Rev. B 93 (8), 085424.

Zhang, D., Wang, Y., Yang, Y., 2019. Design, performance, and application of thermoelectric nanogenerators. Small 15 (32), 1805241.

Zhang, F., et al., 2015. Flexible and self-powered temperature–pressure dual-parameter sensors using microstructure-frame-supported organic thermoelectric materials. Nat. Commun. 6 (1), 8356.

Zhou, A., et al., 2010. Improved thermoelectric performance of higher manganese silicides with Ge additions. J. Electron. Mater. 39 (9), 2002–2007.

Zhu, C., et al., 2017. Multiple nanostructures in high performance Cu2S0. 5Te0. 5 thermoelectric materials. Ceram. Int. 43 (10), 7866–7869.

Index

Note: Page numbers followed by "*f*" indicate figures and "*t*" indicate tables.

A

AACVD. *See* Aerosol assisted chemical vapor deposition (AACVD)
Ab initio molecular dynamics simulations, 14
Acetone, 125, 152, 241
ACI. *See* Alternating cation among interlayer (ACI)
Activated carbon (AC), 200
Addition, 52
Adsorption-based sensing, 236
Adsorption-based solute sensing, 235
Aerosol assisted chemical vapor deposition (AACVD), 179
AFM. *See* Atomic force microscopy (AFM)
ALD. *See* Atomic layer deposition (ALD)
AlN. *See* Aluminum nitride (AlN)
AlSb. *See* Aluminum antimonide (AlSb)
Alternating cation among interlayer (ACI), 149
Aluminum (Al), 118
Aluminum antimonide (AlSb), 116
    applications, 119
    low-buckled, 117*f*
    properties, 116–118
    synthesis and development, 118–119
Aluminum nitride (AlN), 113, 115*t*
    applications, 116
    crystal phases and definition of lattice parameters, 116*f*
    properties, 113–114
    synthesis and development, 114–116
Aminopropyl triethoxysilane (APTES), 259
Ammonia (NH_3), 279, 281*f*
Angle-resolved electronic measurements, 16
Anion vacancies, 36
Antibacterial activity of NL2DMs, 307–310
    copper sulfide nanoparticles, 309*f*
Antimonene, 328
Antimony (Sb), 118
APTES. *See* Aminopropyl triethoxysilane (APTES)
Arsenene, 14, 328–329
Artificial intelligence, 334
Artificial scaffolds, 315
As-based non-layered 2D materials, 13–14
ASC. *See* Asymmetric supercapacitor (ASC)

Aspect ratio of liquid-exfoliated nanosheets, 4–5
Asymmetric supercapacitor (ASC), 200
Atomic force microscopy (AFM), 5, 45
Atomic layer deposition (ALD), 92, 259
Atomic thin crystals, 298

B

B-type natriuretic peptide (BNP), 264
Barium titanate (BT), 319
BAs. *See* Boron arsenide (BAs)
Bethe Salpeter equation (BSE), 118
Biaxial strain, 114
Bimetallic oxide ($NiCo_2O_4$), 38
Bimetallic sulphides, 200
Biomedical applications, 90–91
Bionic robotics, 334
Biosensing. *See also* Sensing
    non-layered 2D materials based biosensing platforms, 255–367
    technologies, 255
Biosensors, 90, 255
    devices based on 2D non-layered composition, 304–307
Black phosphorene, 28
Black phosphorus, 219, 327
Blue phosphorene, 172
BN. *See* Boron nitride (BN)
BNP. *See* B-type natriuretic peptide (BNP)
Boltzmann transport equation (BTE), 133
Boron arsenide (BAs), 110
    applications, 112–113
    experimentally measured physical properties, 112*t*
    properties, 110–111
    synthesis and development, 111
Boron carbide (B_4C), 5
Boron nitride (BN), 103–107, 281–282
    applications, 107
    optimized structure for primary crystalline BN phases, 104*f*
    properties, 103–105
    synthesis and development, 105–106
Boron phosphide (BP), 107
    applications, 110
    electronic band structures, 109*f*

339

optimized crystal structures, 108f
properties, 107–108
synthesis and development, 109–110
Boron-based non-layered 2D materials, 4–7
Bottom-up approach, 43–44
Bovine serum albumin (BSA), 238
BP. *See* Boron phosphide (BP)
BSA. *See* Bovine serum albumin (BSA)
BSE. *See* Bethe Salpeter equation (BSE)
BT. *See* Barium titanate (BT)
BT InAs. *See* Truncated bulk InAs (BT InAs)
BTE. *See* Boltzmann transport equation (BTE)
Bulk GaN, 3

C

c-BN. *See* Cubic boron nitride (c-BN)
c-BP. *See* Cubic boron phosphide (c-BP)
c-GaN. *See* Cubic phase gallium nitride
(c-GaN)
Cabrera–Mott reaction, 44
Cadmium (Cd), 282
Cadmium sulfide (CdS), 177, 179f
growth of, 177–178
Cadmium telluride (CdTe), 19
Calcium digermanide (CaGe$_2$), 10
Calcium titanate (CT), 319
Carbon atoms, 191
Carbon dioxide (CO$_2$), 280
Carbon monoxide (CO), 287
catalysis of, 287–289
CeO$_2$ and Rh catalyst for, 289f
Cardiac Troponin I (cTnI), 259, 265
Catalysis, 88, 285
of CO conversion and organic reactions,
90, 287–289
of water splitting, 89–90, 285–287
Cation vacancies, 36
CBE. *See* Chemical beam epitaxy (CBE)
CBM. *See* Conduction band minimum
(CBM)
CdS. *See* Cadmium sulfide (CdS)
CdTe. *See* Cadmium telluride (CdTe)
Cell adhesion, 315
Cesium (Cs$^+$), 146
Cetyltrimethylammonium bromide
(CTAB), 221, 301
Chalcogens, 17
Chemical beam epitaxy (CBE), 128
Chemical vapor deposition (CVD), 15, 20,
50–52, 72, 75, 78, 92, 105, 118,
122, 170, 220
of 2D In$_2$Se$_3$ nanosheets, 73f

diffusion-controlled CVD setup, 172f
growth of 2D non-layered materials,
170–171
growth of CdS, 177–178
growth of di-indium tri-sulfide,
175–176
growth of hematite, 179–180
growth of lead sulfide, 181–182, 182f,
183f
growth of phosphorene, 172–173
growth of titanium dioxide, 183–184
growth of β-Ga$_2$O$_3$, 176–177
growth of ε-Fe$_2$O$_3$, 180–181
optimised, 223–225, 224f
two-dimensional selenium nanoflakes,
173–175
work, 171–172
Chemiresistive signals, 235
Chemisorption of reactants, 170
Chitosan polymer, 267
Chlorin e6 (Ce6), 312
Cholesterol, 91
CK-MB. *See* Creatine kinase MB (CK-MB)
Close-spaced vapour transport (CSVT),
125
CMOS. *See* Complementary metal-oxide-
semiconductor (CMOS)
Co-precipitation approach, 207
Cobalt (Co), 281
Cobalt (II,III) oxide (Co$_3$O$_4$), 34, 282
crystal structures, 35f
Cobalt monoxide (CoO), 30, 41
crystal structures, 31f
Colloidal NPLs, 156
Colloidal synthesis (CS), 84
Colloidal templated synthesis (CTS), 223
Colloidal-template method, 52
Complementary metal-oxide-semi-
conductor (CMOS), 20
Conduction band minimum (CBM),
117–118, 139
Confined space synthesis (CSS), 84, 221, 301
Controlled drug delivery systems, applica-
tion of non-layered 2D arrays in,
311–314
Conventional energy sources, 324
Cooling applications, 333
Copper sulfide nanoparticles (CuS nano-
particles), 309
Copper(II) oxide (CuO), 30
cubic and monoclinic, 31f
Creatine kinase MB (CK-MB), 264

Crystalline materials, 218
CS. *See* Colloidal synthesis (CS)
CSS. *See* Confined space synthesis (CSS)
CSVT. *See* Close-spaced vapour transport (CSVT)
CT. *See* Calcium titanate (CT)
CTAB. *See* Cetyltrimethylammonium bromide (CTAB)
cTnI. *See* Cardiac Troponin I (cTnI)
CTS. *See* Colloidal templated synthesis (CTS)
Cubic boron arsenide, 110–111
Cubic boron nitride (c-BN), 103–105
  physical and chemical features, 106t
Cubic boron phosphide (c-BP), 107
Cubic phase gallium nitride (c-GaN), 129, 131
CVD. *See* Chemical vapor deposition (CVD)
Cyclic voltammetry (CV) technique, 257
Cysteine, 91

D
DC jet-plasma CVD. *See* Direct-current jet-plasma CVD (DC jet-plasma CVD)
DDCE. *See* Dibenzo thiazolyl dibenzo-18-crown-6 (DDCE)
De-ionized water (DI water), 125
Deformation potential theory, 14
Density functional theory (DFT), 5, 7, 9, 29, 113, 282
Density of states (DOS), 16, 35, 37, 148, 233
DFT. *See* Density functional theory (DFT)
DI water. *See* De-ionized water (DI water)
Di-indium tri-sulfide (In_2S_3), 175, 176f
  growth of, 175–176
Dibenzo thiazolyl dibenzo-18-crown-6 (DDCE), 54
Dielectric confined 2D quantum wells, 157
Dimethyl sulfoxide (DMSO), 151
Dimethylformamide (DMF), 227
Dion Jacobson (DJ), 148
Direct-current jet-plasma CVD (DC jet-plasma CVD), 106
DJ. *See* Dion Jacobson (DJ)
DLHC InAs. *See* Double layer honeycomb InAs (DLHC InAs)
DMF. *See* Dimethylformamide (DMF); N,N-dimethylformamide (DMF)
DMSO. *See* Dimethyl sulfoxide (DMSO)
DOS. *See* Density of states (DOS)

Double layer honeycomb InAs (DLHC InAs), 120, 139
Doxorubicin, 314
Drug delivery systems, 314
Dry methods, 79–83
  2D material families, 80f
  dual-metal precursors CVD growth, 82f

E
E-skin technology, 334
EA. *See* Ethyl acetoacetate (EA)
EC. *See* Electrolysis cell (EC)
ECG. *See* Electrocardiogram (ECG)
ECR-MPCVD. *See* Electron-cyclotron resonance microwave plasma CVD (ECR-MPCVD)
EDLC. *See* Electric double-layer capacitors (EDLC)
EIS. *See* Electrochemical Impedance Spectroscopy (EIS)
EISA. *See* Evaporation-induced self-assembly (EISA)
EL. *See* Electroluminescence (EL)
Electric double-layer capacitors (EDLC), 193–195, 195f
Electric field, 194
Electrocardiogram (ECG), 333
Electrochemical biosensors, 255–261, 258f
Electrochemical Impedance Spectroscopy (EIS), 200, 258
Electrochemical mechanism of non-layered 2D materials, 229–230
Electrochemical sensors, 229–230, 236
Electrode, 202
Electroluminescence (EL), 153, 159
Electrolysis cell (EC), 210
Electron-cyclotron resonance microwave plasma CVD (ECR-MPCVD), 106
Electron-hole pairs, 300
Electronic conductivity, 270
Electronics, 85
Elemental doping, 37–38, 234
Elimination, 52
Energy application, 2D non layered materials for, 193
Energy conversion and storage, 86–88
Energy storage, 213
  2D non-layered materials in, 270–271
Environment applications of non-layered 2D materials
  assessing environmental quality and remediation applications, 279

heavy metals monitoring, 282–283
NH$_3$ and H$_2$S quantification, 279
pH sensing, 283–285
catalysis, 285–289
research hotspots of non-layered 2D
materials, 279f
EPA. *See* US Environmental Protection
Agency (EPA)
ε-Fe$_2$O$_3$, growth of, 180–181
EQE. *See* External quantum efficiency (EQE)
Ethanol detection, 241
Ethyl acetoacetate (EA), 162
Ethylene vinyl acetate (EVA), 283
EVA. *See* Ethylene vinyl acetate (EVA)
Evaporation-induced self-assembly (EISA), 53
Ex-situ atomic force microscopy, 140
Exciton binding energy, 156
Exfoliation techniques, 72
External quantum efficiency (EQE), 159

F

Fabrication methods, 52
for non-layered 2D metal oxides, 43–44
Face-centered cubic (FCC), 104
carbon cells, 104
sublattices, 30
Fast Fourier transform (FFT), 71
FC. *See* Fuel cell (FC)
FCC. *See* Face-centered cubic (FCC)
Ferrocene methanol (Fc), 257
FETs. *See* Field-effect transistors (FETs)
FFT. *See* Fast Fourier transform (FFT)
Field-effect transistors (FETs), 85, 146, 264,
305
First-principles calculations, 14
Fluorescence resonance energy transfer
(FRET), 231
Fluorescence resonance mechanism of
non-layered 2D materials, 231–232
Fluorescence sensors, 231
Fluorescent signals, 235–236
Fool's gold or pyrite, 69
Formadinium (FA$^+$), 146
Formamidinium lead iodide (FAPbI$_3$), 192
FRET. *See* Fluorescence resonance energy
transfer (FRET)
Fuel cell (FC), 209–212
schematic illustration of SOFC, 210f
2D materials performance for, 212t
types of, 211f

G

GaAs. *See* Gallium arsenide (GaAs)
Gallium antimonide (GaSb), 138–140
applications, 140
properties, 138–140
synthesis and development, 140
Gallium arsenide (GaAs), 134, 134f, 135f,
136f, 192
applications, 138
electronic band structure, 137f
properties, 134–137
synthesis and development, 138
Gallium nitride (GaN), 129, 129f, 130f
applications, 133
properties, 129–131
synthesis and development, 131–133
Gallium oxide (β-Ga$_2$O$_3$), 176, 177f, 178f
growth of, 176–177
GaN. *See* Gallium nitride (GaN)
Gas injection method, 45
Gas sensing, non-layered 2D materials for,
238–242, 240f
GaSb. *See* Gallium antimonide (GaSb)
Gauge factor (GF), 227
GCE. *See* Glassy carbon electrodes (GCE)
GeH. *See* Germanane (GeH)
Germanane (GeH), 10
Germanene, 10, 328
atomic structure of, 329f
Germanium (Ge), 12, 300
germanium-based non-layered 2D
materials, 10–12
Germanium nanoplates (Ge-NPts), 12
GF. *See* Gauge factor (GF)
Glassy carbon electrodes (GCE), 282, 307
Glucose, 91
Glucose oxidase (GOx), 257, 267, 307
Glutathione (GSH), 91
GM units. *See* Goeppert-Mayer units
(GM units)
GO. *See* Graphene oxide (GO)
Goeppert-Mayer units (GM units), 158
Gold (Au), 222
GOx. *See* Glucose oxidase (GOx)
Grapevine virus A-type (GVA), 262
Graphene, 2, 14, 190, 219, 327
Graphene oxide (GO), 53, 84, 222, 254,
302
Group IVA elements, 28
GSH. *See* Glutathione (GSH)
GVA. *See* Grapevine virus A-type (GVA)

Index

H
4H polytypes, 126
h-BN. *See* Hexagonal boron nitride (h-BN)
Hard-template method, 52
Hb. *See* Hemoglobin (Hb)
hcp. *See* Hexagonal close-packed (hcp)
Heavy metals, 282
 monitoring, 282–283
Helmholtz layer, 194
Hematite, growth of, 179–180, 180*f*
Hemoglobin (Hb), 259
HER. *See* Hydrogen evolution reaction (HER)
Hexagonal boron nitride (h-BN), 28, 104
Hexagonal close-packed (hcp), 219
Hexagonal InP sheet (HInPS), 123
Hexagonal WO_3 nanosheet (h-WO_3), 280
Heyd-Scuseria-Ernzerhof functional (HSE functional), 127
HI. *See* Hot injection (HI)
High electron mobility, 331
High-resolution scanning electron microscopy (HRSEM), 131
High-Temperature High-Pressure synthesis technique (HTHP synthesis technique), 105
HInPS. *See* Hexagonal InP sheet (HInPS)
Hot injection (HI), 150–151
HRSEM. *See* High-resolution scanning electron microscopy (HRSEM)
HSE functional. *See* Heyd-Scuseria-Ernzerhof functional (HSE functional)
HTHP synthesis technique. *See* High-Temperature High-Pressure synthesis technique (HTHP synthesis technique)
HVPE. *See* Hydride vapor phase epitaxy (HVPE)
Hybrid capacitor 196–200, 197*f*
Hydride vapor phase epitaxy (HVPE), 125, 131
Hydrogen evolution reaction (HER), 42
Hydrogen sulfide (H_2S), 279, 281*f*
Hydrothermal approach, 48–50, 49*f*, 223
Hydrothermal synthesis method, 44, 83

I
IBAD. *See* Ion beam-assisted deposition (IBAD)
ICP-CVD. *See* Inductively coupled plasma CVD (ICP-CVD)
IEF. *See* Internal electric field (IEF); Internal electrostatic field (IEF)
III-V semiconductors, 102, 192, 331–332
III–VI compounds, 28
ILE. *See* Ionic layer epitaxy (ILE)
Immunosensor, 260
In-situ electron diffraction and photoemission spectroscopy, 140
InAs. *See* Indium arsenide (InAs)
Incident polarization photoelectric measurements, 16
Indium antimonide (InSb), 126
 applications, 128–129
 properties, 126–127
 synthesis and development, 128
 unit cell structures, 127*f*
Indium arsenide (InAs), 119, 120*f*
 applications, 122
 properties, 119–122
 synthesis and development, 122
Indium iodide (InI_3), 72
Indium oxide (In_2O_3), 192
Indium phosphide (InP), 123, 124*f*, 192
 applications, 125–126
 properties, 123–124
 synthesis and development, 124–125
Indium(III) sulfide (In_2S_3), 67
Inductively coupled plasma CVD (ICP-CVD), 106
InI_3. *See* Indium iodide (InI_3)
InP. *See* Indium phosphide (InP)
InSb. *See* Indium antimonide (InSb)
Internal electric field (IEF), 132
Internal electrostatic field (IEF), 132
Internet of things based (IOT-based) applications, 334
Ion beam-assisted deposition (IBAD), 106
Ion-sensitive field-effect transistor (ISFET), 285
Ionic layer epitaxy (ILE), 36
IPA. *See* Isopropyl alcohol (IPA)
Iron disulfide (FeS_2), 69
ISFET. *See* Ion-sensitive field-effect transistor (ISFET)
Isomerization reaction, 52
Isopropyl alcohol (IPA), 125

J
Jahn–Teller distortion, 41
JFETs. *See* Junction field effect transistor (JFETs)
Junction field effect transistor (JFETs), 85

L

L2DMs. *See* Layered 2DMs (L2DMs)
LARP. *See* Ligand assisted reprecipitation (LARP)
Layered 2DMs (L2DMs), 298
Layered materials, 148
Lead (Pb), 282
Lead oxide (PbO), 243
Lead sulfide (PbS), 181, 182*f*, 183*f*, 302
  growth of, 181–182
Lead zirconate titanate (PZT), 243
LEDs. *See* Light-emitting diodes (LEDs)
LEED. *See* Low electron energy diffraction (LEED)
LIB. *See* Lithium-ion batteries (LIB)
Ligand assisted reprecipitation (LARP), 151–154
Light-emitting diodes (LEDs), 4, 146, 158
Limit of detection (LOD), 259, 282
Liquid exfoliation, 43
Liquid metal approach, 44–47, 47*t*
Liquid metals synthesis, 225–227, 226*f*
Liquid phase epitaxy (LPE), 124
Lithium-ion batteries (LIB), 200–205, 201*f*
  2D nanostructure performance in, 204*t*
LO phonon mode. *See* Raman longitudinal optical phonon mode (LO phonon mode)
LOD. *See* Limit of detection (LOD)
Low electron energy diffraction (LEED), 128
LPE. *See* Liquid phase epitaxy (LPE)

M

Manganese (Mn), 281
Manganese dioxide (MnO$_2$), 33
  crystal structures, 34*f*
Mass-selected ion beam deposition, 106
Materials, 219
MBE. *See* Molecular beam epitaxy (MBE)
MD. *See* Molecular dynamics (MD)
MDCs. *See* Metal dichalcogenides (MDCs)
Mechanical exfoliation, 43
Mechanical sensing, non-layered 2D materials for, 242–243, 244*f*
MEEG. *See* Migration-enhanced encapsulated growth (MEEG)
Memory resistors, 267
Memristors based 2D non layered material, 267–270
  Au/SnSe/NSTO artificial electronic synapse mimics, 269*f*
Mercury (Hg), 282

Metal cation defects, 239
Metal chalcogenides, 192, 330–331
Metal dichalcogenides (MDCs), 64
Metal organic frameworks (MOFs), 254
Metal organic vapor phase epitaxy (MOVPE), 128
Metal oxides (MOXs), 28, 192, 219, 329–330
  metal oxide-based biosensors, 304
  semiconductor, 256
Metal selenides (MSe$_2$), 71–72
Metal tellurides (MTe$_2$), 72–73
Metal-organic chemical vapor deposition (MOCVD), 79, 124, 131, 138
Metal–insulator–semiconductor field-effect transistors (MISFETs), 85
Metalloid arsenic (As), 282
Metalloids, 328–329
Methane (CH$_4$), 280
Methyl ammonium lead iodide (MAPbI$_3$), 192
Methylammonium (MA$^+$), 146
Microfluidic channels, 305
MicroRNA (miRNA), 304
Microwave-assisted synthesis in ionic liquids, 17
Migration-enhanced encapsulated growth (MEEG), 132
miRNA. *See* MicroRNA (miRNA)
MISFETs. *See* Metal–insulator–semiconductor field-effect transistors (MISFETs)
Mixed metal dichalcogenides, 74
ML GaN. *See* Monolayer GaN (ML GaN)
MOCVD. *See* Metal-organic chemical vapor deposition (MOCVD)
MOFs. *See* Metal organic frameworks (MOFs)
Molecular beam epitaxy (MBE), 79–80, 106, 118, 124, 128, 131–132, 138, 140, 171
Molecular dynamics (MD), 29
Molybdenum dioxides (MoO$_2$), 51, 51*f*, 271
Molybdenum disulphide (MoS$_2$), 64, 192
Molybdenum ditelluride (MoTe$_2$), 72
Monolayer GaN (ML GaN), 133
Mott insulator, 29
MOVPE. *See* Metal organic vapor phase epitaxy (MOVPE)
MOXs. *See* Metal oxides (MOXs)
Multiphoton absorption process, 158

Index

N

N,N-dimethylformamide (DMF), 151
NaHTe. *See* Sodium hydrogen telluride (NaHTe)
Nano-FET based biosensor, 264–267
  flexible and wearable In_2O_3 FET biosensor, 266f
Nanoarchitecture designs for sensing applications, 232–235
Nanofabrication techniques, 218
Nanomaterials, 189
Nanoparticles, 305, 307
Nanoplatelets (NPls), 149, 150f
Nanoscale, 190, 256
Nanosheets (NSs), 149, 202
Nanotechnology, 254
NBE. *See* Near band emission (NBE)
Near band emission (NBE), 262
Nernstian value, 285
Nitride compounds, 102
Nitrous oxide (N_2O), 280
NL-TMCs. *See* Non-layered transition metal carbides (NL-TMCs)
NL2DMs. *See* Non-layered 2DMs (NL2DMs)
Non-layered 2-D metalloids, 328
Non-layered 2D arrays in controlled drug delivery systems, application of, 311–314
Non-layered 2D materials, 260, 286, 327–328
  based biosensing platforms, 255
  electrochemical biosensors, 255–261
  for gas sensing, 238–242, 240f
  III–V Semiconductors, 331–332
  liquid metals synthesis, 225–227
  for mechanical sensing, 242–243, 244f
  metal chalcogenides, 330–331
  metal oxides, 329–330
  metalloids, 328–329
  nano-FET based biosensor, 264–267
  non-layered 2D materials-based sensing application, 280
  optical biosensors, 262–263
  optimised chemical vapour deposition, 223–225
  sensing applications of non-layered 2D materials, 232–245
  sensing mechanisms of, 227–232
  for solute sensing, 235–238, 236f
  structure of, 219–220
  synthesis of, 219–221

  thermoelectric applications, 332–335
  thermoelectric devices and materials, 325–327
  thermoelectric effect, 324–325
  wet chemical synthesis, 221–223
Non-layered 2D metal dichalcogenides
  applications, 84–91
  chemical properties, 78
  defects and grain boundaries, 75
  electronic properties, 76–77
  importance of studying, 66–67
  mechanical properties, 77
  monolayer and few-layer structures, 74–75
  properties, 76–78
  structural characteristics, 74–76
  structural phase transitions, 75–76
  structure of different types, 67–74
  synthesis, 78–84
Non-layered 2D metal oxides, properties of, 38–43
Non-layered 2D nanomaterials, 220, 232
Non-layered 2D sulfides, 308
Non-layered 2DMs (NL2DMs), 298
  antibacterial activity of, 307–310
  application of non-layered 2D arrays in controlled drug delivery systems, 311–314
  biosensors devices based on, 304–307
  lucrative features and healthcare opportunities revealed by, 299f
  PCL-based bone scaffold, 317f
  recent advancement in synthesis and structure of, 299–303
  2D non-layered structures in tissue engineering, 315–319
Non-layered III-V compounds, 103
  AlN, 113–116
  AlSb, 116–119
  BAs, 110–113
  BN, 103–107
  BP, 107–110
  GaAs, 134–138
  GaN, 129–133
  GaSb, 138–140
  InAs, 119–122
  InP, 123–126
  InSb, 126–129
Non-layered materials, 219, 256, 278
Non-layered metal dichalcogenides, 71
  metal selenides, 71–72
  metal tellurides, 72–73

Non-layered metal selenides, 71–72
Non-layered structures, 300
Non-layered tin thiohypodiphosphate, 287
Non-layered transition metal carbides (NL-TMCs), 87, 88*f*
Non-layered transition metal dichalcogenides, 67
  2D non-layered CdS, 68–69
  2D non-layered In_2S_3, 67–68, 67*f*
  (2D) non-layered iron disulfide (FeS_2), 69–71, 70*f*
Non-layered two-dimensional materials, 332
Non-layered two-dimensional metalloids, 3–4
  As-based non-layered 2D materials, 13–14
  atomic geometries, 8*f*
  boron-based non-layered 2D materials, 4–7
  formation energies of terminations, 6*f*
  future perspectives, 19–20
  germanium-based non-layered 2D materials, 10–12
  pure arsenene and arsenene doped with boron, 15*f*
  Sb-based non-layered 2D materials, 14–17
  Si-based non-layered 2D materials, 7–10
  Te-based non-layered 2D materials, 17–19
Non-radiative energy transfer process, 232
Nonlinear optics, 158
NPls. *See* Nanoplatelets (NPls)
NSs. *See* Nanosheets (NSs)

O

OA. *See* Oleic acid (OA)
1-octadecene (ODE), 150
Octadecyl trichlorosilane (OTS), 83
OER. *See* Oxygen evolution reactions (OER)
OLA. *See* Oleyl amine (OLA)
Oleic acid (OA), 150
Oleyl amine (OLA), 150
One dimension (1D), 148
  nanostructures, 14
Optical biosensors, 262–263
  photoluminescence immunosensor based 2D ZnO, 263*f*
Optoelectronics, 85–86
  mechanism of non-layered 2D materials, 230
Organic reactions, catalysis of, 287–289
Organic-inorganic perovskites, 192

ORR. *See* Oxygen reduction reactions (ORR)
OTS. *See* Octadecyl trichlorosilane (OTS)
Oxides, 315
Oxygen evolution reactions (OER), 38
Oxygen reduction reactions (ORR), 39

P

Partial density of states (PDOS), 118
PBE functional. *See* Perdew-Burke-Ernzerhof functional (PBE functional)
PbS. *See* Lead sulfide (PbS)
PCE. *See* Power conversion efficiency (PCE)
PCL. *See* Poly(ε-caprolactone) (PCL)
PDLLA. *See* Poly(D,L-lactic acid) (PDLLA)
PDMS. *See* Polydimethylsiloxane (PDMS)
PDOS. *See* Partial density of states (PDOS); Projected density of states (PDOS)
PDs. *See* Photodetectors (PDs)
PEC-type photodetector. *See* Photoelectrochemical-type photodetector (PEC-type photodetector)
PEI. *See* Polyethylenimine (PEI)
Peltier coefficient, 325
Peltier effect, 324, 325*f*
Perdew-Burke-Ernzerhof functional (PBE functional), 127
Perovskite solar cells (PSCs), 88
Perovskites, 145
  materials, 318
  nanomaterials, 310
  perovskite-based electrochemical biosensors, 306
PET. *See* Polyethylene terephthalate (PET)
PF. *See* Power factor (PF)
pH sensing, 283–285
  pH sensors characterization and performances, 284*f*
Phosphorene, 174*f*
  applications, 173*f*
  growth of, 172–173
Photocatalytic water splitting, 89, 285
Photodetectors (PDs), 4, 158, 243–245
Photoelectrochemical-type photodetector (PEC-type photodetector), 19
Photoluminescence quantum yield efficiency (PLQY), 150, 157
Photovoltaics (PV), 158
Physical vapor deposition (PVD), 105, 171

Index 347

Piezoelectric mechanism of non-layered 2D materials, 229
Piezoelectricity, 229, 242
Piezoresistive mechanism of non-layered 2D materials, 227–228
Pit generation, 234
Plasma treatment, 13
Plasmonic signals, 235
PLD. *See* Pulsed laser deposition (PLD)
PLQY. *See* Photoluminescence quantum yield efficiency (PLQY)
POC. *See* Point-of-care (POC)
Point defects, 75
Point-of-care (POC), 255
Poly(D,L-lactic acid) (PDLLA), 318
Poly(ε-caprolactone) (PCL), 318
Polydimethylsiloxane (PDMS), 45–46
Polyethylene terephthalate (PET), 161
Polyethylenimine (PEI), 160
Polyvinylpyrrolidone (PVP), 19, 84, 221, 301
Pore generation, 234
Power conversion efficiency (PCE), 163
Power factor (PF), 124
Power generation, 332–333
Projected density of states (PDOS), 121, 130
PSCs. *See* Perovskite solar cells (PSCs)
Pseudocapacitor, 195–196, 196*f*
Pulsed laser deposition (PLD), 106
PV. *See* Photovoltaics (PV)
PVD. *See* Physical vapor deposition (PVD)
PVP. *See* Polyvinylpyrrolidone (PVP)
Pyrite (FeS$_2$), 87

Q

QFEG. *See* Quasi freestanding epitaxial graphene (QFEG)
Quantum confinement effect, 155
Quantum hall effect, 148
Quasi freestanding epitaxial graphene (QFEG), 133

R

r-BN. *See* Rhombohedral boron nitride (r-BN)
Radiation sensing, 243–245
Radio frequency sputtering (RF sputtering), 106
Radio-frequency plasma CVD (RF-CVD), 106
Raman longitudinal optical phonon mode (LO phonon mode), 131

Raman spectroscopy, 131
results, 203
Reduced graphene oxide (rGO), 43, 280
Remediation applications, 279–285
Resistance Temperature Detectors (RTDs), 231
RF sputtering. *See* Radio frequency sputtering (RF sputtering)
RF-CVD. *See* Radio-frequency plasma CVD (RF-CVD)
rGO. *See* Reduced graphene oxide (rGO)
rh-BP. *See* Rhombohedral boron phosphide (rh-BP)
Rhombohedral boron nitride (r-BN), 104
Rhombohedral boron phosphide (rh-BP), 107
Rock-salt aluminum nitride (rs-AlN), 113
RP. *See* Ruddlesden popper (RP)
rs-AlN. *See* Rock-salt aluminum nitride (rs-AlN)
RTDs. *See* Resistance Temperature Detectors (RTDs)
Ruddlesden popper (RP), 148
Ruthenium oxides (RuO$_2$), 312
Ruthenium sulfide nanoparticles (RuS$_2$-NPs), 312

S

Salmonella typhimurium, 308
Salt crystals, 53
Scanning electron microscopy (SEM), 140
Scanning tunnelling microscopy (STM), 7, 11, 11*f*
SCs. *See* Supercapacitors (SCs)
SECS. *See* Surface-energy-controlled synthesis (SECS)
Seebeck coefficient, 124, 324
Seebeck effect, 324, 325*f*
Selenium (Se), 17, 18*f*, 19, 72, 268, 300
Selenium nanoflakes (SeNFs), 173–175
Self-limited epitaxial growth, 81, 303
Self-powered sensors, 334
Self-supportive integrated thick electrodes, 203
SEM. *See* Scanning electron microscopy (SEM)
SeNFs. *See* Selenium nanoflakes (SeNFs)
Sensing
applications of non-layered 2D materials, 232
nanoarchitecture designs for, 232–235

non-layered 2D materials for gas sensing, 238–242
non-layered 2D materials for mechanical sensing, 242–243
non-layered 2D materials for solute sensing, 235–238
photodetectors and radiation sensing, 243–245
materials, 227
mechanisms of non-layered 2D materials, 227
electrochemical mechanism, 229–230
fluorescence resonance mechanism, 231–232
optoelectronic mechanism, 230
piezoelectric mechanism, 229
piezoresistive mechanism, 227–228
thermoresistive mechanism, 231
Sensors-based oxides, 280
Si-based non-layered 2D materials, 7–10
SIBs. See Sodium-ion batteries (SIBs)
Silicene, 10
Single layer honeycomb InAs (SLHC InAs), 121
Single layer materials, 192
Single-layered tin(IV) dioxide (SnO_2), 288
nanosheets, 42
Single-walled carbon nanotubes (SWCNTs), 267
Skin disease, 283
SLHC InAs. See Single layer honeycomb InAs (SLHC InAs)
SOC. See Spin-orbit coupling (SOC)
Sodium hydrogen telluride (NaHTe), 19
Sodium ions (Na^+), 205
Sodium yttrium fluoride ($NaYF_4$), 237
Sodium-based layered electrode materials, 206
Sodium-ion batteries (SIBs), 205–209, 206f
2D nanostructure performance in, 208t
Soft-template method, 52
Solar cells, 4
Sol–gel method, 203
Solute sensing, non-layered 2D materials for, 235–238, 236f
Solution-based approaches, 92
Solution-phase approaches, 17
Solvothermal synthesis technique, 49, 223
Space technologies, 333
Spin-orbit coupling (SOC), 118–119, 127
Spinel-type 2D metal oxide nanosheets, 48
Spintronics, 148

ST. See Strontium titanate (ST)
Staphylococcus aureus, 308
STM. See Scanning tunnelling microscopy (STM)
Strain or external electric fields, 14
Strontium titanate (ST), 319
Strontium titanate ferrite (STFx), 310
Substitution, 52
Sulfur passivation, 13
Supercapacitors (SCs), 193, 197
classification of, 194f
EDLC, 193–195
hybrid capacitor 196–200
pseudocapacitor, 195–196
Ragone plot, 194f
2D non layered materials in, 198–199t
Surface morphology, 285
Surface-energy-controlled synthesis (SECS), 84, 221
SWCNTs. See Single-walled carbon nanotubes (SWCNTs)

T

T4 PNK. See T4 Polynucleotide Kinase (T4 PNK)
T4 Polynucleotide Kinase (T4 PNK), 257
Targeted delivery, 311
TDS. See Template-directed synthesis (TDS)
Te-based non-layered 2D materials, 17–19
TEA. See Triethylamine (TEA)
Teflon, 83
TEG. See Thermoelectric generators (TEG)
Tellurium (Te), 17–18, 18f, 19, 300
TEM. See Transmission electron microscopy (TEM)
Template-assisted synthesis, 44, 52–54
Template-directed synthesis (TDS), 84, 222, 302
TENG. See Triboelectric nanogenerator (TENG)
TF-VLS. See Thin-film vapor-liquid solid (TF-VLS)
TG. See Transient grating (TG)
Thermal decomposition strategy, 40
Thermistors, 231
Thermoelectric (TE)
applications of non-layered 2-D materials, 332
applications, 334–335
cooling applications, 333
power generation, 332–333

Index 349

self-powered sensors, 334
space technologies, 333
wearable medical applications,
333–334
current, 334
devices and materials of non-layered 2-
D materials, 325–327, 326f
materials, 124
effect of non-layered 2-D materials,
324–325
Thermoelectric generators (TEG), 325, 334
Thermoresistive mechanism of non-layered
2D materials, 231
Thickness control, 35–36, 232
Thin-film vapor-liquid solid (TF-VLS), 125
Three dimensions (3D), 191
bulk materials, 190
chemical bonding crystal structure, 278
directions, 219
network, 318
perovskites, 146, 147f
Tin (Sn), 268
Tin oxide (SnO_2), 192
Tin selenide (SnSe), 268
Tissue engineering, NL2DMs 2D
non-layered structures in, 315–319
Titanium dioxide (TiO_2), 31, 183
crystal structures, 32f
growth of, 183–184
nanosheets, 42
Titanium tetraisopropoxide (TTIP), 183
TMCs. See Transition metal chalcogenides
(TMCs)
TMDCs. See Transition metal dichalco-
genides (TMDCs)
TMOs. See Transition MOXs (TMOs)
TOPO. See Trioctylphosphine oxide
(TOPO)
TPB. See Triple phase boundary (TPB)
Transient grating (TG), 113
Transition metal chalcogenides (TMCs),
229, 242
Transition metal dichalcogenides
(TMDCs), 14, 28, 67, 71, 74,
219–220, 298, 327
Transition metal oxides, 203
Transition metal selenides (TMSes), 175
Transition MOXs (TMOs), 38
Transmission electron microscopy (TEM),
5, 107, 126
Triboelectric nanogenerator (TENG), 283
Triethylamine (TEA), 41

Trioctylphosphine oxide (TOPO), 160
Triple phase boundary (TPB), 211
Truncated bulk InAs (BT InAs), 121
TTIP. See Titanium tetraisopropoxide
(TTIP)
Tungsten diselenide (WSe_2), 64, 192
Tungsten disulphide (WS_2), 192
Tungsten ditelluride (WTe_2), 72
Tungsten oxide (WOx)-based materials 284
Tungsten trioxide (WO_3), 259
Two dimension (2D), 148
heterostructure assembly, 234
layered materials, 191
metalloids, 4
nanomaterials, 28, 190, 213
nanosheets, 218
nanostructures, 14
non-layered CdS, 68–69, 69f
non-layered In_2S_3, 67–68, 67f
non-layered iron disulfide (FeS_2),
69–71, 70f
non-layered materials, 10, 191–192,
255, 272, 278, 285
applications of, 192–193
challenges and future perspectives,
272–273
for energy application, 193
in energy storage, 270–271
growth of, 170–171, 171f
memristors based, 267–270
non-layered 2D materials based biosensing platforms, 255–267
non-layered structures in tissue engineering, NL2DMs, 315–319
perovskites, 148, 153f
applications, 158–163
methods, 154–155
morphology of $CsPbBr_3$, 152f
PL results, 157f
structure and physical properties,
155–158
synthesis, 149–155
synthesis of $Cs_3Bi_2Br_9$ NPls, 163f
selenium nanoflakes, 173–175, 175f
thermoelectric materials (2-D thermoelectric materials), 327
Two dimension electron gas (2DEG), 148
Two dimension metal dichalcogenides
(MDCs), 64, 65f
applications of non-layered 2D metal
dichalcogenides, 84–91
challenges and future directions, 91–93

characterization techniques, 91
current challenges in research of, 91–92
fundamental understanding, 92
future research directions, 92–93
importance of studying non-layered 2D
    metal dichalcogenides, 66–67
integration with other materials, 91
properties of non-layered 2D metal
    dichalcogenides, 76–78
stability and degradation, 91
structural characteristics of non-layered
    2D metal dichalcogenides, 74–76
structure of different types of non-
    layered 2D metal dichalcogenides,
    67–74
synthesis
    of non-layered 2D metal dichalco-
        genides, 78–84
    and scalability, 91
Two-dimensional materials (2DMs), 2,
    190–193, 218, 228, 254, 270, 278,
    285, 298, 327–328
material-based gas sensors, 238
types of, 161f
Two-dimensional metal oxides (MOXs), 28
advantages and disadvantages of 2D
    non-layered synthesis approaches,
    55t
challenges and future perspectives,
    54–56
crystal structures of NiO, 30f
CVD, 50–52
fabrication approaches for non-layered
    2D metal oxides, 43–44
hydrothermal approach, 48–50
liquid metal approach, 44–47, 47t
properties of non-layered 2D metal
    oxides, 38–43
structural arrangements and influence on
    properties, 35–38
structural characteristics of non-layered,
    29
template-assisted approach, 52–54
ultrathin sheets with bonding patterns,
    29–34
Two-dimensional non-layered sheets
    (2DNLSs), 13

U

Ultrafast open space transformation, 234
Ultrathin 2D nanosheets, 232

Ultrathin sheets with bonding patterns,
    29–34
Ultraviolet-visible (UV–vis) spectroscopy,
    211, 262
Universal dual-metal precursors method, 81
US Environmental Protection Agency
    (EPA), 282

V

Vacancy engineering, 234
Vacancy manipulation, 36–37
Vacuum ultraviolet (VUV), 244
Vacuum-tube hot-pressing technique, 12
Valence band maximum (VBM), 117–118,
    130, 139, 234
Van der Waals epitaxy (vdWE), 17–18, 20,
    79, 171, 223, 302
Vapor phase growth, 17
Vapor-phase deposition synthesis, 44
VBM. See Valence band maximum (VBM)
vdWE. See Van der Waals epitaxy (vdWE)
Very low electron energy diffraction
    (VLEED), 128
VLEED. See Very low electron energy
    diffraction (VLEED)
VOCs. See Volatile organic compounds
    (VOCs)
Volatile organic compounds (VOCs), 239
Voltammetric sensors, 230
VUV. See Vacuum ultraviolet (VUV)
V–VI compounds, 28

W

w-BN. See Wurtzite boron nitride (w-BN)
w-BP. See Wurtzite boron phosphide
    (w-BP)
Water splitting, catalysis of, 89–90, 285–287
Wearable medical applications, 333
Wearable thermoelectric devices, 333–334
Wet chemical synthesis, 221–223, 222f, 301
Wet chemistry methods, 83–84
Wiedemann–Franz law, 326
Wurtzite (WZ), 126, 134
Wurtzite boron nitride (w-BN), 104
Wurtzite boron phosphide (w-BP), 107
WZ. See Wurtzite (WZ)

X

X-ray diffraction (XRD), 126, 131, 203
X-ray photoelectron spectroscopy (XPS),
    181, 203, 211

Index

XPS. *See* X-ray photoelectron spectroscopy (XPS)

XRD. *See* X-ray diffraction (XRD)

Z

ZB. *See* Zinc blende (ZB)

Zero dimension (0D), 148

Zigzag InP nanoribbons (ZInPNRs), 123

Zinc blende (ZB), 118, 126, 134

Zinc oxide (ZnO), 32, 192, 256
  crystal structures, 33*f*

nanostructures, 263

synthesis, structural and morphological characterization, 46*f*

Zinc oxide nanosheets–regenerated cellulose (ZNSRC), 40–41

Zinc telluride (ZnTe), 19

Zincblende aluminum nitride (zb-AlN), 113–114

ZInPNRs. *See* Zigzag InP nanoribbons (ZInPNRs)

ZNSRC. *See* Zinc oxide nanosheets-regenerated cellulose (ZNSRC)

Printed in the United States
by Baker & Taylor Publisher Services